A NEW LOOK AT
GEOMETRY

IRVING ADLER
With Diagrams by Ruth Adler

Introduction to the Dover Edition by Peter Ruane

DOVER PUBLICATIONS, INC.
Mineola, New York

The English translation of Gauss's letter to Taurinus is by Professor Harold E. Wolfe, and is reproduced from his book *Non-Euclidean Geometry* with the permission of Professor Wolfe and the publishers, Holt, Rinehart & Winston.

Copyright

Bibliographical Note

This Dover edition, first published in 2012, is an unabridged republication of the work originally published by The John Day Company, New York, in 1966. A new Introduction to the Dover edition, written by Peter Ruane, has been specially prepared for the present volume.

Library of Congress Cataloging-in-Publication Data

Adler, Irving.
    A new look at geometry / Irving Adler with diagrams by Ruth Adler ; introduction to the Dover edition by Peter Ruane. — Dover ed.
      p. cm.
    Originally published: New York : John Day Co., 1966.
    Includes bibliographical references and index.
    ISBN-13: 978-0-486-49851-5
    ISBN-10: 0-486-49851-4
    1. Geometry. I. Title.

QA445.A3 2012
516—dc23

                                      2012013992

Manufactured in the United States by Courier Corporation
49851401
www.doverpublications.com

Contents

Introduction to the Dover Edition

When it was first published in the USA by The John Day Company in 1966, this was, to my knowledge, the first book to provide an accessibly broad description of classical and modern geometries. Since then, similar books on geometry have appeared, but they assume a higher level of mathematical expertise on the part of the reader, and the scope of their contents is somewhat narrower. Consequently, I believe this book retains its uniqueness in the above sense. On the other hand, I have never understood why it became out of print, while many lesser texts remained in circulation.

Suited to readers of high school standard and above, *A New Look at Geometry* describes the emergence of geometrical ideas from the ancient Egyptians to the work of Bolyai, Lobachevsky, and Felix Klein. One of the book's central themes is the relationship between physical space and the geometries that represent it. In other words, it relates Euclidean geometry to Newtonian physics, and it explains the relevance of non-Euclidean geometry to relativity.

The work of Descartes and Fermat form the context for the development of another of the book's major themes—the relationship between algebra and geometry. Coordinates, vector algebra, isometries, and symmetry groups are shown to be effective geometric tools and, following subsequent introductions to non-Euclidean and projective geometry, the importance of Klein's Erlanger Programme is considered. However, because this is an introductory survey of geometries that is intended for non-specialists, Irving Adler has avoided undue reference to mathematical technicalities.

The central chapters explain how the theory of parallels, the theory of curved surfaces and the "geometry of position" jointly form the basis of modern geometry. Saccheri, Lambert, and Legendre are the main characters in the drama of Euclid's parallel postulate, while the work of Gauss, Bolyai, and Lobachevsky characterize its denouement. The relevance of Hilbert's axioms to Euclidean and non-Euclidean geometry is also clarified. Projective geometry is approached algebraically and synthetically, and the chapter "Calculus and Geometry" offers a highly intuitive survey of elementary differential geometry that leads to the notion of Gaussian curvature and the concept of a manifold.

A New Look at Geometry was written on the basis of sound educational principles that are clearly spelled out in the first chapter ("One Book and Three Metaphors"). In practice, this means that each of the book's important geometric themes are introduced in such a way that anyone with a knowledge of high school mathematics can gain at least a basic understanding of them. It also means that there is no fragmentation of the book's contents, because each of the different geometries is shown to be interconnected. However, although Irving Adler states that this book is not a history of geometry, the all-pervading historical observations add to the coherence of the narrative.

Fortunately, nothing about this book has changed since it first appeared and, in this respect, one should apply the dictum "If it ain't broke, don't fix it." What has changed, however, is the world around it. That is to say, compared to the late 1960s, there are now far more books on particular geometrical themes that have been written for school and university usage. For example, there are numerous texts on non-Euclidean geometry for use at the lower undergraduate level, and differential geometry has long been a standard subject at later stages. Other publications deal with the foundations of geometry, and thereby expose the logical flaws in Euclid. However, at a very accessible level, Irving Adler provides an introduction to—and an overview—of all such aspects of geometry. And yet, despite the clarity of the exposition, and the carefully considered presentation of its contents, there is no hint of oversimplification. As such, this book constitutes a serious challenge to the reader.

It is obvious that Irving Adler has a deep understanding of geometry and its history. Moreover, it is clear that, in his day, he was a first-rate teacher, and I would imagine that he was able to teach across a wide range of abilities. Consistent with this belief is the fact that he is the author of fifty-six books (some under the pen name Robert Irving). These are about mathematics, science, and education; but he is also the co-author of thirty others, for both children and adults. His books have been published in thirty-one countries in nineteen different languages.

PETER RUANE
February 2012

A NEW LOOK AT
GEOMETRY

(See page 288)

1

One Book and Three Metaphors

The purposes of this book may be expressed in terms of three metaphors.

The Gem

From the moment he is born a child begins to explore his environment by seeing and hearing it, by moving around in it, and by touching and manipulating things. Out of these explorations there crystallize his first primitive notions of space and the existence of objects in space. Through his daily experience he acquires conceptions of size and shape and distance. When he goes to school he refines these conceptions by absorbing into his experience some of the ideas about space created by inventive minds in the past. At first he studies geometry informally, with emphasis on mensuration. Later, in high school, he gets a brief introduction to Euclidean geometry as a deductive mathematical system. If he goes on to college he learns of new aspects of geometry presented under the titles "Analytic Geometry" and "Calculus." He may hear hints, too, of other mysterious divisions of geometry that only the specialist penetrates: non-Euclidean Geometry, Projective Geometry, Topology, and others. If he is one of the lucky specialists, he discovers that geometry is a many-faceted gem cut and polished from the raw material of our daily experience. One purpose of this book is to permit the reader to share the pleasure of the specialist as he turns this gem in the light and catches the brilliant flashes of color reflected from its facets.

The Valley and the Mountain

Geometry today consists of many subdivisions. There are synthetic geometry, analytic geometry, and differential geometry. There are Euclidean geometry, hyperbolic geometry, and elliptic geometry. There are also metric geometry, affine geometry, projective geometry, and other branches besides. The subdivisions of geometry have been compared to the distinguishable regions within a complex landscape. Most of these regions are in a valley. An explorer who is deep within one region can easily lose sight of the fact that the other regions exist. At a boundary where one region touches another he can see the fact that the regions are related to each other. But seeing the regions pair by pair does not suffice to reveal the pattern of this relationship. There is a path from the valley that leads up the side of a mountain to a clearing at the top. The explorer who reaches this clearing suddenly sees the whole valley laid out before his eyes. From his height at the top of the mountain he can see all the regions of the valley and the pattern that they form. A second purpose of this book is to lead the reader from region to region in the valley, where he can savor the special beauties for which each is famous, and then to take him up to the top of the mountain where he can see the grand design of the valley in all its breathtaking splendor.

The Motion Picture Film

A motion picture theatre tries to interest the passerby in the film that is being shown by putting on display selected still photographs from the film. The passerby, looking at these "stills," sees people in frozen attitudes of action. However, he knows that each of these pictures is but one of many frames on the film; that these frames form a time sequence; and that if he enters the theatre to see the pictures flashed on the screen in quick succession he will see the action and movement by which the story of the film unfolds. The many subdivisions of geometry are like the still photographs of a motion picture film. If we view them in sequence, they, too, tell a story, the story of the evolution of geometry through five thousand years of history. A third purpose of this book is to show the reader the motion picture as well as the "stills," so that he may see the exciting story of geometry evolving.

Neither Fish Nor Fowl

The form of the book, determined by its threefold purpose, is a compromise between exposition and narration. There is much geometry in the book, but the book is not a textbook of geometry. The sequence in which ideas are developed is approximately chronological, but the book is not a history of geometry. Frequently, when we encounter an idea in an ancient setting, we shall view it with hindsight from the modern point of view, in order to see the full range of its implications. The book is organized around a few basic themes: 1) The relationship between physical space and mathematical space, and our changing conceptions of each. 2) The relationship between algebra and geometry, and how this relationship has changed in the course of time. 3) The story of how three separate streams of thought, the theory of parallels, the theory of curved surfaces, and the geometry of position converged to form one integrated whole. 4) The crystallization of the ideas which will permit us to answer the questions, "What is a space?" and "What is a geometry?"

2

Geometry Before Euclid

The Measurement of Physical Objects

Geometry in its earliest form, as developed in ancient Babylonia and Egypt, was concerned with the measurement of physical objects. It dealt with such practical problems as finding the length of a piece of cloth, the area of a field, or the volume of a basket.

There are three basic steps that are involved in making a measurement: 1) selection of a unit; 2) repetition of the unit; 3) counting the number of times that the unit is repeated. For example, to measure the area of a floor, we may choose a particular square tile as unit, and we count the number of such tiles that must be put side by side in order to cover the floor. The first significant results in geometry were short-cuts for carrying out the third step. For example, if a floor is covered by 5 rows of tiles and there are 3 tiles in each row, it is not necessary to count the tiles one by one to find the area of the floor. It suffices to multiply the numbers 3 and 5. This fact was already known to the priests of ancient Babylonia over five thousand years ago. Though they had no algebraic symbolism with which to express it, they were familiar with the formula for the area of a rectangle, $A = hb$, where h is the length of the height of the rectangle, and b is the length of its base. The Babylonians also knew the analogous formula for the volume of a prism or cylinder, $V = hB$, where h is the length of the height of the solid, and B is the area of its base.

The Use of Averages

The Babylonians tried to derive from the formula $A = hb$ a more general rule for computing the area of a quadrilateral from the lengths of its four sides. We have no record of the reasoning that they used, but it has been surmised by some historians that it may have taken the following form: "To find the area of a

12

quadrilateral, first replace it by a rectangle with approximately the same area. If the pairs of opposite sides of the quadrilateral have lengths a, a', and b, b' respectively, use the average of a

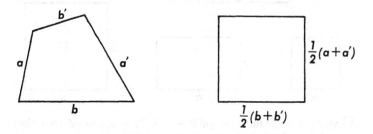

and a' as the height of the rectangle, and use the average of b and b' as the base of the rectangle. Then, using the formula for the area of a rectangle, we get $A = [\frac{1}{2}(a + a')][\frac{1}{2}(b + b')]$, or $A = \frac{1}{4}(a + a')(b + b')$." In any case, the latter formula is the one that was used in Babylonia about 3000 B.C. Unfortunately it gives a correct result only when the quadrilateral is a rectangle.

Similar reasoning may have been used to derive the Babylonian formula for the volume of a basket whose height is h and whose upper and lower bases have areas B and B' respectively. If the volume of the basket is assumed to be equal to that of a cylinder

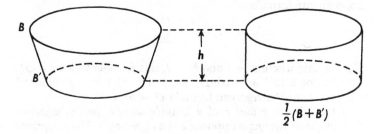

of the same height whose base has an area equal to the average of B and B', then we get the Babylonian formula for the volume of the basket, $V = h[\frac{1}{2}(B + B')] = \frac{1}{2}h(B + B')$. Unfortunately, this formula is correct only if $B = B'$.

The Egyptians who lived one thousand years later were more ingenious and more successful in using the averaging principle. To compute the volume of a frustum of a square pyramid whose height is h and whose bases have edges a and b respectively, they used the correct formula $V = \frac{1}{3}h(a^2 + ab + b^2)$ instead of the

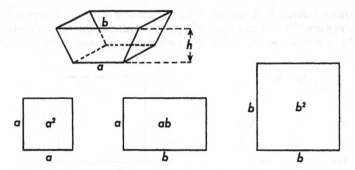

Babylonian formula $V = \frac{1}{2}h(B + B')$. It is surmised that they derived their formula by equating the frustum to a prism whose height is h and whose base area is the average of three areas, namely a^2, b^2, and ab. The third of these areas, ab, is itself a kind of average between a^2 and b^2, known as their *geometric mean*. It is the area of a rectangle whose height is a and whose base is b.

The Sides of a Triangle

The Babylonians and the Egyptians who followed them were aware of the fact that the length of the hypotenuse of a right triangle is related to the lengths of the other two sides. The Babylonians computed the hypotenuse c by means of the approximate formula

$$c = a + \frac{b^2}{2a}.$$

For the case where $a = 4$ and $b = 3$, this formula yields $c = 5\frac{1}{8}$. This is not a bad approximation to the value $c = 5$ obtained from the later Pythagorean formula $c^2 = a^2 + b^2$.

If the sides a, b and c of a triangle satisfy the Pythagorean formula, then the angle opposite c is a right angle. The Egyptians were aware of at least a special case of this rule, and used it in their technique for constructing a right angle. The surveyors of that time, known as "rope-stretchers," laid out a right angle with

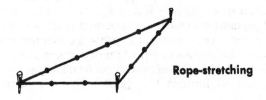

Rope-stretching

the help of a rope that was divided into equal segments by a series of knots. They used the rope to form a triangle whose sides had the ratio 3 to 4 to 5. The angle opposite the longest side was the sought-for right angle.

The Concept of Physical Space

When the Babylonians or Egyptians computed a volume, it was always the volume of a particular physical object, such as a basketful of grain or a block of stone. Long experience with such computations led to the emergence of a new idea, that the volume was not a property of the grain as such, or of the stone as such, but of the space occupied by the grain or stone. This space may be thought of as a container that may be filled with any substance whatever. The same space that contains a basketful of grain may be filled with a basketful of sand instead. The same space that is occupied by a block of stone may be occupied by a block of wood instead. On the basis of this idea, geometry becomes the study of the *space* occupied by physical objects, rather than a study of the particular physical objects themselves. This space, an abstraction from our experience with physical objects, is an aspect of the physical world in which we live, so we refer to it as *physical space*.

The first person to formulate the abstract idea of physical space was the Greek merchant, mathematician and philosopher Thales of Miletus (640–546 B.C.). During his active days as a merchant, Thales had often traveled to Egypt. There he learned the geometry of the Egyptians, and brought it back to Greece. When he retired he devoted his time to studying and teaching mathematics and philosophy. Thales knew enough about astronomy to be able to predict the solar eclipse of 585 B.C.

Deductive Proof

Geometry as the study of physical space is a physical science. Therefore there are two distinct ways in which its propositions can be proved. They can be proved empirically, by experiments in which particular configurations are observed and measured. Or they may be proved by showing that they are logical consequences of other propositions that have already been proved. Thales is credited with the first significant use of the latter type of proof. Among the propositions that were discovered and proved by Thales are these:

The base angles of an isosceles triangle are equal.

An angle inscribed in a semicircle is a right angle.
A diameter of a circle bisects the circle.

Numbers and Space

In the measurement of lengths, areas or volumes, numbers are
intimately related to space. In this sense we may say that, from
its inception, geometry, the study of space, was fused with
arithmetic and algebra, the study of numbers. This fusion was
first intensified and then undermined by the work of Pythagoras
and his followers.

Pythagoras (about 548 B.C. to 495 B.C.) was a pupil of Thales.
Like his master, he traveled in Egypt and absorbed the geometric
knowledge of the Egyptian priests. In 529 B.C. he settled in
Crotona, in southern Italy, where he became a celebrated teacher
of mathematics and philosophy. His pupils organized a religious
society known as the Order of the Pythagoreans. In the course
of their scientific-philosophic-religious speculations, the Pythagoreans made many important discoveries in geometry. They
followed the custom of attributing all these discoveries to
Pythagoras himself. Consequently we cannot be sure which of
these discoveries were made by Pythagoras and which were made
by his pupils.

In the thinking of Pythagoras, the fusion of numbers and space
took on a peculiar one-sided form. To Pythagoras, whole numbers
were not merely an aspect of space. They were the essence of
space and of all things in the universe. The number *one*, from
which all other whole numbers could be generated by addition,
was the essence of divinity. The even numbers *two*, *four*, *six*,
eight, etc., contained the female principle. The odd numbers,
three, *five*, *seven*, etc., contained the male principle. *Five* stood
for marriage, since it was the result of uniting the first even with
the first odd number. *Six* was the cause of cold, *seven* of health,
and *eight* of love.

Somewhat less mystical, and more meaningful mathematically,
was the linking of numbers with shapes. To Pythagoras, the
numbers *one*, *three*, *six*, *ten*, and so on, obtained by taking the
sums 1, $1 + 2$, $1 + 2 + 3$, $1 + 2 + 3 + 4$, etc., are *triangular
numbers*, and the numbers *one*, *four*, *nine*, and so on, obtained
by taking the products 1×1, 2×2, 3×3, etc., are *square
numbers*. This linking of numbers and shape helps to reveal many
interesting number relationships.*

---

* For details, see Chapter III of *Magic House of Numbers*, by the same
author, The John Day Company, New York, 1957.

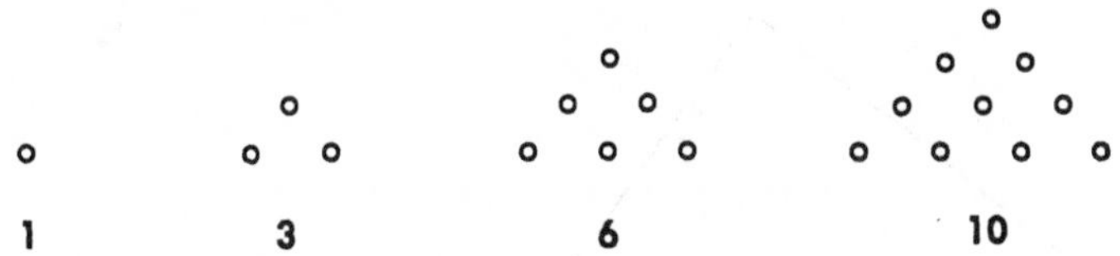

Triangular numbers

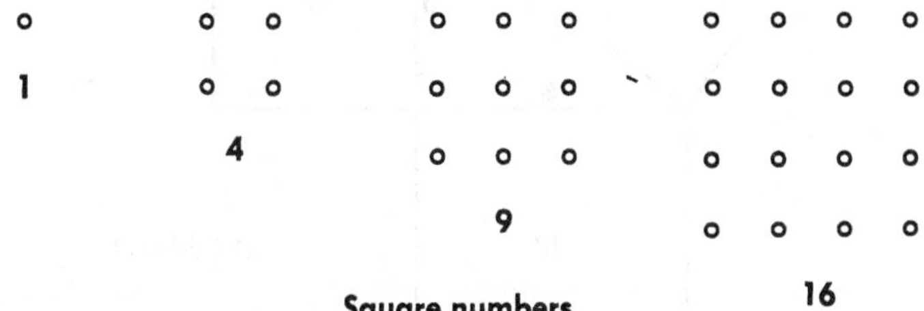

Square numbers

The Pythagoreans made many important discoveries in geometry. We shall examine now a few of the discoveries that play significant roles in the later evolution of geometric ideas.

The Pythagorean Theorem

The name "Pythagorean Theorem" is reserved for the rule that describes how the lengths of the sides of a right triangle are related. In contemporary textbooks of geometry the rule is usually given as, "The square of the hypotenuse of a right triangle is equal to the sum of the squares of the legs." If the legs of the right triangle have lengths a and b respectively, and the hypotenuse has length c, then the rule is expressed by the equation $a^2 + b^2 = c^2$. The rule did not take this algebraic form for Pythagoras, because algebra had not yet been invented. For Pythagoras the rule was a purely geometric one concerning the areas of squares, and was expressed in this form: "The square on the hypotenuse of a right triangle is equal to the sum of the squares on the legs." The squares referred to are shown in the next diagram. There are many ways of proving this theorem. In the book, *The Pythagorean Proposition*, by Elisha S. Loomis, 256 different proofs are given! Diagrams I, II and III contain the essence of one proof that goes directly to the heart of the geometric content of the theorem. In diagram I we see a right

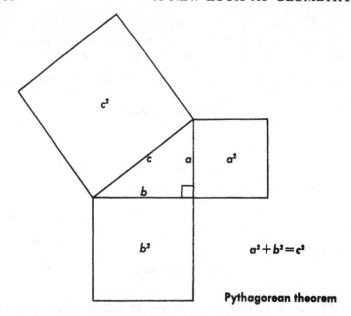

$$a^2 + b^2 = c^2$$

Pythagorean theorem

triangle and the square on each leg. In diagram II, three more triangles have been adjoined to the figure to fill out a large square whose side has length $a + b$. If we denote by T the area of the triangle, then the area of this square, as shown in diagram

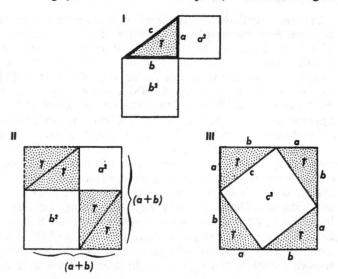

II, is $a^2 + b^2 + 4T$. In diagram III we see another way of dissecting this large square into parts. A triangle congruent to the original one is cut off at each corner of the square. The remaining piece is clearly a square whose side has length c. The area of the large square, as shown in diagram III, is $c^2 + 4T$. Equating these two expressions for the same area, we get the equation $a^2 + b^2 + 4T = c^2 + 4T$. If we remove the four triangles in each of the diagrams II and III, the areas that remain are equal. That is, $a^2 + b^2 = c^2$.

Space-filling Figures

The Pythagoreans initiated the theory of space-filling figures. The main problem of this theory is to find figures that can be repeated, like tiles on a floor, to fill out a plane. Three simple solutions to the problem are shown below. In the first one, the figure that is repeated is an equilateral triangle; in the second one, it is a square; in the third one, it is a regular hexagon. In all three solutions, the figure that is repeated is a regular polygon. It is interesting that these are the only solutions to the problem

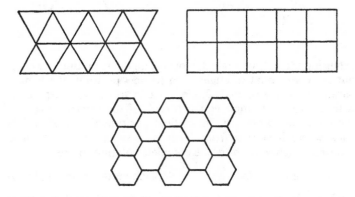

in which the figure that is repeated is a regular polygon. To be able to prove this assertion, let us first review quickly some elementary facts about regular polygons.

A regular polygon is one that has equal sides and equal angles. To draw a regular polygon, it suffices to divide a circle into three or more equal parts and join the successive points of division. Therefore a regular polygon can be drawn with n sides for any integral value of n greater than or equal to 3. Let us call a

regular polygon with *n* sides a regular *n*-gon. It is obvious that
a regular *n*-gon has *n* angles.

It is easy to calculate the number of degrees in each angle of a
regular *n*-gon by first calculating the number of degrees in each
exterior angle formed by extending one side. Let us denote by *x*
the number of degrees in the exterior angle. To calculate *x* we
take one exterior angle at each vertex of the *n*-gon, as shown in
the diagram below, and then add them up. To add the angles,
we use a hand of a clock in this way: Start with the hand placed
parallel to the horizontal side of angle 1. Rotate the hand
counterclockwise until it has swept out an angle equal to angle 1.
The hand will end up parallel to the other side of angle 1, whose

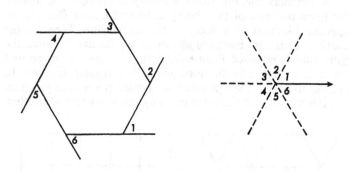

extension is a side of angle 2. Now rotate the hand until it sweeps
out an angle equal to angle 2. In its new position, the hand is
parallel to a side of angle 3. Continue in this way, sweeping out
with the hand an angle equal to each of the exterior angles of
the *n*-gon in succession. In its final position the hand will be
back where it started from. That is, *n* successive rotations of
x degrees each add up to one complete rotation, or 360 degrees.
Then, since $nx = 360$, $x = \dfrac{360}{n}$. At each vertex of the *n*-gon, an
angle of the *n*-gon and the exterior angle next to it add up to
180 degrees. Consequently the number of degrees in each angle
of a regular *n*-gon is $180 - \dfrac{360}{n}$, or $180\left(1 - \dfrac{2}{n}\right)$.

Now we are ready to consider the problem of finding all the
values of *n* for which repetitions of a regular *n*-gon can fill out a
plane. Let *p* be the number of *n*-gons that occur at each vertex.
Then there will be *p* angles at each vertex, and since they fill out
the space in the plane around the vertex, their sum is 360 degrees.

Consequently the number of degrees in each of them is $\frac{360}{p}$. But we know already that the number of degrees in each angle of a regular n-gon is $180\left(1 - \frac{2}{n}\right)$. Equating these two expressions, and dividing by 180, we get the equation;

(1) $$1 - \frac{2}{n} = \frac{2}{p}, \text{ or}$$

(2) $$1 - \frac{2}{n} - \frac{2}{p} = 0.$$

If we multiply equation (2) by np, we get

(3) $$np - 2p - 2n = 0.$$

If we add 4 to both sides of equation (3) we get

(4) $$np - 2p - 2n + 4 = 4.$$

Factoring the left-hand side of equation (4) we get

(5) $$(n - 2)(p - 2) = 4.$$

Since $n - 2$ and $p - 2$ are whole numbers, and their product is 4, we get all possible values of n and p by equating the pair $n - 2$, $p - 2$ to all possible pairs of whole numbers whose product is 4. These possible pairs are $4, 1; 2, 2;$ and $1, 4$. Consequently there are only three solutions, given by the three pairs of equations:

$$\begin{cases} n - 2 = 4 \\ p - 2 = 1 \end{cases} \qquad \begin{cases} n - 2 = 2 \\ p - 2 = 2 \end{cases} \qquad \begin{cases} n - 2 = 1 \\ p - 2 = 4 \end{cases}$$

The first pair yields the solution $n = 6, p = 3$. The second pair yields the solution $n = 4, p = 4$. The third pair yields the solution $n = 3, p = 6$. Therefore there are only three ways of filling a plane by repeating a regular n-gon: use equal regular 6-gons (regular hexagons) with three at each vertex, or equal regular 4-gons (squares) with four at each vertex, or equal regular 3-gons (equilateral triangles) with six at each vertex.

The Regular Solids

The three-dimensional analogue of a regular polygon is a regular solid. A regular solid is a polyhedron whose faces are congruent regular polygons arranged so that the same number of faces occurs at each vertex. The ancient Egyptians knew of three regular solids: the regular tetrahedron, in which there are four faces, each of which is an equilateral triangle, and there are three

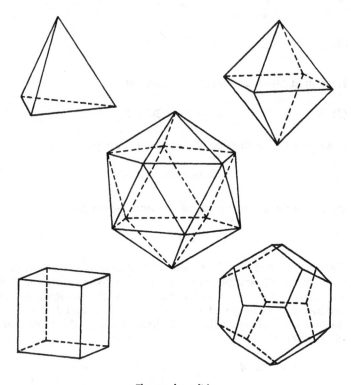

The regular solids

faces at each vertex; the regular hexahedron (cube), in which there are six faces, each of which is a square, and there are three faces at each vertex; and the regular octahedron, in which there are eight faces, each of which is an equilateral triangle, and there are four faces at each vertex. The Pythagoreans discovered two others: the regular dodecahedron, in which there are twelve faces, each of which is a regular pentagon, and there are three faces at each vertex; and the regular icosahedron, in which there are twenty faces, each of which is an equilateral triangle, and there are five faces at each vertex. A pattern for making each of these five regular solids is shown below. To make one of the solids, first draw a network of regular polygons as shown in the pattern. Cut along the outside edges only, and fold along the other edges to form the polyhedron. Join adjacent faces with tape where necessary.

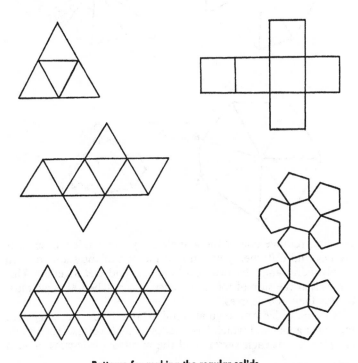

Patterns for making the regular solids

The five regular solids described above are the only ones that are possible. To prove this fact, observe first that the sum of the face angles at a vertex of a polyhedron cannot be arbitrarily large. To see this fact intuitively, place a dot on a piece of paper, and draw lines radiating from the dot to divide the 360 degrees around the dot into adjacent angles. (In the diagram below the dot is labeled O, and the lines radiate from O.) Make a crease along each line, and then try to fold the paper into a polyhedral angle with edges along these creases. You will find that you cannot do so, because the angles will not fold out of the plane of the paper unless at least one of them buckles. However, if

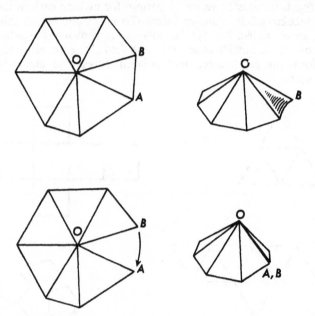

you first remove one of these angles, say, the angle whose sides are OA and OB, then you can fold the rest of the sheet to form a polyhedral angle by bringing lines OA and OB together. This shows that the sum of the face angles of a polyhedral angle must be less than 360 degrees.

Now consider any regular solid whose faces are congruent regular n-gons, and which has p faces at each vertex. Then there are p angles at each vertex, and the number of degrees in each angle is $180\left(1 - \dfrac{2}{n}\right)$. Consequently,

$$180\left(1 - \frac{2}{n}\right)p < 360.$$

Dividing both sides by $180p$, we get the inequality

(6) $$1 - \frac{2}{n} < \frac{2}{p}.$$

Then, by the same sequence of steps used to transform equation (1) into equation (5) on page 21, we can transform (6) into the equivalent inequality

(7) $$(n - 2)(p - 2) < 4.$$

This inequality shows us that $n - 2$ and $p - 2$, which we know to be positive integers, have a product that is less than 4. The only pairs of positive integers that have this property are $(1, 1)$; $(2, 1)$; $(1, 2)$; $(3, 1)$; and $(1, 3)$. Therefore n and p must satisfy one of the following five pairs of equations:

$$\text{I. } \begin{cases} n - 2 = 1 \\ p - 2 = 1 \end{cases} \qquad \text{II. } \begin{cases} n - 2 = 2 \\ p - 2 = 1 \end{cases} \qquad \text{III. } \begin{cases} n - 2 = 1 \\ p - 2 = 2 \end{cases}$$

$$\text{IV. } \begin{cases} n - 2 = 3 \\ p - 2 = 1 \end{cases} \qquad \text{V. } \begin{cases} n - 2 = 1 \\ p - 2 = 3 \end{cases}$$

The first pair yields the solution $n = 3, p = 3$, which corresponds to the regular tetrahedron, whose faces are regular 3-gons (equilateral triangles), and which has 3 faces at each vertex. The second pair yields the solution $n = 4, p = 3$, which corresponds to the cube, whose faces are regular 4-gons (squares), and which has 3 faces at each vertex. The third pair yields the regular octahedron, with $n = 3$, $p = 4$. The fourth pair yields the regular dodecahedron, with $n = 5, p = 3$. The fifth pair yields the regular icosahedron with $n = 3$, $p = 5$. Consequently the five regular solids that were known to the Pythagoreans are the only ones that are possible.

The Stuff of the Elements

The five regular solids played a key role in the cosmology of Pythagoras. The universe, according to Pythagoras, is completely enclosed in the heavenly sphere, and consists of fire, air, water, and earth. Fire, he taught, is composed of units, each of which is a regular tetrahedron; air is composed of octahedra; water is composed of icosahedra; and earth is composed of cubes. The heavenly sphere is represented in the Pythagorean doctrine by a regular dodecahedron inscribed in it. The dodecahedron, discovered by Pythagoras, and used as a symbol of the heavenly sphere, was the prime secret of the Order of the Pythagoreans. Hippasus, a member of the Order, violated his oath of membership by revealing this secret to outsiders. When he died in a

shipwreck, the Pythagoreans were sure that his death was divine
punishment for his violation of his oath.

Euler's Formula

Imagine a regular polyhedron whose faces are made of rubber
sheets. By stretching each face, we can deform the polyhedron
into a sphere. Any polyhedron that can be deformed into a
sphere in this way is called a *simple* polyhedron. There is a
formula, discovered by Leonhard Euler (1707–1783), that relates
to each other the number of vertices, the number of faces, and
the number of edges of a simple polyhedron. To derive this
formula, imagine a simple polyhedron deformed into a sphere,
so that each face becomes a region of the sphere, bounded by a

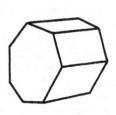 

spherical polygon, and each edge becomes a boundary line on
the sphere between two adjacent regions. We can obtain the same
network of edges on the sphere by first drawing one polygon,
and then adding, one at a time, additional edges. At each stage
of the construction let us find the value of $V - E + F$, where V

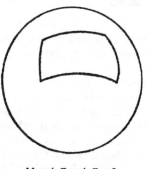

$$V = 4, E = 4, F = 2$$
$$V - E + F = 2$$

is the number of vertices, E is the number of edges, and F is the number of separate regions on the sphere. At the beginning, when we have only a polygon drawn on the sphere, $V = E$, and $F = 2$. Therefore the initial value of $V - E + F$ is 2. When we add new edges to the network, we can do so in two ways: 1) We can put in a new vertex not already present, and draw an edge that joins it to an old vertex, or 2) we can draw an edge that joins two vertices that are already there. (See diagrams I and II below.) In the first case, V is increased by 1, E is increased by 1, and F is unchanged. As a result, $V - E + F$ remains unchanged. In the second case, V is unchanged, E is increased by 1, and F is increased by 1. In this case, too, $V - E + F$ is unchanged. Consequently, when we finally obtain the network of edges

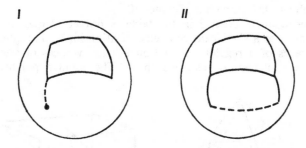

formed by the simple polyhedron, the value of $V - E + F$ is the same as its initial value 2. But F now is the number of faces of the simple polyhedron. This proves that in a simple polyhedron the number of vertices V, the number of edges E, and the number of faces F are related by the formula

(8) $$V - E + F = 2.$$

The values of V, E, and F for each of the regular polyhedra are shown in the table below.

Regular Polyhedron	V	E	F
tetrahedron	4	6	4
hexahedron	8	12	6
octahedron	6	12	8
dodecahedron	20	30	12
icosahedron	12	30	20

This table reveals an interesting symmetry relationship that connects each regular polyhedron with another. The fifth line of the table indicates that the following statement is true: *There is a regular polyhedron that has 12 vertices, 30 edges, and 20 faces.* If we interchange the words *vertices* and *faces* in this statement, we get another statement called its *dual: There is a regular polyhedron that has 12 faces, 30 edges, and 20 vertices.* This statement is also true because it summarizes the information in the fourth line of the table. The corresponding statements that express the information given in the second and the third lines are related to each other in the same way, that is, each is the dual of the other. The corresponding statement that expresses the information given in the first line is its own dual. What we have observed here is a foretaste of a more general symmetry known as *duality* which we shall discuss in Chapter 10.

The duality revealed in the table is not accidental. It corresponds to a significant geometric relationship: The centers of the faces of a regular polyhedron are the vertices of another regular polyhedron inscribed in it. The regular polyhedron

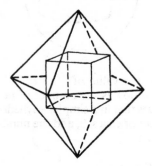 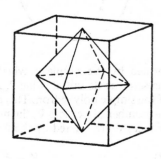

inscribed in another is called its *dual*. The dual of the regular icosahedron is the regular dodecahedron, and vice versa. The dual of the regular octahedron is the regular hexahedron, and vice versa. The regular tetrahedron is its own dual.

The Angles of a Triangle

According to Proclus (410–485), the Pythagoreans discovered the theorem, "The sum of the angles of a triangle is two right angles." The proof they gave for this theorem is the familiar one commonly found in contemporary geometry textbooks. Since we shall have occasion to discuss this theorem and its proof later, we reproduce the proof here:

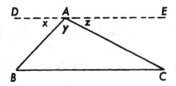

Given: *triangle ABC*
Prove: $\angle B + \angle A + \angle C = 2$ *right angles*

Proof: Through A draw DE parallel to BC. $\angle x + \angle y + \angle z = 2$ *right angles*. When two parallel lines are cut by a transversal, the alternate interior angles are equal. Hence $\angle x = \angle B$, and $\angle z = \angle C$. Moreover, $\angle y = \angle A$. Therefore $\angle B + \angle A + \angle C = 2$ *right angles*.

The Pythagorean Crisis

It was the Pythagorean view that underlying all spatial relations were whole numbers or at least the ratios of whole numbers. This view was undermined by Pythagoras himself when he discovered the *irrational*. The discovery arose in connection with the problem of comparing the length of one line with the length of another.

Suppose when we compare a short line r with a longer line s we find that r fits exactly a whole number of times into s. Then we say that r is a *measure* of s, and s is a multiple of r. For

$$s = 3r \qquad \frac{s}{r} = \frac{3}{1}$$

example, if r fits into s three times, $s = 3r$, and $s/r = 3/1$. Moreover, if r is chosen as the unit of length, then the length of s is $3 = 3/1$. If r does not fit exactly a whole number of times into s, it may be possible to find a smaller length t that fits a whole number of times into both r and s. In that case we say that t is a *common measure* of r and s. Then the ratio s/r may still be expressed as a ratio of whole numbers. For example, suppose t fits twice into r, and it fits seven times into s. Then

$r = 2t$, $s = 7t$, and $s/r = 7/2$. If r is the unit of length, then the length of s is $7/2$.

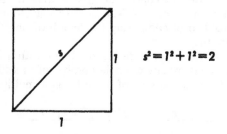

$$r = 2t \qquad s = 7t \qquad \frac{s}{r} = \frac{7}{2}$$

The Pythagoreans believed that every pair (r, s) of lengths was like one of the two pairs used as examples above: either one length was a measure of the other, or the two lengths had a common measure. This belief implied that if r was the unit of length, then the length of s could be expressed as a ratio of whole numbers. This belief was shattered when Pythagoras found a pair of lengths that do not have a common measure, namely, the side and the diagonal of a square.

We present here the proof that the side r and diagonal s of a square do not have a common measure. Two adjacent sides and a diagonal of a square form a right triangle. Let us take the side r of the square as the unit of length. Then, by the Pythagorean Theorem, $s^2 = 1^2 + 1^2 = 2$. If r and s have a

$$s^2 = 1^2 + 1^2 = 2$$

common measure, then the length of s may be written in the form m/n where m and n are whole numbers. We shall see, however, that this is impossible.

Suppose $s = m/n$. Since every fraction can be reduced to lowest terms, let us assume that this has already been done, so that m/n is in lowest terms. Then it is clear that m and n are not both even numbers. (An even number is one that is double another whole number.) If $s = m/n$, then $m^2/n^2 = s^2 = 2$, and $m^2 = 2n^2$. That is, m^2 must be an even number. This implies that m *is an even number*, because if m were odd, m^2 would also be odd. Consequently there exists a whole number k such that $m = 2k$.

Then $m^2 = 4k^2$. But $m^2 = 2n^2$. Equating these two expressions for m^2, we see that $2n^2 = 4k^2$, or $n^2 = 2k^2$. That is, n^2 must be even, and therefore n *must be even*. Notice that while m and n are not both even, we have been compelled to say that they are both even. We were led into this absurdity as a result of assuming that the length s may be expressed as a ratio of whole numbers. Therefore this assumption must be false, and we conclude that the side and diagonal of a square do not have a common measure.

Greek arithmetic was not sufficiently developed to provide a way of expressing by numbers the ratio of two lengths that do not have a common measure. Thus the unity of arithmetic and geometry was broken. Geometry and arithmetic went their separate ways for about two thousand years until an improved arithmetic made it possible to restore the unity of arithmetic and geometry in a new form.

Digression on Arithmetic and Algebra

The number system we use today is not something that sprang full-grown, like Athene, from the brow of Zeus. It is the product of a long process of growth. At first the only numbers known were the counting numbers, 0, 1, 2, 3, and so on. Then fractions were introduced, to make it possible to express, at least approximately, lengths that are not whole number multiples of the unit of length. Irrational numbers were invented to represent lengths that cannot be expressed by fractions. Negative numbers were brought in to permit the subtraction of any two numbers. So-called "imaginary" numbers were created to guarantee that every algebraic equation has a solution.*

To see some features of the number system that have a bearing on geometric concepts that we discuss later, we shall retrace the steps in the growth of the number system. However, we shall depart from the actual historical sequence by introducing negative numbers early, before we introduce fractions. We shall find it convenient and enlightening to use a pictorial representation of numbers, first as points on a line, and finally as points on a plane.

The Rational Number System

We begin by drawing a straight line that extends endlessly to the right and to the left. We choose any point on the line and

---

* For a step-by-step expansion of the number system, see *The New Mathematics,* by the same author, The John Day Company, New York, 1958.

call it 0. We choose a unit of length, and locate to the right of 0 points whose distances from 0 are 1 unit, 2 units, 3 units, etc. We call these points 1, 2, 3, etc. Similarly we locate points that

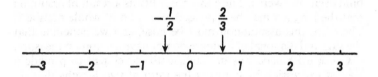

are 1 unit, 2 units, 3 units, and so on, to the left of 0, and call these points -1, -2, -3, etc. Now we divide each unit interval into halves, thirds, fourths, etc., thus introducing more points of division on the line. If m is any counting number, and n is any counting number except 0, we use the labels $\dfrac{m}{n}$ and $\dfrac{-m}{n}$ respectively for the points to the right and left of 0 at a distance of $\dfrac{m}{n}$ units. The points labeled in this way constitute the *rational number system*. Each rational number is represented by many equivalent fractions. Rational numbers may be added and multiplied in accordance with the familiar rules $\dfrac{a}{b} + \dfrac{c}{d} = \dfrac{ad + bc}{bd}$, $\dfrac{a}{b} \cdot \dfrac{c}{d} = \dfrac{ac}{bd}$.

Fields

The operations addition and multiplication in the rational number system have the following properties: If x, y and z are any rational numbers,

1. $x + (y + z) = (x + y) + z$. (Associative Law of Addition)
2. $x + y = y + x$. (Commutative Law of Addition)
3. There exists a member *0* such that, for every x in the system, $x + 0 = 0 + x = x$.
4. For each x, there exists a member denoted by $-x$ that has the property $x + (-x) = (-x) + x = 0$.
5. $x(yz) = (xy)z$. (Associative Law of Multiplication)
6. $xy = yx$. (Commutative Law of Multiplication)
7. There exists a member *1* such that, for every x in the system, $x \cdot 1 = 1 \cdot x = x$.

8. If $x \neq 0$, there exists a member denoted by $\left(\dfrac{1}{x}\right)$ that has the property $x\left(\dfrac{1}{x}\right) = \left(\dfrac{1}{x}\right) x = 1$.

9. $x(y + z) = xy + xz$.   (Distributive Law)

Any system of elements for which two operations, addition and multiplication, are defined that satisfy conditions 1 to 9 is known as a *field*.

The Real Number System

The construction shown below locates a point P to the right of O whose distance from O is the length of the diagonal of a square whose side is 1 unit. As Pythagoras discovered, there is no rational number that represents this length. Therefore the scheme outlined above by which every rational number is used as a label

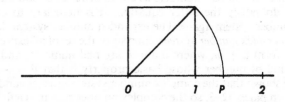

for a point on the line leaves some points, like P, without labels. These points, which are not labeled by any rational number, are known as *irrational points*. It is necessary to expand the number system in order to obtain numbers that can serve as labels for these points. The expanded number system, in which there is a number for every point on the line, and a point for every number, is called the *real* number system. A simple way to carry out the expansion is to define the real number system as the set of all non-terminating decimals, with the understanding that some pairs of non-terminating decimals, like .5000 . . . and .4999 . . . represent the same point. Addition and multiplication of real numbers are defined by means of sums and products of longer and longer terminating decimals. It can be shown that the real number system satisfies conditions 1 to 9 on pages 32–3, so the real number system is a field. All the rational numbers are contained within the real number system in the guise of repeating decimals.*

---

* For details, see Chapter V of *The New Mathematics*.

Every point r in the real number system divides the rational numbers into two parts A and B, where A includes all the rational points that are to the left of r, and B includes all the remaining rational points. Then B consists of all the rational points that are to the right of r, as well as r itself if r is rational. Every point in A is to the left of every point of B. If x is in A, then $x < r$ (where $<$ means "less than", or "to the left of"). If x is in B, then $x > r$, (where $>$ means "greater than" or "to right of"), or $x = r$. The last case can occur only if r itself is a rational number. Such a division of the rational numbers into two parts is called a *Dedekind cut*, after Richard Dedekind (1831–1916) who used these cuts for introducing an alternate way of constructing the real number system.

The Complex Number System

The equation $x^2 - 1 = 0$ is satisfied by two real numbers, namely, 1 and -1. On the other hand, the equation $x^2 + 1 = 0$ is satisfied by no real numbers at all. In order to obtain numbers which do satisfy the latter equation, it is necessary to expand the number system again. The expanded number system, known as the *complex number system*, consists of the set of all expressions of the form $a + bi$, where a and b are real numbers, and i is a new kind of number whose basic property is that $i^2 = -1$.* It can be shown that the complex number system satisfies conditions 1 to 9 on pages 32–3, so the complex number system is a field. All the real numbers are contained in the complex number system in the form $a + 0i$. That is, the real number a occurs in the complex number system as the complex number $a + 0i$.

To obtain a pictorial representation of the complex number system, we use the set of all points in a plane. First draw a horizontal line in the plane, and associate the real number system with the points on the line, as we have already done. We call this line the *axis of real numbers*. To associate a point of the plane with every complex number $a + bi$, we use this procedure: At the point a on the axis of real numbers draw the line that is perpendicular to the axis of real numbers. Associate a real number scale with the points on this vertical line in such a way that the unit of length on it is the same as on the axis of real numbers, and the positive half of the scale is above the axis of real numbers. Locate on the vertical line the point whose scale number on this line is b. We use this point to represent the

---

* For a systematic construction of the complex number system, see Chapter VIII of *The New Mathematics*.

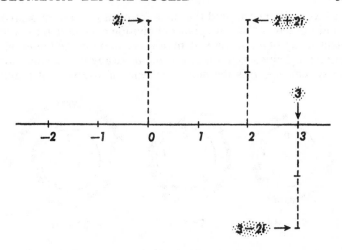

complex number $a + bi$. In the diagram above we show the points that represent the numbers 3, $2i$, $2 + 2i$, and $3 - 2i$

Finite Fields

The rational number system, the real number system, and the complex number system are all fields. Each of these three fields contains infinitely many members. It is an interesting fact, first discovered by Evariste Galois (1811–1832) that there are also fields that contain only finitely many members. These are referred to as *finite fields*. To construct a simple example of a finite field, follow this procedure. Draw a circle and divide it into five equal

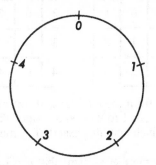

arcs. Take one of these arcs as unit of arc length on the circle. Number the points of division 0, 1, 2, 3, and 4, as shown in the diagram. Then the number associated with a point indicates its clockwise distance around the circle from the *0* point, measured in units of length. The five points of division constitute the entire membership of a number system in which addition and multiplication are defined as follows: If *a* and *b* are any two members of the system, $a + b$ is the point you reach if you start at *0* and

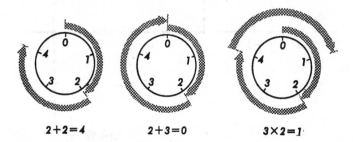

| 2+2=4 | 2+3=0 | 3×2=1 |

then move clockwise around the circle first *a* units and then *b* units of arc length; and *ab* is the point you reach if you start at *0* and move clockwise *a* times in succession a distance of *b* units of arc length. All possible sums and products obtained in this way are recorded in the addition and multiplication tables shown below:

+	0	1	2	3	4
0	0	1	2	3	4
1	1	2	3	4	0
2	2	3	4	0	1
3	3	4	0	1	2
4	4	0	1	2	3

×	0	1	2	3	4
0	0	0	0	0	0
1	0	1	2	3	4
2	0	2	4	1	3
3	0	3	1	4	2
4	0	4	3	2	1

It is easy to verify from these tables that this number system satisfies all the conditions 1 to 9 listed on pages 32–3, and hence is a field. This field, which contains exactly five members, is denoted by the symbol F_5.

There are many number systems that can be constructed in a similar manner. If *n* is any positive integer greater than or equal to 2, divide a circle into *n* equal arcs. Number the points of division *0, 1, . . . , n − 1*, and define addition and multiplication as we have done in F_5. The number systems obtained in this way are called *modular number systems* (popularly known as *clockface number systems*). Every modular number system satisfies conditions 1 to 7 and condition 9 of pages 32–3. Condition 8 is also satisfied if and only if n is a prime number. In that case, and only in that case, the modular number system is a field. If p is a prime whole number, we denote by F_p the field obtained in this way that contains exactly p members.

It is possible that part of a field may itself be a field. For example, the rational number system is part of the field of real numbers, and the rational number system is itself a field. It is shown in the theory of fields that among all the fields that lie within a given field there is a smallest field that is contained in every one of them. Moreover, this smallest field is either the rational number system or one of the fields F_p. These smallest possible fields are called *prime fields*. The prime field contained in a finite field cannot be the rational number system, since the rational number system contains infinitely many members. Therefore the prime field in any finite field must be one of the fields F_p. It is proved in the theory of fields that if the prime field contained in a finite field is F_p, then the number of members in that finite field is p^n, where n is some positive integer. Moreover, for every prime number p and every positive integer n there is a finite field with exactly p^n members, and it contains the prime field F_p. Thus, there are finite fields with 2 members, 4 members, 8 members, etc., and each of these finite fields contains the prime field F_2; there are finite fields with 3 members, 9 members, 27 members, etc., and each of these finite fields contains the prime field F_3; there are finite fields with 5 members, 25 members, 125 members, etc., and each of these finite fields contains the prime field F_5; and so on.

We shall encounter these finite fields again in Chapter 10.

The Concept of a Group

A field is a set of objects for which two operations, + (plus) and · (times), are defined, satisfying conditions 1 to 9 of pages 32–3. If we consider only one of these operations at a time, we shall observe a somewhat simpler algebraic structure that plays a very important part in the later development of geometric ideas.

Let us consider first those properties of a field F that relate to the operation + (plus) alone. If x and y are in F there is a member of F denoted by $x + y$, and the operation + in F

satisfies conditions 1 to 4 of page 32. We repeat these conditions
here for convenience:

1. $x + (y + z) = (x + y) + z$.
2. $x + y = y + x$.
3. There exists a member 0 such that, for every x in the system,
   $x + 0 = 0 + x = x$.
4. For each x there exists a member denoted by $-x$ that has the
   property $x + (-x) = (-x) + x = 0$.

Let us consider next those properties of a field that relate to
the operation · (times) alone. Because the number 0 is an
exception for condition 8, let us exclude 0 from consideration.
Denote by F* the set of all non-zero members of F. If x and y
are in F* there is a member of F* denoted by $x \cdot y$, and the
operation · in F* satisfies conditions 5 to 8 of pages 32–3. We
repeat these conditions with a slight modification of 8 made
possible by the fact that we are now talking about F* instead
of F. (We also show each multiplication sign explicitly.)

5. $x \cdot (y \cdot z) = (x \cdot y) \cdot z$.
6. $x \cdot y = y \cdot x$.
7. There exists a member 1 such that, for every x in the system,
   $x \cdot 1 = 1 \cdot x = x$.
8. For each x there exists a member denoted by $\left(\dfrac{1}{x}\right)$ that has the
   property $x \cdot \left(\dfrac{1}{x}\right) = \left(\dfrac{1}{x}\right) \cdot x = 1$.

A comparison of the properties of $+$ in F with those of · in
F* shows a great similarity between them. Condition 5 is like
condition 1, except that · takes the place of $+$ in it. Similarly,
condition 6 is like condition 2. Condition 7 is like condition 3,
with the number 1 playing the same role with respect to multiplication that the number 0 plays with respect to addition.
Condition 8 is like condition 4, with the number $\left(\dfrac{1}{x}\right)$ related to
multiplication and 1 in the same way that the number $(-x)$ is
related to addition and 0.

To emphasize this similarity let us rewrite properties 1 to 4
of F with a change of notation. Write ∘ (for operation) instead
of $+$. Write e instead of 0, and write (*inverse of x*) instead of $-x$.
Then conditions 1 to 4 take new forms A, B, C, D, listed below.

A. $x \circ (y \circ z) = (x \circ y) \circ z.$
B. $x \circ y = y \circ x.$
C. There exists a member e such that, for every x in the system, $x \circ e = e \circ x = x.$
D. For each x there exists a member denoted by (*inverse of x*) that has the property $x \circ (\textit{inverse of } x) = (\textit{inverse of } x) \circ x = e.$

Now rewrite properties 5 to 8 of F* with a similar change of notation. Write $\circ$ instead of $\cdot$, write e instead of 1, and write (*inverse of x*) instead of $\left(\dfrac{1}{x}\right)$. Then the conditions 5 to 8 take the same form A, B, C and D. A set of objects in which an operation $\circ$ is defined that satisfies conditions A, C and D is called a *group*. If condition B is also satisfied, the group is said to be *commutative*. The member e of the group that has the special property described in condition C is called the *identity element* of the group. On the basis of this definition we can say that the membership of a field F is a group with respect to the operation $+$. The identity element of this group is 0, and every member x of the group has an inverse, denoted by $(-x)$, with respect to the operation $+$. Similarly, we can say that the membership of F* is a group with respect to the operation $\cdot$. The identity element of this group is 1, and every member x of the group has an inverse, denoted by $\left(\dfrac{1}{x}\right)$, with respect to the operation

The algebraic structure called a group is singled out for special attention because it occurs in many places in mathematics. We shall encounter it again on page 57 when we discuss some mathematical ideas related to the concept of motion.

Mathematical Space

The physical space occupied by a physical object is part of the physical world. Consequently, geometry as the study of physical space is really a branch of the science of physics. Geometry became a branch of mathematics only when it was transformed into the study of *mathematical space*, a conceptual structure obtained through the idealization of physical space. This transformation took place in intimate connection with the development of the philosophy of Plato (431 B.C. to 351 B.C.).

Plato believed, as did Heraclitus before him (around 513 B.C.), that the universe we observe with our senses is in a state of perpetual change. Consequently all things in it are transitory.

He was impressed, on the other hand, by the permanence of ideas, such as those expressed by common nouns. The senses, he believed, are deceptive, but we can attain truth that is independent of sense perception by reasoning in the realm of ideas. As he expressed it in the dialogue called *Timaeus*, "That which is apprehended by intelligence and reason is always in the same state; but that which is conceived by opinion with the help of sensation and without reason, is always in a process of becoming and perishing and never really is." He concluded then that there are three different kinds of being. The first is the *idea*, which is a pattern for things in the world of sense perception. It is immutable, uncreated, and indestructible. The second is the *object of sense perception*, a mere imitation of the pattern. It is created, always in motion, and perishable. The third is space; "it is the receptacle, and in a manner the nurse, of all generation." The object is "that which is in process of generation;" space is "that in which the generation takes place;" the idea is "that of which the thing generated is a resemblance." Thus Plato made two abstractions from an object such as a cylindrical block of stone. One of these is the space occupied by the cylindrical block of stone, the "receptacle" in which the block is generated. The other is the cylindrical form of the block. The first of these abstractions is what we have referred to as physical space. The second of these abstractions belongs to mathematical space. In Plato's view, mathematical space, which is the object of study of the geometer, belongs to the realm of ideas, and not to the realm of sense perception. As he expressed it in *The Republic*, although the geometers "use visible figures and argue about them, they are not thinking about these figures but of those things which the figures represent; thus it is the square in itself and the diameter in itself which are the matter of their arguments, not that which they draw; similarly, when they model or draw objects, which may themselves have images in shadows or in water, they use them in turn as images, endeavoring to see those absolute objects which cannot be seen otherwise than by thought."

There is a characteristic method of procedure in the study of mathematical space. Here is how Plato described it in *The Republic*: "I think you know that those who deal with geometrics and calculations and such matters take for granted the odd and the even, figures, three kinds of angles and other things cognate to these in each field of inquiry; assuming these things to be known, they make them hypotheses, and henceforth regard it as unnecessary to give any explanation of them either to themselves or to others, treating them as if they were manifest to all; setting out from these hypotheses, they go at once through the remainder of the argument until they arrive with perfect consistency at the goal to which their inquiry was directed." In modern terminology we say that the mathematician begins with assumptions concerning certain undefined terms, and then draws conclusions from these assumptions by deductive reasoning. This method was developed to a high level by Euclid, in his book, *The Elements*, which we discuss in the next chapter. We shall see in Chapters 3, 7 and 8 how, as the method was refined and perfected, our understanding of the significance of the assumptions in a mathematical system has changed.

The Theory of Ratio

The concept of ratio is easy to define for two magnitudes of the same kind that have a common measure. Suppose, for example, two line segments a and b have a common measure c. That is, there exist positive integers m and n such that $a = mc$, and $b = nc$. Then we say that the segments a and b have the ratio m *to* n. In modern notation, we represent this ratio by the fraction m/n. Using this definition of ratio, there is an obvious and easy way to determine when two pairs of segments have the same ratio: If the ratio of a to b is the fraction m/n, and the ratio of c to d is the fraction p/q, then the ordered pairs (a, b) and (c, d) are in the same ratio if and only if the fractions m/n and p/q are equivalent. (An ordered pair is a pair for which we specify by the order in which we write its members that one of them is first and the other one is second.)

The discovery by Pythagoras that some magnitudes of the same kind do not have a common measure undermined this easy approach to the concept of ratio. It became necessary to redefine the concept of ratio in such a way that it would make sense for any two magnitudes of the same kind, whether or not they have a common measure. This was done brilliantly by Eudoxus (408 B.C. to 355 B.C.), one of Plato's pupils. Here is the essential content of Eudoxus's definition, expressed in modern language and notation:

I. Two magnitudes a and b have a ratio if there exists a positive integer m such that $ma > b$, and there exists a positive integer n such that $nb > a$.

II. If (a, b) and (c, d) are two ordered pairs of magnitudes that have a ratio, then they have the same ratio if for any choice of positive integers m and n,

either 1) $ma > nb$, and $mc > nd$,
or   2) $ma = nb$, and $mc = nd$,
or   3) $ma < nb$, and $mc < nd$.

The theory of ratio presented in Euclid's *Elements* was developed on the basis of this definition. It is worth while for us to pause here for some discussion of the implications of this definition.

The condition stated in part I of the definition serves a three-fold purpose: a) It requires that magnitudes be of the same kind in order to have a ratio. Thus, if a is a length, and b is a length, some multiple of a is greater than b, and vice versa, so a pair of lengths has a ratio. But if a is a length and b is an area, no multiple of a is greater than b, and the ordered pair (a, b) does not have a ratio. b) It excludes from consideration the cases where a or b have zero magnitude. This is equivalent to the rule in arithmetic that we never divide by zero. c) It excludes from consideration systems of ordered magnitudes in which one element may be "infinitesimal" compared to another, that is, systems in which a may be so small that no finite multiple of a is greater than b. An example of such an ordered system is the set of all polynomials $a_n x^n + \cdots + a_1 x + a_0$ with real coefficients, in which the order relation $>$ is defined as follows: To determine which of two polynomials is the greater, first write them both as polynomials of the same degree by supplying some zero coefficients where necessary. Then starting with the highest power of x, and proceeding through descending powers of x, compare the polynomials term by term until you reach the first term in which the polynomials differ. Then that polynomial is greater which has the greater coefficient in this term. For example, to compare $3x^2 + 2x - 1$ with $3x^2 + 5x - 4$, compare in succession the terms $3x^2$ and $3x^2$, then $2x$ and $5x$, then -1 and -4. The first pair in which the coefficients differ is $2x$ and $5x$. Since $5 > 2$, we say that $3x^2 + 5x - 4 > 3x^2 + 2x - 1$. In this system the element x is "infinitesimal" compared to x^2, because no matter how large a positive integer n may be, $nx < x^2$ (that is, $0x^2 + nx < 1x^2$).

The assumption that lengths of line segments satisfy condition I is commonly called the *axiom of Archimedes*. It really should be called the axiom of Eudoxus, since Eudoxus was the first to state it explicitly. An ordered system is called *Archimedean* or *non-Archimedean* according as it does or does not satisfy this axiom. The system of polynomials with real coefficients, with the order relation defined above, is non-Archimedean. It is a sign of Eudoxus's profound mathematical insight that he did not take it for granted that every system of ordered magnitudes satisfies condition I, or, as we say today, is Archimedean.

To see the significance of part II of Eudoxus's definition, let *a*, *b*, *c* and *d* be positive real numbers, suppose that the ordered pairs (*a*, *b*) and (*c*, *d*) have the same ratio, and let (*m*, *n*) be any ordered pair of positive integers. Let B be the set of all positive rational numbers *m*/*n* such that the ordered pair (*m*, *n*) satisfies condition 1) or condition 2). Let A be the set of all positive rational numbers *m*/*n* such that the ordered pair (*m*, *n*) satisfies condition 3). Then the separation of all positive rational numbers into the two sets A and B is a Dedekind cut, and Eudoxus's definition says in effect that the ordered pairs (*a*, *b*) and (*c*, *d*) have the same ratio if *b*/*a* and *d*/*c* define the same Dedekind cut. It is another sign of the profundity of Eudoxus's thinking that he anticipated by over two thousand years the basic concept that underlies the modern theory of the real number system.

Plenum or Void?

Ever since the emergence of the concept of physical space, physicists and philosophers have had to consider the question, "What is the relationship between space and the matter that is in it?" The thought of ancient Greece produced two diametrically opposite answers to this question. One answer, given by Democritus (about 420 B.C.) and his disciple Epicurus (about 340 B.C. to 270 B.C.) is that space is a *void*, in which the atoms that constitute all material bodies are in constant motion. Space, to them, is like the stage on which the drama of material existence is enacted, and the atoms are the actors. The Epicurean viewpoint implies that the properties of space are independent of the atoms that move in it, just as the properties of a stage are independent of the actors on it.

Plato's answer to the question is entirely different. We have seen that Plato thought of physical space as "the receptacle, and in a manner the nurse, of all generation." But it was not an empty receptacle. The process of generation was one of the acquisition of form, in imitation of the eternal forms found in the realm of ideas. But the substance that took on the form was already present in space. Space was a substrate common to all material bodies. In fact Plato accepted the view that the units of earth, water, air and fire were regular solids (see page 25), and hence were portions of space that differed only in form. From this point of view space is a *plenum*, that is, it is inseparably associated with material substance.

The idea of empty space was also rejected by Aristotle (385 B.C. to 322 B.C.). He did not accept Plato's view that space was the material of which bodies are made, but he thought of it as a plenum nevertheless, inseparable from material bodies. Aristotle's notion of space is expressed in his definition of *place* as the limit between the surrounding and the surrounded body. Thus, position, in Aristotle's view, is purely relative to material bodies.

The conflict between these two opposing views of space as *plenum* and space as *void* has extended through the centuries. Since it concerns the nature of physical space, it has followed the winds of change in the science of physics. We shall return to this question again in Chapter 7, where we discuss the Newtonian conception of space, and in Chapter 11, where we discuss the Einsteinian conception of space.

Finite or Infinite?

The early Greek philosophers also gave thought to the question, "Is space finite or infinite?" Two different answers were given to this question, too. Anaximander, a disciple of Thales, expressed the view that all things originate in an infinite ether. The earth, he said, is a cylindrical body floating freely in the infinite ether. Similarly, Melissus of Samos (about 440 B.C.) said that *being* is infinite in time and infinite in space. On the other hand, Parmenides (about 500 B.C.) expressed the view that the universe is finite: The universe, he said, is composed of a series of concentric spheres contained within an outermost sphere that is solid, dark, and cold. Similarly, Plato and Aristotle asserted that the body of the universe is a sphere, and hence finite.

The word "infinite" was used rather loosely by the philosophers and mathematicians of ancient Greece. They used it to mean both "infinite in extent" and "unbounded." This confusion in meanings was eliminated for the first time by Bernhard Riemann (1826–1866) when he pointed out that infinite in extent and unbounded are two different concepts, and that an unbounded space need not be infinite in extent. For example a straight line is both unbounded and infinite in extent. It is unbounded because, as you move along the line, no matter what point you reach, there are more points of the line beyond that point. A circle on the other hand, though it is finite in extent, is also unbounded. That is, as you move along the circle, no matter what point you reach, there are more points of the circle beyond that point. If we separate these two possible meanings of the word "infinite," the question "Is space finite or infinite?" becomes two separate questions: "Is space finite or infinite in extent?" and "Is space bounded or unbounded?" We shall return to these questions again in Chapters 7, 8 and 11.

There are two other possible meanings of the word *infinite*. It is sometimes used in the sense of *infinitely many*, and it has also been used, at times, in the sense of *infinitely divisible*. Confusion of these many meanings of the word *infinite* has led to some interesting paradoxes. We present below some of the famous paradoxes of Zeno, a pupil of Parmenides. Then, after a brief excursion into the modern theory of the infinite, in which we disentangle some of the different meanings of the word *infinite*, we show how these paradoxes are resolved.

Paradoxes of Zeno

Zeno (about 450 B.C.) was one of the philosophers who believed that motion is illusory. To support this belief he devised four ingenious arguments purporting to show that the concept of motion is self-contradictory. These arguments are known as the *Dichotomy*, the *Achilles*, the *Arrow*, and the *Stadium*. We give here only the *Dichotomy* and the *Achilles*. (For the *Arrow*, see page 266.)

Dichotomy. Suppose a body is to move from A to B. Before it can reach B, it must first reach the half-way point B_1; before it can reach B_1, it must first reach the half-way point B_2; before it can reach B_2, it must first reach the half-way point B_3; and so on, *ad infinitum*. Thus the motion from A to B cannot even begin.

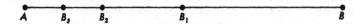

Achilles. Suppose a tortoise is moving along a straight line from P_1, and Achilles, starting from a position behind the tortoise, moves twice as fast in the same direction. By the time Achilles reaches P_1, the tortoise has advanced to P_2; by the time Achilles reaches P_2, the tortoise has advanced to P_3; by the time Achilles reaches P_3, the tortoise has advanced to P_4; and so on, *ad infinitum*. Thus the tortoise is always ahead of Achilles.

Although he moves faster than the tortoise, Achilles never catches up with him.

Finite and Infinite Sets

In order to resolve these paradoxes it is necessary first to develop a precise definition of the concept "infinitely many." This can be done with the help of some elementary notions from the theory of sets developed by Georg Cantor (1845–1918) and others.

A set is any collection of objects, real or conceptual, with a definite membership that is specified either by listing its members or by stating a rule by which it is possible to determine whether or not a given object is a member of the set. Here are some familiar examples of sets: 1) The set of digits of the Arabic system of numerals. The members of this set are the digits *0, 1, 2, 3, 4, 5, 6, 7, 8,* and *9.* 2) The set of positive integers. The members of this set are *1, 2, 3, . . .*, where the three dots indicate that the list of members is to be continued in an obvious way. 3) The set of even positive integers. The members of this set are *2, 4, 6, . . .* 4) The set of positive rational numbers. 5) The set of real numbers between 0 and 1.

One set is said to be a *subset* of another if every member of the first set is also a member of the second set. For example, the set of even positive integers is a subset of the set of positive integers, since the numbers *2, 4, 6, . . .* are included among the numbers *1, 2, 3, . . .* The set of integers is a subset of the set of rational numbers, and the latter, in turn, is a subset of the set of real numbers. Every set is a subset of itself. A subset of a given set is called a *proper subset* if it does not contain all the members of the given set. Thus, the set of even positive integers is a proper subset of the set of positive integers, since the even positive integers do not include the integers *1, 3, 5, . . .*

When can we say that one set has "as many" members as another set? This question is answered with the help of the concept of a one-to-one correspondence. Before defining the concept of a one-to-one correspondence we first introduce the concept of a function.

If A and B are sets, a *function on A to B* is an assignment of one and only one member of B to each member of A. For example, if A is the set of people in a family and B is the set of days of the year, there is a function which we may call the birthday function, which assigns to each member of the family the day when he celebrates the anniversary of his birth. A function on A to B is essentially a set of ordered pairs, in which the first member of an ordered pair is a member of A and the second member of the ordered pair is the member of B that is assigned to the first member, and in which every member of A occurs once and only once as a first member of an ordered pair. Thus, if, in a family of three, father, mother and child were born on March 1, February 10, and April 20 respectively, the birthday function for this family is the set of three ordered pairs {(father, March 1), (mother, February 10), (child, April 20)}.

A function on a set A to a set B is said to be *one-to-one* if no two members of A have the same member of B assigned to them. If a one-to-one function on A to B is viewed as a set of ordered pairs, no member of B occurs more than once as a second member of an ordered pair. A function on a set A to a set B is said to be *onto* if every member of B is assigned to at least one member of A. If an onto function on A to B is viewed as a set of ordered pairs, every member of B occurs at least once as a second member of an ordered pair.

A function on A to B is called a *one-to-one correspondence between A and B* if the function is both one-to-one and onto. If a one-to-one correspondence between A and B is viewed as a set of ordered pairs, each member of A occurs once and only once as a first member of an ordered pair, and each member of B occurs once and only once as a second member of an ordered pair. Thus, in a one-to-one correspondence between two sets there is a pairing of the members of one set with the members of the other set in such a way that each member of one set is paired with one and only one member of the other set. For example, the pairing indicated below by the double-headed arrows establishes a one-to-one correspondence between the set of vowels *a*, *e*, *i*, *o* and *u* and the set of positive integers that are less than 6:

$$
\begin{array}{ccccc}
a & e & i & o & u \\
\updownarrow & \updownarrow & \updownarrow & \updownarrow & \updownarrow \\
1 & 2 & 3 & 4 & 5
\end{array}
$$

Two sets are called *equivalent* if they can be put into one-to-one correspondence. Thus the set of vowels *a*, *e*, *i*, *o* and *u* is equivalent to the set whose members are 1, 2, 3, 4, and 5. All sets may be grouped into families of equivalent sets. There is a *cardinal number* associated with each such family. For our purposes it will suffice to think of each cardinal number as a symbol used to identify a family of equivalent sets. The symbol associated with the family of sets each of which contains only a single member is 1. The symbol associated with the family of all pairs is 2. The symbol associated with the family of all triples is 3, etc. The symbol associated with the family of sets that are equivalent to the set of positive integers is $\aleph_0$ (aleph-null). The cardinal number of a set is the symbol associated with the family of equivalent sets to which it belongs. Consequently two sets that are equivalent have the same cardinal number. The cardinal number of a set is what we use to answer the question "How many members are there in the set?"

We are now ready to make a precise distinction between finite and infinite sets. A set is said to be *infinite* if it is equivalent to one of its proper subsets.

For example, the one-to-one correspondence indicated below, where each positive integer n is paired with $2n$, shows that the set of positive integers is equivalent to the set of even positive integers, which is one of its proper subsets:

$$\begin{array}{ccccccc} 1 & 2 & 3 & \ldots & n & \ldots \\ \updownarrow & \updownarrow & \updownarrow & & \updownarrow & \\ 2 & 4 & 6 & \ldots & 2n & \ldots \end{array}$$

Consequently the set of positive integers is infinite.

If a set is not infinite, it is said to be *finite*. A finite set is said to have *finitely many* members, and its cardinal number is called a *finite number*. An infinite set is said to have *infinitely many* members, and its cardinal number is called an *infinite number*. The number 5 is an example of a finite number. The number $\aleph_0$ is an example of an infinite number.

The Points on a Line

Let us designate by c the cardinal number of the set of all real numbers. When we introduce a number scale on a straight line, as we did on page 31, the scale establishes a one-to-one correspondence between the real number system and the set of points on the line. Consequently c is also the cardinal number of the set of points on the line. We shall show that *c* is infinite by showing that the set of all points on the line is equivalent to one of its proper subsets, namely the set of points between 0 and 1. Make a duplicate copy of the line segment between 0 and 1, bend it to make an angle at its midpoint, and then place it above the number line in the position shown in the diagram below, so

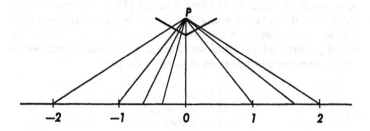

that the endpoints of the segment are at the same height above the number line, and the midpoint of the segment is directly above the zero point of the number line. Let *P* be the point half-way between the endpoints of the segment in this position. Through *P* draw a straight line through each point of the bent segment, and extend it until it meets the number line. The lines drawn in this way join each point of the bent line segment with one and only one point of the number line, and vice versa. Thus they establish a one-to-one correspondence between the set of all real numbers and the subset that contains only the real numbers between 0 and 1. Consequently the set of real numbers between 0 and 1 is equivalent to the set of all real numbers, and its cardinal number is *c*.

A procedure like the one described above can be used to set up a one-to-one correspondence between the set of all points on a line and the set of points between the ends of *any* line segment. Consequently the cardinal number of the set of points between the ends of any line segment is *c*. That is, all line segments, no matter how long they are, contain the same number of points.

Different Infinities

An infinite set, by definition, is equivalent to one of its proper subsets, and the latter is necessarily infinite, too. This fact suggests the question, "Are all infinite sets equivalent to each other?" The answer to this question turns out to be "No." In fact, the set of real numbers between 0 and 1 is not equivalent to the set of positive integers. That is, the infinite cardinal numbers c and $\aleph_0$ are different. To prove this fact we show that it is impossible to have a one-to-one correspondence between the set of positive integers and the set of all real numbers between 0 and 1.

We note first that every real number between 0 and 1 can be represented by a non-terminating decimal of the form $.a_1a_2a_3 \ldots$, where a_1 is the digit in the first decimal place, a_2 is the digit in the second decimal place, a_3 is the digit in the third decimal place, and so on. If there is a one-to-one correspondence between the set of positive integers and some set of real numbers between 0 and 1, then we can put it on display as follows:

$$1 \leftrightarrow .a_1a_2a_3 \ldots$$
$$2 \leftrightarrow .b_1b_2b_3 \ldots$$
$$3 \leftrightarrow .c_1c_2c_3 \ldots$$
etc.

We now show that the set of real numbers shown on the right-hand side of this table cannot include *all* real numbers between 0 and 1. We do so by producing one that is not in the table. Let x_1 be a digit that is different from a_1, 0, and 9; let x_2 be a digit that is different from b_2, 0, and 9; let x_3 be a digit that is different from c_3, 0, and 9; and so on. Then consider the non-terminating decimal $.x_1x_2x_3 \ldots$. It represents a real number between 0 and 1. Since none of its digits are zeros or nines, it is the only non-terminating decimal that represents that number. (If two non-terminating decimals represent the same real number, as for example 2.1999 . . . and 2.2000 . . . , one has a long string of nines in it, and the other has a long string of zeros in it.) It is not the same as $.a_1a_2a_3 \ldots$, because it differs from it in the first decimal place. It is not the same as $.b_1b_2b_3 \ldots$, because it differs from it in the second decimal place. It is not the same as $.c_1c_2c_3 \ldots$, because it differs from it in the third decimal place. In general it is not the same as the real number listed on the nth line of the table because it differs from it in the nth decimal place. Consequently the number $.x_1x_2x_3 \ldots$ differs from all the numbers listed on the right-hand side of the table. Therefore the set of real numbers listed on the right-hand side does not include all real numbers between 0 and 1.

A set whose cardinal number is $\aleph_0$ is said to be *denumerable* or *denumerably infinite*. We have shown that the set of all real numbers is not denumerable.

Finite or Infinite Extent

By introducing the concepts of an infinite set and the cardinal number of an infinite set we have given a precise meaning to the expression "infinitely many." Now we shall give a precise meaning to the term "of infinite extent" as it is applied to a straight line. The line we have been discussing is one that can be provided with a number scale by setting up a one-to-one correspondence between the points of the line and the system of real numbers. Since this is essentially the kind of straight line that occurs in the geometry of Euclid, let us refer to it from now on as a *Euclidean line*. When a Euclidean line has been provided with a number scale, every point on the line is associated with a definite real number. This makes it possible to associate with every line segment a real number known as its *length*, in the following way: If the real number associated with a point P is p, and the real number associated with a point Q is q, the length of the segment PQ is $p - q$ or $q - p$, whichever is positive. If we think of the segment as being movable, like a stick, we can also determine its length in this way: Place the segment on the positive half of the line so that one end of the segment is at the zero point of the scale. Then the length of the segment is the real number associated with the point where the other end of the segment falls.

Let b be the length of a segment. The number 1 is the length of the unit segment on the scale. According to the axiom of Archimedes, there is a positive integer n such that $n \cdot 1 > b$. Consequently, if we take n unit segments and place them end to end over the segment of length b, starting at one end of the segment, then the last of these units will extend beyond the other end. In other words, the segment of length b can be completely covered by a set of n units. For example, a segment of length π can be completely covered by a set of 4 units. Since the number of units in this covering set is finite, we say that the covered segment has a *finite length*, or is *finite in extent*. On the other hand, no finite set of units will suffice to cover an entire Euclidean line. An infinite number ($\aleph_0$) of units is needed to cover the whole line. For this reason we say that a Euclidean line is *infinite in length*, or is *of infinite extent*. Similarly a Euclidean plane is of infinite extent because an infinite number of unit squares are needed to cover it, and Euclidean space of three dimensions is of infinite extent because an infinite number of unit cubes are needed to cover it.

Both Finite and Infinite

There is a sense in which a line segment is finite: it is finite as a length, because it can be covered by a finite number of unit segments. There is also a sense in which a line segment is infinite: it is infinite as a set of points, because it contains an infinite number of points. This two-fold aspect of a segment may seem like a paradox, but the paradox is more apparent than real. The finiteness of the length of a segment is in no way contradicted by the infinity of the number of points it contains.

The Dichotomy

One of the consequences of the fact that a line segment of finite length contains an infinite number of points is that the segment is infinitely divisible, that is, it may be divided into an infinite number of segments. Zeno's dichotomy, for example, decomposes the segment AB into the sequence of segments B_1B, preceded by B_2B_1, preceded by B_3B_2, etc.

Zeno's argument points out that for a body to move from A to B, it must pass through in reverse order the points $B_1, B_2, B_3, \ldots$. That is, it must pass through an infinite number of points. There is a definite instant at which the body would be at each of these points. Let us say the body reaches B_1 at time t_1, reaches B_2 at time t_2, etc. Then t_2 precedes t_1, t_3 precedes t_2, and so on. Thus to reach the time when the motion begins, we must go backwards in time through an infinite sequence of instants, $t_1, t_2, t_3, \ldots$. Zeno concludes from this fact that the starting time is infinitely remote. This conclusion is false, however, because just as there are infinitely many points in a line segment of finite length, there are infinitely many instants in a finite time interval. Suppose, for example, that the length of AB is one unit, and that the body, moving at a uniform speed of one unit per hour, reaches B at 3:00 o'clock. Then t_1 is 2:30, t_2 is 2:15, t_3 is 2:07$\frac{1}{2}$, etc. The entire infinite sequence $t_1, t_2, t_3, \ldots$ is contained in the finite interval between 2:00 o'clock and 3:00 o'clock. The starting time is not infinitely remote. In fact, it is precisely 2:00 o'clock.

An Infinite Series

To establish beyond any doubt that the sequence of instants $t_1, t_2, t_3, \ldots$ is contained in a finite interval, let us re-examine the Dichotomy from a slightly different point of view. By reckoning backwards from the instant that the body arrives at point B, let us try to compute the length of time that the body is in motion. It takes the body $\frac{1}{2}$ hour to move from B_1 to B. It takes the body $\frac{1}{4}$ hour to move from B_2 to B_1. It takes the body $\frac{1}{8}$ hour to move from B_3 to B_2, and so on. The total number of hours that the body is in motion before it reaches B may therefore be represented by the infinite series

$$\frac{1}{2} + \frac{1}{4} + \frac{1}{8} + \frac{1}{16} \ldots,$$

in which each term is one-half of the term that precedes it. The essential question we have to face is this: Is it possible for an infinite series of terms to have a finite sum? Before we can answer this question we must define carefully what is meant by the "sum" of an infinite series of terms. To do so, we first compute the sums that we can get by using only a finite number of terms in succession of the series. We call these the *partial sums* of the series. If we take only the first term, the partial sum is $\frac{1}{2}$. If we take only the first two terms, the partial sum is $\frac{3}{4}$. If we take only the first three terms, the partial sum is $\frac{7}{8}$, and so on. Thus we get a sequence of partial sums,

$$\frac{1}{2}, \frac{3}{4}, \frac{7}{8}, \frac{15}{16}, \ldots .$$

If we denote by S_n the nth partial sum, we can compute it by using the well-known formula for the sum of a geometric progression:

$$S_n = \frac{a - ar^n}{1 - r}.$$

In this case, $a = \frac{1}{2}$ and $r = \frac{1}{2}$. Making these substitutions, we find that $S_n = 1 - (\frac{1}{2})^n$. This formula shows that S_n is always less than 1, that for large values of n the value of S_n is almost equal to 1, and, in fact, we can make the value of S_n as close to 1 as we please by taking a large enough value of n. Under these conditions we say that S_n approaches 1 as a limit as n increases "toward infinity." It is this limit approached by S_n that is used as the "sum" of the infinite series $\frac{1}{2} + \frac{1}{4} + \ldots$.

In general, suppose we have an infinite series

$$a_1 + a_2 + a_3 + \ldots .$$

We form the partial sums $S_1 = a_1$, $S_2 = a_1 + a_2$, $S_3 = a_1 + a_2 + a_3$, etc. Then we consider the sequence of partial sums,

$$S_1, S_2, S_3, \ldots .$$

If there exists a number S such that S_n approaches S as a limit as n increases toward infinity, we say that S is the sum of the infinite series, and we write

$$S = a_1 + a_2 + a_3 + \dots \; .$$

Whenever this limit S exists, the infinite series has a finite sum.

Since the series $\frac{1}{2} + \frac{1}{4} + \dots$ has a finite sum equal to 1, the body was in motion for exactly 1 hour before it reached B. Consequently, the sequence of instants t_1, t_2, t_3, . . . is contained in a finite interval, and the starting time was 2:00 o'clock, as we asserted in the preceding paragraph.

Achilles

The concept of the sum of an infinite series makes it possible for us to analyze the Achilles paradox. Let us assume that Achilles starts from a position that is 1 mile behind the tortoise. Let us assume, too, that Achilles moves at a speed of 1 mile per hour, while the tortoise moves at a speed of one-tenth of a mile per hour. Then Achilles reaches P_1 after 1 hour. Meanwhile the tortoise advances to P_2 a distance of .1 mile. It takes Achilles .1 hour to go from P_1 to P_2. Meanwhile the tortoise advances to P_3 a distance of .01 mile. It takes Achilles .01 hour to cover this distance, and so on. Thus the total number of hours that elapse before Achilles overtakes the tortoise is $1 + .1 + .01 + .001 + .0001 + \dots$. Zeno's assertion that Achilles never catches up with the tortoise would be correct only if this infinite series had an infinite sum. However, it has a finite sum. In fact the sum is $1.1111 \dots = \frac{10}{9}$. That is, it takes Achilles one and one-ninth hours to catch up with the tortoise.

The Method of Exhaustions

When the sum S of an infinite series $a_1 + a_2 + a_3 + \dots$ exists, it is obtained as the limit of the sequence of partial sums S_1, S_2, S_3, . . . , which are successively better and better approximations to the value of S. This is an example of a fundamental technique that plays an important role in contemporary mathematics, the technique of obtaining a sought-for number as the limit of a sequence of approximations. Although this technique is completely modern in its spirit of logical rigor, it is also very ancient. It is essentially the same as the *method of exhaustions* introduced into geometry by Eudoxus. Eudoxus used this method, for example, to prove the theorem that a line that is parallel to one side of a triangle divides the other two sides proportionally. He first proved it for the case where the ratio of the segments of one side is a rational number. He then extended the proof to the case where the ratio of the segments is an irrational number by using in effect the fact that an irrational number can be approached as the limit of a sequence of rational approximations to it. He used the method, too, to prove some theorems concerning the perimeter and area of a circle. In these cases, he first proved the theorem for a regular polygon. He then extended the proof so that it applied to a circle by using the fact that the perimeter and area of a circle are the limits respectively of the perimeter and areas of a sequence of inscribed polygons with more and more sides. Eudoxus's anticipation by two thousand years of the modern method of limits is a third example of his genius as a mathematician. The ideas of Eudoxus, which have stood so well the test of time and searching criticism, entitle him to be rated as one of the truly great mathematicians of all time.

Transformations

The elementary concepts of geometry are abstractions derived from our experience with physical objects. A *point* is an abstraction from an object of negligible size, such as a dot on a piece of paper or a speck of dust. A *line* is an abstraction from a taut string, imagined to be endless. A *plane* is an abstraction from a flat table top, imagined to be endless. Another concept that has played an important part in the evolution of geometry, the concept of a *transformation*, is an abstraction from the motion of a physical object from one place to another.

Imagine an endless stiff wire resting along a straight line. If we slide the wire along the line, every point of the wire undergoes a change of position. As a result the motion associates with each point P on the line, which was initially occupied by a particular point of the wire, the new point P' to which that point of the wire has been moved. This association of each point of the line with another point of the line is called a *transformation* of the line. It is clear from what we said on page 47 that a transformation of a line is a function *on* the set of points of the line *to* the same set of points.

Although the essence of the transformation is the mere association of each point with another, we find it convenient to use a metaphor for describing it. We dispense with the stiff wire altogether and picture the transformation as consisting of a motion of the line itself that carries each point P to a new position P'. If the line moves as if it were a stiff wire, the length of any segment of the line is unchanged by the motion. In this case the transformation is called a *rigid motion*. There are other possible transformations of the line besides a rigid motion. If the line moves as if it were stretched like a rubber band, the motion effects a transformation of the line in which the lengths of segments are not preserved.

In a rigid motion of a line along itself, no two points of the line are carried to the same new position. A transformation that has this property is said to be *one-to-one*. If P' is any point on the line, a rigid motion of the line carries some point P to P'. A transformation that has this property is said to be *onto*. A stretching of a line is also a one-to-one onto transformation. Nearly all the transformations that we shall encounter later in this book are one-to-one onto transformations. A significant feature of a one-to-one onto transformation is that it is reversible, that is, if a one-to-one onto transformation moves every point P to a new position P', there is another transformation that moves every point P' back to its original position P.

The concept of a transformation is easily extended to a plane and to a three-dimensional space. Just as a transformation of a line may be pictured metaphorically as a "motion" that carries each point of the line to a new position on the line, a transformation of a plane may be pictured as a "motion" that carries each point of the plane to a new position on the plane, and a transformation of (three-dimensional) space may be pictured as a "motion" that carries each point of space to a new position.

Continuous Transformations

One of the features of both rigid motions and stretches is that they carry points that are near each other into new positions that are near each other. Transformations that have this property are said to be *continuous*. This definition of a continuous transformation is somewhat crude, but it is adequate for our purposes now. (A better definition is given on page 319.) Nearly all the transformations we shall encounter in this book are continuous.

Groups of Transformations

Consider the set of all one-to-one onto transformations of a plane. We shall use lower case letters in italics, such as x, y, and z, to designate each of these transformations. Suppose a transformation x moves each point P to a new position P', and another transformation y moves each point P' to a new position P''. If we first perform the transformation x and then perform the transformation y, the combined effect of the two transformations is itself a transformation that moves each point P to P''. When we perform two transformations x and y in succession, we say that we are "multiplying" them. We call the resulting transformation that moves each point P to P'' the *product* of x and y, and we designate it by $x \circ y$. Since x and y are one-to-one and onto, the transformation $x \circ y$ is also one-to-one and onto. We pause to consider some properties of the set of all one-to-one onto transformations of the plane in relation to the multiplication operation $\circ$ that signifies performing two transformations in succession.

Suppose x moves P to P', y moves P' to P'', and z moves P'' to P'''. Then $y \circ z$ moves P' to P''', and $x \circ y$ moves P to P''. The transformation $x \circ (y \circ z)$ is the combined effect of first performing x and then performing $y \circ z$. Since x moves P to P', and $y \circ z$ moves P' to P''', $x \circ (y \circ z)$ moves P to P'''. The transformation $(x \circ y) \circ z$ is the combined effect of first performing $x \circ y$ and then performing z. Since $x \circ y$ moves P to P'', and z moves P'' to P''', $(x \circ y) \circ z$ moves P to P'''. That is, $x \circ (y \circ z)$ and $(x \circ y) \circ z$ have the same effect, or $x \circ (y \circ z) = (x \circ y) \circ z$. That is, the set of all one-to-one onto transformations of the plane satisfies condition A of page 39.

Let us designate by e the transformation that "moves" each point P of the plane into itself, that is, which does not move it at all. Then it is easy to verify that if x is any one-to-one onto transformation of the plane, $x \circ e = e \circ x = x$. That is, condition C of page 39 is satisfied.

Let x be any one-to-one onto transformation of the plane that takes P to P'. Since the transformation is one-to-one and onto it is reversible. That is, there is another transformation that carries every point P' back to its original position P. Let us call this transformation the *inverse of x*. Since x carries P to P', and *inverse of x* carries P' to P, the product $x \circ$ (*inverse of x*) carries P to P, that is $x \circ$ (*inverse of x*) $= e$. Similarly, (*inverse of x*) $\circ x = e$. That is, condition D of page 39 is also satisfied.

Since conditions A, C, and D of page 39 are all satisfied, the set of all one-to-one onto transformations of a plane is a group with respect to the multiplication operation $\circ$ which we have defined. In general, the set of all one-to-one onto transformations of any space form a group with respect to the multiplication operation $\circ$ that signifies performing two transformations in succession. Certain subsets of the set of all one-to-one onto transformations of the space also form a group. These are known as *subgroups* of the group of one-to-one onto transformations of the space. We shall encounter such subgroups in later chapters.

For the sake of brevity in writing about transformations, it is customary to omit the multiplication sign ∘ whenever a product is written. In this notation, we write xy instead of $x \circ y$. It is also customary to use the symbol x^{-1} for *inverse of x*. Since the identity transformation e behaves with respect to the multiplication of transformations the way the number 1 behaves with respect to the multiplication of numbers, it is convenient to use the symbol 1 instead of e for the identity transformation. In this modified notation, conditions A, C, and D of page 39 take this form:

A. $x(yz) = (xy)z$.
C. There exists a member 1, the identity transformation, such that for every x in the system, $x1, = 1x = x$.
D. For each x there exists a member denoted by x^{-1} that has the property $xx^{-1} = x^{-1}x = 1$.

EXERCISES FOR CHAPTER 2

1. *The averaging principle.* Let the bases of a trapezoid have lengths b and b' respectively, and let the height of the trapezoid be h. Assuming that the area of the trapezoid is equal to the area of a rectangle of the same height whose base is the average of b and b', derive a formula for the area of the trapezoid. Compare this formula with the one that is derived in Euclidean geometry.

2. *Rope stretching.* Tie knots at equal intervals along a string until you have twelve equal intervals. Use the knotted string to form a triangle whose sides contain exactly three, four, and five intervals respectively. With a carpenter's square verify that the angle opposite the longest side is a right angle.

3. *Empirical observations with the help of paper-folding.* a) On a sheet of paper, draw two equal straight line segments PA and PB from the same point P. Then draw the line segment AB, to form the isosceles triangle PAB. Now fold the paper to make the equal segments PA and PB coincide. Verify that when this is done the base angles of triangle PAB will also coincide. What geometry theorem of Thales is verified by this exercise?

b) Draw a circle on a sheet of paper. Then fold the paper so that the crease passes through the center of the circle. Verify that after the paper is folded, the two arcs of the circle that lie on opposite sides of the crease coincide. What geometry theorem of Thales is verified by this exercise?

c) Using the circle drawn for exercise 3 b), take any point on the circle and draw lines from that point to the two points where the crease meets the circle. Using a carpenter's square, verify that the angle formed by these lines is a right angle. What geometry theorem of Thales is verified by this exercise?

4. *Triangular numbers.* The first three triangular numbers are $1, 1 + 2 = 3$, and $1 + 2 + 3 = 6$. Find the next seven triangular numbers. Add each triangular number to the next larger triangular number. (For example, $1 + 3 = 4$, $3 + 6 = 9$.) What kind of number is each of the sums obtained in this way?

5. *The Pythagorean Theorem.* In diagram III on page 18, prove that the four triangles cut off at the corners of the large square are congruent to each other. Then prove that the unshaded quadrilateral that remains has four equal sides and four right angles.

6. *The regular solids.* a) Follow the directions given on page 23 to make each of the five regular solids.

b) Prove that inequality (6) on page 24 implies inequality (7).

7. *Euler's formula.* Using models of the regular solids, count the vertices, edges and faces in each to verify the values of V, E and F shown in the table on page 27. Verify that in each case $V - E + F = 2$.

8. Every even integer has the form $2n$, and every odd integer has the form $2n + 1$, where n is an integer. Prove that the square of an odd integer is an odd integer.

9. *Rational numbers.* a) The number $-x$ that has the property that $x + (-x) = 0$ is called the negative of x. What is the negative of 2? of 5? of -3? of 0?

b) If $x \neq 0$, the number $\dfrac{1}{x}$ that has the property that $x \cdot \dfrac{1}{x} = 1$ is called the reciprocal of x. What is the reciprocal of $\frac{2}{3}$? of 5? of -1? of 1?

10. *Non-terminating decimals.* The non-terminating repeating decimal $.333\ldots$ represents the rational number $\frac{1}{3}$. The non-terminating repeating decimal $.222\ldots$ represents the rational number $\frac{2}{9}$. By adding $\frac{1}{3}$ and $\frac{2}{9}$ and converting the sum into a non-terminating decimal, verify that $.333\ldots + .222\ldots = .555\ldots$.

11. a) Verify that the real number 1 satisfies the equation $x^2 - 1 = 0$.

b) Verify that the real number -1 also satisfies the same equation.

c) If r is a real number, why is it impossible for r to satisfy the equation $x^2 + 1 = 0$?

12. *Finite fields.* Divide a circle into three equal arcs and number the points of division 0, 1, and 2, going clockwise. Using the definition of addition and multiplication given on page 36, construct the addition and multiplication tables for the modular number system F_3. Use the tables to answer the following questions:

a) What is the negative of 1 in F_3?
b) What is the negative of 2 in F_3?
c) What is the reciprocal of 1 in F_3?
d) What is the reciprocal of 2 in F_3?

13. How many elements are there in the smallest finite field that contains F_7 but is different from F_7?

14. The addition and multiplication tables for the prime field F_2 are given below:

Addition

+	0	1
0	0	1
1	1	0

Multiplication

×	0	1
0	0	0
1	0	1

Use a clock-face diagram to verify these tables.

15. The smallest finite field that contains F_2 but is different from F_2 has four elements 0, 1, a and b with the following addition and multiplication tables:

Addition

+	0	1	a	b
0	0	1	a	b
1	1	0	b	a
a	a	b	0	1
b	b	a	1	0

Multiplication

×	0	1	a	b
0	0	0	0	0
1	0	1	a	b
a	0	a	b	1
b	0	b	1	a

a) Verify that a satisfies the equation $x^2 + x + 1 = 0$.
b) Verify that b satisfies the equation $x^2 + x + 1 = 0$.

c) Verify that every member of this field of four elements satisfies the equation $x^4 - x = 0$.

16. If A is the set of 31 days of the month of January in a particular year, and B is the set of seven days of the week, there is a function defined by the calendar for that year which assigns to each day of January the day of the week on which it falls.
a) Is this function one-to-one? b) Is it onto?

17. The institution of marriage defines a function on the set of all married men to the set of all women, by assigning one wife to each married man, provided that no man is guilty of bigamy.
a) Under what conditions would this function be one-to-one?
b) Under what conditions would this function be onto?

18. *Finite and infinite sets.* Which of these sets is finite and which is infinite?

a) the set of eyes in your face;
b) the set of hairs on your head;
c) the set of positive integers that are integral multiples of 3;
d) the set of points on a Euclidean line.

19. If a line segment is one foot long,
a) What is the number of inches in its length?
b) What is the number of points between the ends of the line segment?
c) Are these numbers finite or infinite?

20. *The sum of an infinite series.*

a) Use the formula $S_n = \dfrac{a - ar^n}{1 - r}$ to find the nth partial sum of the infinite series $\frac{1}{3} + \frac{1}{9} + \frac{1}{27} + \frac{1}{81} + \ldots$ in which each term is one third of the term that precedes it. What is the limit approached by S_n as n increases toward infinity? What is the sum of the infinite series $\frac{1}{3} + \frac{1}{9} + \ldots$?

b) Consider the infinite series $\frac{1}{2} + \frac{1}{3} + \frac{1}{4} + \frac{1}{5} + \ldots$ where the term following $\dfrac{1}{n}$ is $\dfrac{1}{n+1}$. Consider the partial sums S_n where n is one less than a power of two:

$$S_1 = \tfrac{1}{2}, \ S_3 = \tfrac{1}{2} + \tfrac{1}{3} + \tfrac{1}{4},$$
$$S_7 = \tfrac{1}{2} + \tfrac{1}{3} + \tfrac{1}{4} + \tfrac{1}{5} + \tfrac{1}{6} + \tfrac{1}{7} + \tfrac{1}{8}, \text{ etc.}$$

Note that $S_3 = \tfrac{1}{2} + (\tfrac{1}{3} + \tfrac{1}{4}) > \tfrac{1}{2} + (\tfrac{1}{4} + \tfrac{1}{4}) = 2(\tfrac{1}{2})$

$$S_7 = \tfrac{1}{2} + (\tfrac{1}{3} + \tfrac{1}{4}) + (\tfrac{1}{5} + \tfrac{1}{6} + \tfrac{1}{7} + \tfrac{1}{8})$$
$$> \tfrac{1}{2} + (\tfrac{1}{4} + \tfrac{1}{4}) + (\tfrac{1}{8} + \tfrac{1}{8} + \tfrac{1}{8} + \tfrac{1}{8}) = 3(\tfrac{1}{2}), \text{ etc.}$$

Thus

$$S_3 > 2(\tfrac{1}{2}), \; S_7 > 3(\tfrac{1}{2}), \; S_{15} > 4(\tfrac{1}{2}), \text{ etc.}$$

Does S_n approach a limit as n increases toward infinity? Does the infinite series $\tfrac{1}{2} + \tfrac{1}{3} + \tfrac{1}{4} + \ldots$ have a sum?

21. *Groups of transformations.* Let P be a fixed point of a plane. Denote by A, B, C and D the clockwise rotation of the plane about P through 90°, 180°, 270°, and 360° respectively. The set of transformations consisting of A, B, C and D alone is a subgroup of the group of all one-to-one and onto transformations of the plane.

a) Copy the multiplication table below and complete it by entering in each vacant space the transformation that results when you first perform the transformation listed at the left of the row that the space is in, and then perform the transformation listed at the top of the column that the space is in. (For example, a 90° rotation clockwise followed by a 180° rotation clockwise is equivalent to a 270° rotation clockwise. Hence AB = C, and we write C in the space that is in the A row and the B column.)

	A	B	C	D
A		C		
B				
C				
D				

b) Which transformation is the identity transformation?
c) What is the inverse of A? of B? of C? of D?

3

Euclid's Geometry

Euclid and His Purpose

Euclid (about 300 B.C.) was a teacher at the great library in Alexandria, Egypt, during the reign of the first Ptolemy. He was the most successful textbook writer of all time. His book, *The Elements*, which includes a systematic presentation of much of the geometric knowledge of his day, has been the basis for the teaching of geometry for over two thousand years.

The Elements was not intended to be an encyclopedia of geometry. It had a more limited and specialized purpose: it was to be used to prepare students for philosophical studies in the spirit of Plato. This purpose determined many of the specific features of the content and form of the book.

Because the book was an introduction via geometry to the realm of pure ideas, it dealt with abstract mathematical space rather than with physical space. Consequently no consideration is given anywhere in the book to the practical applications of geometry. This emphasis on theory to the exclusion of practice is the theme of an anecdote about Euclid told by Stobaeus. A student, after learning the first theorem of *The Elements*, asked Euclid "What shall I get by learning these things?" Euclid replied by calling his slave and saying, "Give him a coin, since he must make gain out of what he learns."

Because of the limited purpose of the book, it does not include all of the geometry known in Euclid's time. For example, it does not include anything on the theory of conic sections and higher curves (see page 123), although Euclid himself had written a book on conic sections.

The special purpose of the book is reflected in the choice of content for the final and climactic chapter of the book. This chapter (Book XIII of *The Elements*) deals with the construction of the five regular solids, identified by the Pythagoreans and by Plato with the four elements and the Universe. (See page 25.)

Since the book deals with pure ideas and their relationships with each other, it gives no clues whatever to the way in which these ideas were discovered. Instead it concentrates exclusively on the derivation of some ideas from others, in the manner described by Plato in the statement quoted on page 40. That is, the book organizes the propositions of geometry into a deductive system in which nearly all the propositions are proved by logical deduction from a small set of primitive propositions. The exclusive emphasis of the book on deductive reasoning has made it notoriously difficult for school boys and kings. Proclus reports that King Ptolemy once asked Euclid if there was in geometry any shorter way than that of *The Elements*. Euclid replied, "There is no royal road to geometry."

A Deductive System

Euclid's *Elements* was the first attempt to organize an entire branch of human knowledge into a deductive system. A deductive system is a collection of propositions in which an attempt is made, to the extent that it is possible, to define each term that is used, and to validate each proposition by proving it by logical deduction from other propositions in the system. As every high school student knows, it is not possible to prove *every* proposition in a deductive system. Suppose certain propositions a, b, and c, are proved by deduction from propositions d, e, and f. If propositions d, e, and f are proved, and there is no circular reasoning, then they must be deduced from some other propositions, say g and h. If propositions g and h are proved, and there is no circular reasoning, then they must be deduced from still other propositions, say j, k and l, and so on. This argument leads to a chain of sets of propositions in which each set of propositions is deduced from the next set of propositions in the chain. In order to avoid an infinite regression, the chain must come to an end. Then the propositions in the last link of the chain are not proved by deduction from any other propositions in the system. These unproved propositions, which serve as the foundation from which all the other propositions are derived, are known as *axioms*. The propositions that are deduced from the axioms are known as *theorems*.

A deductive system may be compared to the structure of a tree. The axioms of the system are like the trunk of the tree. Propositions deduced directly from the axioms are like branches supported by the trunk. Propositions deduced from these propositions are like twigs supported by the branches. Propositions deduced from the latter propositions are like stems supported by the twigs, etc. The theorems of a deductive system are like the stems, twigs and branches, each of which is supported by some other part of the tree. The axioms of a deductive system are like the trunk of the tree, which is itself unsupported, but supports everything else on the tree.

Just as it is not possible to prove every proposition in a deductive system, it is also not possible to define every term that is used in the system. In order to define a term *a*, it is necessary to use certain other terms, say *b* and *c*. In order to define terms *b* and *c*, it is necessary to use other terms, say *d*, *e*, and *f*, etc. Since an infinite regression must be avoided, we are ultimately led back in this way to terms that are not defined with the help of other terms. These are known as *undefined terms:* The undefined terms occur in the axioms. The axioms are assertions about the undefined terms, and give them whatever meaning they have in the deductive system. Thus, a deductive system begins with certain undefined terms, and certain axioms about them and then, on this foundation, defines all other terms used, and proves all other propositions that are asserted.

Euclid's feat in organizing geometry into a deductive system was an outstanding scientific accomplishment. It inspired other scientists and even philosophers to try to organize the results of their own investigations in the same way. Thus, Newton, in his great and elegant work *Principia*, derived theorems of mechanics from a small set of axioms. Similarly, Spinoza organized his thoughts on *Ethics* in the same form. Today all branches of pure mathematics are organized into deductive systems, in which theorems are deduced from axioms about undefined terms.

The Contents of *The Elements*

There are thirteen chapters in *The Elements*. The first four chapters, Books I to IV, are concerned with the concepts of line segment, angle, congruence, and area, and apply these to the study of triangles, parallelograms, circles, regular polygons, etc. The work on polygons is based mostly on the investigations of the Pythagoreans. The chapter on circles incorporates the results of the research of Hippocrates.

Books V and VI contain the theory of ratio and proportion developed by Eudoxus, and apply this theory to the study of similar figures. As we have seen, the Eudoxian theory of ratio anticipates the modern theory of the real number system.

Books VII, VIII and IX are devoted to the theory of numbers, largely Pythagorean in origin. Book X contains a geometric classification of irrational numbers that are expressible in terms of square roots.

Book XI consists of some elementary solid geometry.

Book XII develops Eudoxus's method of exhaustions, which anticipates the modern theory of limits. Euclid uses the method to prove that the areas of two circles are to each other as the squares of their radii.

Book XIII is devoted to the construction of the regular solids.

Euclid's Axioms

Euclid begins *The Elements* with some definitions of basic terms, and with ten axioms from which all his theorems are to be derived. He calls five of the axioms *postulates*, and calls the other five *common notions*. Today we attach no significance to the distinction he made between them, but merely class them all as *assumptions*. These are the ten assumptions as translated by Sir Thomas L. Heath in his famous annotated edition of *The Elements*:

Postulates
1. To draw a straight line from any point to any point.
2. To produce a finite straight line continuously in a straight line.
3. To describe a circle with any center and distance.
4. That all right angles are equal to one another.
5. That, if a straight line falling on two straight lines make the interior angles on the same side less than two right angles, the two straight lines, if produced indefinitely, meet on that side on which are the angles less than the two right angles.

Common notions
1. Things which are equal to the same thing are also equal to one another.
2. If equals be added to equals, the wholes are equal.
3. If equals be subtracted from equals, the remainders are equal.
4. Things which coincide with one another are equal to one another.
5. The whole is greater than the part.

The Theorems of Book I

We shall find it convenient to list here for future reference the 48 propositions that occur in Book I of *The Elements*, as translated by Heath. Some of the propositions are *constructions* that are carried out, and the rest are *theorems* that are proved.

1. On a given finite straight line to construct an equilateral triangle.
2. To place at a given point (as an extremity) a straight line equal to a given straight line.
3. Given two unequal straight lines, to cut off from the greater a straight line equal to the less.
4. If two triangles have the two sides equal to two sides respectively, and have the angles contained by the equal straight lines equal, they will also have the base equal to the base, the triangle will be equal to the triangle, and the remaining angles will be equal to the remaining angles respectively, namely those which the equal sides subtend.
5. In isosceles triangles the angles at the base are equal to one another, and, if the equal straight lines be produced further, the angles under the base will be equal to one another.
6. If in a triangle two angles be equal to one another, the sides which subtend the equal angles will also be equal to one another.
7. Given two straight lines constructed on a straight line (from its extremities) and meeting in a point, there cannot be constructed on the same straight line (from its extremities), and on the same side of it, two other straight lines meeting in another point equal to the former two respectively, namely each to that which has the same extremity with it.
8. If two triangles have the two sides equal to two sides respectively, and have also the base equal to the base, they will also have the angles equal which are contained by the equal straight lines.
9. To bisect a given rectilineal angle.
10. To bisect a given finite straight line.
11. To draw a straight line at right angles to a given straight line from a given point on it.
12. To a given infinite straight line, from a given point which is not on it, to draw a perpendicular straight line.
13. If a straight line set up on a straight line makes angles, it will make either two right angles or angles equal to two right angles.

14. If with any straight line, and at a point on it, two straight lines not lying on the same side make the adjacent angles equal to two right angles, the two straight lines will be in a straight line with one another.

15. If two straight lines cut one another, they make the vertical angles equal to one another.

16. In any triangle, if one of the sides be produced, the exterior angle is greater than either of the interior and opposite angles.

17. In any triangle two angles taken together in any manner are less than two right angles.

18. In any triangle the greater side subtends the greater angle.

19. In any triangle the greater angle is subtended by the greater side.

20. In any triangle two sides taken together in any manner are greater than the remaining one.

21. If on one of the sides of a triangle, from its extremities, there be constructed two straight lines meeting within the triangle, the straight lines so constructed will be less than the remaining two sides of the triangle, but will contain a greater angle.

22. Out of three straight lines, which are equal to three given straight lines, to construct a triangle: thus it is necessary that two of the straight lines taken together in any manner should be greater than the remaining one.

23. On a given straight line and at a point on it to construct a rectilineal angle equal to a given rectilineal angle.

24. If two triangles have the two sides equal to two sides respectively, but have the one of the angles contained by the equal straight lines greater than the other, they will also have the base greater than the base.

25. If two triangles have the two sides equal to two sides respectively, but have the base greater than the base, they will also have the one of the angles contained by the equal straight lines greater than the other.

26. If two triangles have the two angles equal to two angles respectively, and one side equal to one side, namely, either the side adjoining the equal angles, or that subtending one of the equal angles, they will also have the remaining sides equal to the remaining sides and the remaining angle to the remaining angle.

27. If a straight line falling on two straight lines make the alternate angles equal to one another, the straight lines will be parallel to one another.

28. If a straight line falling on two straight lines make the ex-

terior angle equal to the interior and opposite angle on the same side, or the interior angles on the same side equal to two right angles, the straight lines will be parallel to one another.

29. A straight line falling on parallel straight lines makes the alternate angles equal to one another, the exterior angle equal to the interior and opposite angle, and the interior angles on the same side equal to two right angles.

30. Straight lines parallel to the same straight line are also parallel to one another.

31. Through a given point to draw a straight line parallel to a given straight line.

32. In any triangle, if one of the sides be produced, the exterior angle is equal to the two interior and opposite angles, and the three interior angles of the triangle are equal to two right angles.

33. The straight lines joining equal and parallel straight lines (at the extremities which are) in the same directions (respectively) are themselves also equal and parallel.

34. In parallelogramic areas the opposite sides and angles are equal to one another, and the diameter bisects the areas.

35. Parallelograms which are on the same base and in the same parallels are equal to one another.

36. Parallelograms which are on equal bases and in the same parallels are equal to one another.

37. Triangles which are on the same base and in the same parallels are equal to one another.

38. Triangles which are on equal bases and in the same parallels are equal to one another.

39. Equal triangles which are on the same base and on the same side are also in the same parallels.

40. Equal triangles which are on equal bases and on the same side are also in the same parallels.

41. If a parallelogram have the same base with a triangle and be in the same parallels, the parallelogram is double of the triangle.

42. To construct, in a given rectilineal angle, a parallogram equal to a given triangle.

43. In any parallelogram the complements of the parallelograms about the diameter are equal to one another.

44. To a given straight line to apply, in a given rectilineal angle, a parallelogram equal to a given triangle.

45. To construct, in a given rectilineal angle, a parallelogram equal to a given rectilineal figure.

46. On a given straight line to describe a square.
47. In a right-angled triangle the square on the side subtending the right angle is equal to the squares on the sides containing the right angle.
48. If in a triangle the square on one of the sides be equal to the squares on the remaining two sides of the triangle, the angle contained by the remaining two sides of the triangle is right.

Defects of Euclid's Axioms

Euclid had set himself the goal of deriving all the significant theorems of geometry by logical deduction from his axioms. Although he deserves the highest praise for being the first to undertake this task, it is necessary to say also that he did not fully succeed in carrying it out. Modern critical analysis of *The Elements* reveals that there are serious defects in Euclid's system of axioms. We call attention to some of the most important of these defects in the next few paragraphs.

A Fallacious Proof

There is a gap in Euclid's axioms which serves as an open door through which fallacious proofs can be smuggled into the body of geometry. As an example, we give below a "proof" that purports to show that every triangle is isosceles. See if you can find the error in the "proof."

Let ABC be any triangle. Draw the bisector AD of angle A and draw the perpendicular bisector of side BC. Either AD is perpendicular to BC, or it is not perpendicular to BC. We call the first of these possibilities case I. If AD is not perpendicular to BC, then it intersects the perpendicular bisector of BC in some point O. Either O is inside triangle ABC, or outside triangle ABC, or on side BC. Let us call these possibilities case II, case III and case IV. Diagrams for these four cases are shown on page 71.

Case I. AD is perpendicular to BC. Therefore angle $BDA =$ angle CDA. Since AD bisects angle BAC, angle $BAD =$ angle DAC. Moreover, $AD = AD$. Therefore triangle ABD is congruent to triangle ACD. Hence $AB = AC$, and triangle ABC is isosceles.

Case II. Draw OE perpendicular to AC, and draw OF perpendicular to AB. Triangle AFO is congruent to triangle AEO because angle $FAO =$ angle EAO, angle $OFA =$ angle OEA, and $AO = AO$. Consequently $AF = AE$, and $OF = OE$. Now

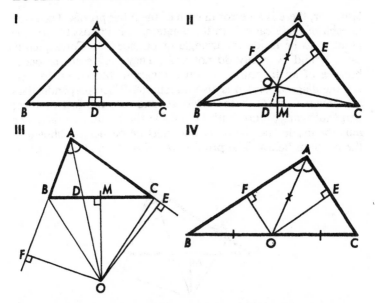

we consider triangles OFB and OEC. In these triangles, angles OFB and OEC are right angles. $OB = OC$, because any point on the perpendicular bisector of a segment is equally distant from the ends of the segment. Moreover, we have already proved that $OF = OE$. Consequently, triangles OFB and OEC are congruent, and $FB = EC$. Adding this equation to the equation $AF = AE$, already established above, we find that $AF + FB = AE + EC$. That is, $AB = AC$.

Case III. As in case II, draw OE perpendicular to AC and draw OF perpendicular to AB. The same argument given in case II establishes that triangle AFO is congruent to triangle AEO, triangle OFB is congruent to triangle OEC, $AF = AE$, and $FB = EC$. Subtracting the last equation from the one before it, we find that $AF - FB = AE - EC$. That is, $AB = AC$.

Case IV. The argument in case IV proceeds exactly as in case II, and leads to the conclusion that $AF + FB = AE + EC$. That is, $AB = AC$.

The Fallacy

There is no error in the proof of case I. In fact, under the conditions of case I, the bisector of angle A coincides with the perpendicular bisector of BC, and triangle ABC is indeed isosceles.

However, there is an error in each of the other proofs. The error in each of these cases is in the diagram. It is impossible for the point O to lie inside the triangle or on side BC. Consequently cases II and IV really do not arise. They only seem to occur because of an inaccurately drawn diagram. Moreover, in case III, the diagram drawn is also inaccurate. When perpendiculars are drawn from O to the sides AC and AB, the feet of these perpendiculars cannot both lie outside these sides. In fact, one must be inside and the other one must be outside, as shown in the diagram below. The proof that $AF = AE$ and $FB = EC$ is

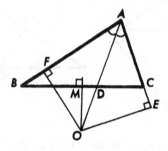

valid. However, $AB = AF + FB$, while $AC = AE - EC = AF - FB$. Then, since AB and AC are equal to the sum and difference respectively of the same two segments, they cannot be equal to each other.

We have asserted that the diagrams originally drawn for cases II, III and IV are wrong. This fact can be verified empirically for physical space by drawing accurately the bisector of angle A, the perpendicular bisector of BC, and the perpendiculars from O to AB and AC. However, such an empirical verification is not possible for mathematical space, which deals with ideal lines and points rather than with diagrams drawn with a pencil. In mathematical space it should be possible to determine, without relying on any drawn diagram, whether or not the relationships represented in the diagram are correct. Unfortunately, Euclid's axioms do not provide the means for making such a determination. To decide whether the diagrams are right or wrong we have to be able to answer these questions: Is the point O inside the triangle, on a side of the triangle, or outside the triangle? Is the point E between A and C or not? Is the point F between A and B or not? These questions concern the *order* in which points are arranged on a line or in a plane. Euclid's axioms cannot provide answers to such questions because they do not include any axioms about order relations. Euclid's failure to provide axioms about order is the gap which made possible the fallacious proof we have just seen. To close this gap, Euclid's axioms have to be revised so that they will include axioms about order relations. This is done in the modern versions of the axioms of Euclidean geometry.

Begging the Question

In his proposition 4 of Book I, Euclid undertakes to prove that two triangles are congruent if two sides and the included angle of one triangle are equal respectively to two sides and the included angle of the other. He does so by first placing one triangle on top of the other. This procedure begs the question, because after he "moves" a triangle it is no longer the same triangle. When he assumes that the triangle in its new position is congruent to the triangle in its original position, he is in effect assuming what he set out to prove. He was forced into this unconscious error because of another gap in his system of axioms. They do not include any axioms about congruence that might serve as a foundation for legitimate proofs about congruence. To remedy this defect, a modern version of the axioms of Euclidean geometry should include axioms about congruence. In the modern axioms listed on page 77 we shall see that proposition 4, which Euclid really assumed although he thought that he proved it, is explicitly assumed as an axiom.

Lines Without Holes

Suppose that point P is inside a circle, point Q is outside the circle, and the line segment PQ is drawn. In such circumstances, Euclid concludes without further ado that the circle intersects the segment in some point R. However, there is nothing in Euclid's axioms that would justify this conclusion. His axioms

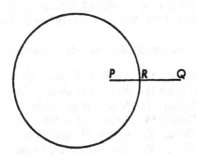

do not exclude the possibility that the segment is full of holes, and that the circle passes through one of the holes, so that it does not intersect the segment at all. In order to close the door against this possibility, it is necessary to include among the axioms an axiom of *continuity*. In the modern axioms listed on page 77, the axiom of Archimedes is used to help give a straight line the continuity properties that correspond to our intuitive notion that a line has no holes in it.

An Infinite Line

In his proof of proposition 16 of Book I, Euclid tacitly assumes that a line is infinite in extent. This assumption is, in effect, another one of his axioms, although he never stated it as such. We shall see later how Riemann, by discarding this assumption, was able to create another geometry different from Euclid's.

Undefined Terms

Although Euclid recognized the need for unproved propositions (the axioms), he failed to recognize the need for undefined terms. He opened his book with attempts to define the primitive concepts of point, line, etc. His attempts did not succeed, because at best they merely substituted other words for the words he tried to define. In modern versions of the axioms of Euclidean geometry, certain terms are frankly taken to be undefined. The properties ascribed to them are only those that are expressed in the axioms.

Hilbert's Axioms

Several different sets of axioms have been drawn up that are designed to eliminate the defects of Euclid's axioms. The best known of these are the axioms of David Hilbert (1862–1943), first presented by him in a series of lectures at the University of Göttingen in 1898–1899. We list them below in slightly modified form.

Hilbert's axioms are divided into six groups, as follows:

Group I.   Axioms of connection
Group II.   Axioms of order
Group III. Axiom of parallels
Group IV.   Axioms of congruence
Group V.   Axiom of continuity
Group VI.   Axiom of completeness

The axioms of connection contain the undefined terms, *point*, *straight line*, and *plane*, relate them to each other, and assert the existence of some points:

I,1. There is a straight line that contains two given distinct points.

I,2. There is at most one straight line that contains two given distinct points.

I,3. There is a plane that contains three given points which do not lie on the same straight line.

I,4. There is at most one plane that contains three given points which do not lie on the same straight line.

I,5. If two points of a straight line lie in a plane, then every point of the straight line lies in the plane.

I,6. If two planes have one point in common, then they have at least a second point in common.

I,7. On every straight line there are at least two points, in every plane there are at least three points that are not on the same straight line, and in space there are at least four points that are not on the same plane.

The axioms of order introduce the undefined term *between* as a relation among points on a line. Axioms of order were first formulated by M. Pasch in 1882.

II,1. If *A*, *B*, and *C* are points of a straight line, and *B* is between *A* and *C*, then *B* is also between *C* and *A*.

II,2. If *A* and *C* are two points of a straight line, then there is at least one point *B* that is between *A* and *C*, and there is at least one point *D* such that *C* is between *A* and *D*.

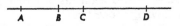

II,3. Of any three points on a straight line there is always one and only one which is between the other two.

II,4. Any four points on a straight line can be labeled *A*, *B*, *C*, and *D* in such a way that *B* is between *A* and *C* and also between *A* and *D*, and that *C* is between *A* and *D* and also between *B* and *D*.

On the basis of these axioms, a *segment AB* is defined as the set of points between the two points A and B. The term *segment* occurs in the next axiom of order:

II,5. Let A, B and C be three points that are not on the same straight line, and let a be a straight line in the plane determined by these points, and not passing through any of them. Then if a passes through a point of the segment AB, it will also pass through either a point of the segment BC or a point of the segment AC. (This axiom is known as the *Pasch Axiom*.)

On the basis of these axioms it is possible to show that a point on a line divides it into two parts which we call *rays* or *half-lines*, and that a line on a plane divides it into two *half-planes*. An *angle* (h, k) is defined as the figure formed by two half-lines h and k that emanate from one point and are on two distinct straight lines. A *polygon* is defined as a closed chain of segments joined end to end. A polygon is *simple* if none of these segments contains an end or an inside point of another one of these segments. In a plane it is then possible to define the *interior* and *exterior* of an angle and the *interior* and the *exterior* of a simple polygon. It can be shown with the help of the Pasch Axiom that if a ray from the vertex of an angle of a triangle lies between the sides of the angle, it must intersect the side opposite the angle.

The axiom of parallels has the form given to it by Proclus and popularized by Playfair (1795). We shall see later that it is equivalent to Euclid's postulate 5.

III. In a plane containing a given straight line and a given point that is not on the line, there is one and only one straight line that does not intersect the given line (Playfair Axiom).

The axioms of congruence introduce the undefined term "congruent" (represented by the usual symbol $\cong$) as a relation that may hold between two segments or two angles, and characterized by the following properties:

IV,1. If AB is a segment, and A' is a point on a line, then there is on that line on a given side of A' one and only one point B' such that $AB \cong A'B'$. $AB \cong AB$.

IV,2. If $AB \cong A'B'$, and $AB \cong A''B''$, then $A'B' \cong A''B''$.

IV,3. Let AB and BC be two segments of a straight line which lie on opposite sides of B, and let $A'B'$ and $B'C'$ be two segments of a straight line that lie on opposite sides of B'. Then if $AB \cong A'B'$, and $BC \cong B'C'$, we also have $AC \cong A'C'$.

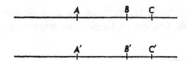

IV,4. If (h, k) is an angle, and a is a line on a plane, and h' is a half-line on a with vertex O, then there is on the plane on a given side of a one and only one half-line k' with vertex O such ·that angle $(h, k) \cong (h', k')$. Angle $(h, k) \cong$ angle (h, k).

IV,5. If angle $(h, k) \cong$ angle (h', k'), and angle $(h, k) \cong$ angle (h'', k''), then angle $(h', k') \cong$ angle (h'', k'').

IV,6. If, in the two triangles ABC and $A'B'C'$, $AB \cong A'B'$, $AC \cong A'C'$, and angle $BAC \cong$ angle $B'A'C'$, then angle $ABC \cong$ angle $A'B'C'$, and angle $ACB \cong$ angle $A'C'B'$.

Notice that axioms IV,2 and IV,5 assert for congruence of segments and congruence of angles the property that Euclid expressed in general form in *common notion 1* (see page 66). Axiom IV, 3 asserts for congruence of segments the property that Euclid expressed in general form in *common notion 2*. Hilbert gives no axiom to assert the same property for congruence of angles, because it can be proved as a theorem from the other axioms. Axiom IV,6 is essentially the same as Euclid's proposition 4 of Book I, which as we have seen, Euclid unconsciously assumed when he tried to prove it by superposition.

The axiom of continuity is the axiom of Archimedes:

V. Let A_1 be any point on a straight line between the points A and B. Take the points $A_2, A_3, A_4, \ldots$ so that A_1 is between A and A_2, A_2 is between A_1 and A_3, A_3 is between A_2 and A_4, etc., and so that

$$AA_1 \cong A_1A_2 \cong A_2A_3 \cong A_3A_4 \cong \ldots .$$

Then there is a positive integer n such that B is between A and A_n.

The axiom of completeness says in effect that the space under consideration is the largest one that satisfies the other axioms.

VI. No additional points, lines or planes can be added to the system without violating one of the axioms in groups I to V.

With the six groups of axioms of Hilbert as a foundation the geometry of Euclid can be developed without any logical flaws.

The geometry derived from these axioms or any equivalent set of axioms is now known as Euclidean geometry.

What Is an Axiom?

The transition from the axioms of Euclid to the axioms of Hilbert for the same system of geometry was a product of over two thousand years of evolution of geometric ideas, some aspects of which are presented in the chapters that follow. Another product of this evolution was a radical change in our understanding of the nature of an axiom in a deductive system. The old view, which was dominant up to the nineteenth century, was that an axiom is a self-evident truth. The modern view is that an axiom is a mere assumption. We shall see later why the old view had to be rejected. We pause now to explore briefly the meaning of the modern view.

In modern mathematics we make a distinction between *pure mathematics* and *applied mathematics*. In pure mathematics, the axioms of a deductive system are arbitrary assumptions, subject to the sole restriction that they must be consistent with each other. The theorems of the system are logical deductions from the axioms. Statements made in pure mathematics all have an "if . . . then" form. They assert that *if* the axioms are true, *then* the theorems are also true. They do not assert that either the axioms or the theorems are true. They merely assert that the axioms *imply* the theorems. In fact, if we change the axioms, we may find that the theorems are changed as a result. For example, the axioms of Euclid as modified by Hilbert imply that the sum of the angles of a triangle is two right angles. However, as we shall see in Chapter 8, the axioms of Lobatchewsky imply that the sum of the angles of a triangle is less than two right angles. Thus, in pure mathematics, we do not assert the truth of the statement that the sum of the angles of a triangle is two right angles. We merely say that the statement is true if Hilbert's axioms are true, but it is false if Lobatchewsky's axioms are true.

We must keep in mind, too, that in pure mathematics, the undefined terms have no meaning other than that given to them by the axioms. For example, in Euclidean geometry as a system of pure mathematics, the word *point* does not refer to a dot on a piece of paper, and the word *line* does not refer to a taut string. Points and lines are merely things that are assumed to be related to each other the way the axioms say they are. In fact, instead of using the word *points*, we might use the expression *things of the first kind*, and instead of using the word *lines*, we might use the expression *things of the second kind.* Then, as pure mathematicians we would assert that if things of the first kind and things of the second kind are related to each other as the axioms say they are, then they are also related to each other as the theorems say they are.

Now it may happen that we may find some specific system of objects, either conceptual or physical, in which the objects have the relations described in the axioms of a deductive system. Then the specific system of objects is called a *concrete representation* of the deductive system. If we interpret the undefined terms of the axioms as referring to the corresponding objects in the concrete representation, then they are true statements about these objects. Then because the axioms are true statements about these objects, the theorems are also true statements about them. When we thus *apply* a deductive system to a concrete system of which it is a mathematical model, we have left the realm of pure mathematics and entered the realm of applied mathematics. For example, when we use Euclidean geometry as an approximate model of physical space, we are in the realm of applied mathematics.

In applied mathematics, where a deductive system is used as a mathematical model of some specific system of objects, each undefined term in the axioms refers to a certain type of object in the specific system. In pure mathematics, however, an undefined term has no specific referent. In this sense, in pure mathematics we do not know what the undefined terms mean. In applied mathematics, the axioms and theorems are true statements about the objects in the specific system to which the mathematical model has been applied. In pure mathematics, however, our statements have a conditional form: we merely assert that the theorems are true *if* the axioms are true. Since we are not really asserting the truth of the axioms, we have no basis for asserting the truth of the theorems. In this sense, in pure mathematics we do not know whether what we say is true. This is why Bertrand Russell defined pure mathematics "as the subject in which we never know what we are talking about, nor whether what we are saying is true."

Actually, the choice of axioms in a system of pure mathematics is not entirely arbitrary. Historically, each significant mathematical system has emerged as an abstraction from some concrete system that was under investigation. However, once the mathematical system has been formulated, it is studied abstractly without regard to the concrete system from which it was derived. This disregard of the original concrete system is not a loss for the mathematician but a gain. By not tying the mathematical system to one particular concrete representation of it, we leave the door open for finding many different concrete representations of the same system. For example, a *field* is an abstract mathematical system characterized by the axioms given on pages 32–3, and the real number system is a concrete representation of a field. By deducing the properties of a field from these axioms alone without reference to the real number system we arrive at conclusions that are equally applicable to the real number system, the rational number system, the complex number system, or a finite field, all of which are different concrete representations of the abstract field structure.

Some Theorems Proved

Three concepts of special significance in Euclidean geometry are those of order, congruence, and parallelism. In later chapters we trace the emergence of geometries in which one or more of these concepts are absent. To make possible some comparisons between the methods and content of Euclidean geometry with the methods and content of the later geometries, we present below proofs of several key theorems of Euclidean geometry:

The medians of a triangle are concurrent.

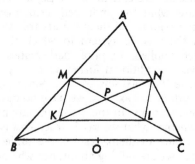

Proof: We first prove that any two medians of a triangle divide each other in the ratio 1:2. In triangle *ABC*, shown above, let *M, N,* and *O* be the midpoints of *AB, AC,* and *BC* respectively. Then the medians are *CM, BN,* and *AO*. (Only *CM* and *BN* are drawn in the diagram.) In triangle *AMC*, the line *BN* passes through none of the vertices, and contains a point of the segment *AC*, but not of the segment *AM*. Hence, by the Pasch Axiom, it must contain a point of the segment *CM*. That is, *BN* and *CM*

intersect at some point *P*. Let *K* and *L* be the midpoints of *BP* and *CP* respectively, and draw *MN, NL, LK,* and *KM.* In triangle *ABC, MN* joins the midpoints of two sides of the triangle. Hence it is parallel to the third side *BC*, and also is equal to half of it. In triangle *BPC, KL* joins the midpoints of two sides of the triangle, so it too is parallel to the third side *BC* and equal to half of it. Since *MN* and *KL* are both parallel to the same line, they are parallel to each other. Since they are both equal to half of the same line segment, they are also equal to each other. Consequently, quadrilateral *MNLK*, in which *MN* and *KL* are both parallel and equal, is a parallelogram. It follows that the diagonals *ML* and *KN* bisect each other. That is, *MP = PL* (which is half of *PC*), and *NP = PK* (which is half of *PB*). Therefore *CM* and *BN* divide each other in the ratio 1:2. By a similar proof, using the median *AO* and *MC*, we can show that *AO* intersects *MC* in a point that divides it in the ratio 1:2. Then this point of intersection must also be *P*. Consequently, all three medians of triangle *ABC* pass through *P*. That is, they are concurrent.

We shall give another proof of this theorem in Chapter 5. In Chapter 10 we shall see that, in a sense that will be explained then, this theorem is not really a theorem of Euclidean geometry, but is a theorem of a more general kind of geometry known as *affine* geometry, in which the concept of congruence is absent, but the concept of parallelism is present.

In a plane, if points A, B and C are on one straight line and points D, E and F are on another straight line, and if AE is parallel to BD and EC is parallel to FB, then AF is parallel to CD. (This theorem is known as *Pappus's Theorem*, after Pappus of Alexandria, who discovered it about A.D. 350.)

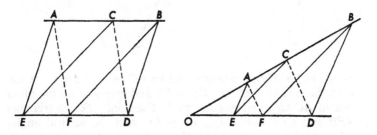

Proof: The lines *AB* and *ED* are either parallel or they intersect at some point *O*. We consider each of these cases separately. If *AB* is parallel to *ED*, then quadrilateral *ABDE* is a parallelogram, since both pairs of its opposite sides are parallel. Consequently

$AB = ED$. Similarly, quadrilateral $CBFE$ is a parallelogram, and $CB = EF$. Subtracting these two equations, we get $AC = FD$. Since AC is also *parallel* to FD, we conclude that quadrilateral $ACDF$ is a parallelogram, and hence AF is parallel to CD. If AB intersects ED at O, then, in triangle OBD, AE is parallel to one side of the triangle, and hence divides the other two sides proportionally. That is, $\dfrac{OA}{OB} = \dfrac{OE}{OD}$. Similarly, in triangle OBF, since EC is parallel to FB, we have $\dfrac{OC}{OB} = \dfrac{OE}{OF}$. Dividing these two equations, we find that $\dfrac{OA}{OC} = \dfrac{OF}{OD}$, that is, in triangle OCD the line AF divides two of the sides proportionally. Therefore it is parallel to the third side CD.

In Chapter 10 we shall prove by other methods a generalization of this theorem. The generalization is a theorem of a more general kind of geometry known as *projective* geometry, in which order relations, congruence, and parallelism are all absent.

In any triangle, if one of the sides be produced, the exterior angle is greater than either of the interior and opposite angles. (This is proposition 16 in Book I of Euclid's *Elements*.)

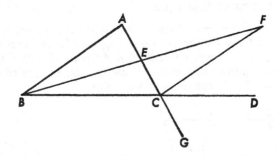

Proof: Let ABC be a triangle, and let BC be extended to D. Let E be the midpoint of AC. Draw BE and extend it on the other side of AC an equal distance to F. Then draw FC. In triangles AEB and CEF, $AE = EC$, $BE = EF$, and the vertical angles AEB and CEF are also equal. Hence the triangles are congruent, and angle $FCA =$ angle A. Since angle DCA is greater than angle FCA, it is also greater than angle A. If AC is extended to G, we can show in the same manner that angle BCG, which is equal to angle DCA, is greater than angle B.

A key step in this proof is the extension of *BE* on the other side of *AC* an equal distance to *F*. The concept of the "other side" of *AC* is based on the axioms of order. The possibility of making *EF* equal to *BE* is assured by Hilbert's axiom IV,1. Together these axioms imply that a line is infinite in extent. (If the line were finite in extent, the line *BE* might not extend far enough on the other side of *AC* to permit making *EF* equal to *BE*.) So the proof is valid only on the assumption that a line is infinite in extent. In Chapter 6 we shall encounter a geometry in which a line is finite in extent. In this geometry, known as *elliptic* geometry, the order axioms do not hold, and the theorem we have just proved does not apply. In fact, we get a foretaste of this other kind of geometry on pages 85 to 92.

If a straight line falling on two straight lines make the alternate (interior) angles equal to one another, the straight lines will be parallel to one another. (This is proposition 27 in Book I of Euclid's *Elements*.)

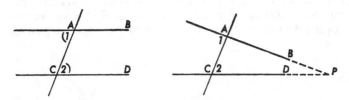

Proof: Let *AB* and *CD* be cut by *AC* so that angle 1 = angle 2, as shown in the diagram. If *AB* and *CD* meet at some point *P*, then, by the theorem of the preceding paragraph, angle 1 is greater than angle 2, contradicting our hypothesis. Therefore *AB* and *CD* cannot meet. That is, they are parallel.

Since this proof makes use of Euclid's proposition 16 of Book I, it, too, depends on the assumption that a line is infinite in extent. Consequently, the theorem need not apply in elliptic geometry, where a line is finite in extent.

The sum of the angles of a triangle is equal to two right angles. (This is proposition 32 in Book I of Euclid's *Elements*.)

Proof: See page 29 for the proof given by the Pythagoreans. Notice that this proof makes use of proposition 29 in Book I of Euclid's *Elements* (the converse of proposition 27). We shall show on page 198 that this proposition is equivalent to the axiom of parallels, and hence cannot apply in either an *elliptic* or a *hyperbolic* geometry in which this axiom does not hold.

In right-angled triangles the square on the side subtending the right angle is equal to the (sum of) the squares on the sides containing the right angle. (This is the *Pythagorean Theorem*, and is proposition 47 in Book I of Euclid's *Elements*.)

Proof: See the proof given on page 17. The proof assumes the existence of a square. We shall find in Chapter 8 that there are no such things as squares in either elliptic or hyperbolic geometry, and that in these geometries the Pythagorean Theorem does not hold, but is replaced by a different theorem about the sides of a right-angled triangle.

Ideal Points

In the proof of Pappus's theorem, we had to use two separate diagrams to illustrate two possibilities: the two given lines might be parallel, or they might intersect at some point O. These two possibilities can be related to each other in the following way. Suppose we have a line a, a point P that is not on a, and a line b that passes through P and intersects a at O. Let us see what happens if we allow O to move to the left, through the positions

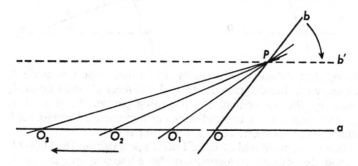

O_1, O_2, O_3 and beyond. The diagram shows that as O moves to the left, the line b rotates clockwise around P, and the angle between a and b becomes progressively smaller. As the point O moves off to infinity, the line b approaches as a limit the line b' that is parallel to a. This fact suggested to Johannes Kepler (1571–1630) the idea of saying that parallel lines meet at a "point at infinity." Actually there is no such thing as a "point at infinity" on a Euclidean line, so the statement that two lines meet at a "point at infinity" must be understood as a metaphor for the statement that the lines are parallel. To permit the use of this metaphor, it is necessary to add to the points of a line an additional point supposed to be "at infinity." All lines in a plane that are parallel to a given direction pass through the same "point at infinity." There is a separate "point at infinity" for each direction in a plane. The points at infinity are referred to as the "ideal" points in the plane. The Euclidean plane, augmented by the addition of ideal points, is called the *extended Euclidean plane*.

At first the ideal points were used only as a convenient metaphor that permitted separate cases such as parallel lines and intersecting lines to be treated as one. However, they have turned out to have a deeper significance, as we shall see in Chapter 10.

Units of Measure

The existence of the congruence relation for line segments makes it possible to introduce the measure known as the *length* of a line segment. To do so, we first choose an arbitrary segment to use as unit of length, and then compare with the unit the segment to be measured. A segment that can be divided into two units is said to have length 2, etc.

The existence of the congruence relation for angles also makes it possible to introduce a measure of angles. This measure, too, is obtained by comparing with some unit angle the angle that is to be measured. However, the choice of the unit angle need not be arbitrary. Euclidean geometry is equipped with a *natural unit* of angle measure, which is easily defined. This natural unit is the *right angle*, defined as the angle formed by two intersecting straight lines that form equal adjacent angles.

There is no way, except by arbitrary choice, of specifying a unit of measure for segments. Consequently, *in Euclidean geometry, while there is a natural unit of angle measure, there is no natural unit of length.*

Bringing Heaven Down to Earth

An important use of geometry in ancient times was for the study of astronomy, since the annual cycle of the seasons is geared to the shifting position of the stars in the sky. To the observer looking at the sky, the sky looks like a sphere of which he is the center. This sphere is called the *celestial sphere*. Half of it is visible above him. The other half, below him, is hidden by the ground. The stars in the sky look like points on the surface of the sphere. Consequently the study of astronomy led directly to the study of *spherical geometry*, which explores the relation-

ships of points, lines, triangles, etc., drawn on the surface of a
sphere. A triangle drawn on the surface of a sphere is called a
spherical triangle. Spherical triangles made their first appearance
in mathematical literature in a book on spherics by Menelaus
(about A.D. 100).

Long after the sky was pictured as a sphere it was recognized
that the earth itself is a sphere. Then spherical geometry, originally developed for the study of the sky, was brought down to
earth as an aid to geodesy and navigation.

Lines and Angles on a Sphere

The shortest path on a plane between two points on the plane
is along the straight line that joins them. The shortest path on a
sphere between two points on the sphere is along the great circle

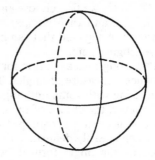

Great circles on a sphere

that joins them. (A great circle is one whose plane passes through
the center of the sphere.) For this reason, great circles play the
same role in spherical geometry that straight lines do in plane

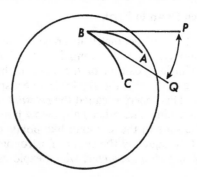

geometry. The angle between two great circles intersecting at a
point is defined to be the angle between the straight lines that are
tangent to them at that point. In the diagram, for example,
where *PB* is tangent to arc *AB* at *B*, and *QB* is tangent to arc *CB*
at *B*, angle *ABC* is defined to be equal to angle *PBQ*. An angle
equal to angle *PBQ*, and hence to spherical angle *ABC* can be
obtained in the following way: the planes that contain arcs *AB*
and *BC* intersect along the line *BO*, where *O* is the center of the
sphere. A plane perpendicular to *BO* intersects these planes in
lines *RS* and *TS*. Angle *RST* is equal to angle *PBQ*.

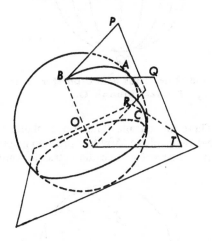

Sphere Versus Plane

A navigator of a ship traveling short distances on the surface
of the earth (less than 200 miles) may safely assume that the
surface of the earth is flat, because in that short distance it
deviates only slightly from a plane. However, if the ship travels
long distances, the navigator must take into account the fact that
the surface of the earth is a sphere. Consequently, while he may
use plane geometry for short distances, he must use spherical
geometry for long distances. It is interesting and instructive to
compare the two geometries to see some of their resemblances
and differences. In making this comparison, when we say "line"
on a plane it will be understood to be a straight line, and when
we say "line" on a sphere it will be understood to be a great
circle.

Connection. On a plane, two points determine one and only
one line. This is also true on a sphere, except when the two points

are the ends of a diameter of the sphere. In that exceptional case, every line through one point also goes through the other.

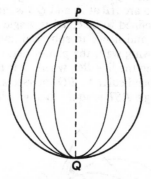

If P and Q are ends of a diameter, every line through P goes through Q.

Parallelism. There are parallel lines on a plane, but there are no parallel lines on a sphere. Any two lines on a sphere intersect, and in fact they intersect at two diametrically opposite points.

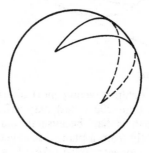

Any two lines on a sphere intersect at two diametrically opposite points.

Order relations. Both on a sphere and on a plane, if a point is chosen on a line, it is possible to move along the line away from the point in exactly two opposite directions, shown by the arrows in the next diagram. However, although in plane geometry a point divides a line into two parts, this is not true in spherical geometry. This is seen most clearly by thinking of the line as a

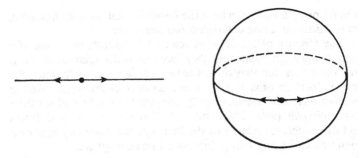

string and the point on the line as a cut in the string. Since a line on a sphere is a closed loop, a single cut does not divide it into two pieces. Because of this fact, the order relation that exists among points on a line in a plane does not hold for points on a line on a sphere. For example, on a plane, if *A*, *B*, and *C* are three distinct points on a line, then one and only one of them is between the other two. This statement does not apply to three points on a line of a sphere. For example, in the diagram below, on the straight line *AC*, *B* is between *A* and *C*. *A* is not between *B* and *C*, and *C* is not between *A* and *B*. These relations can be observed in the fact that if you move from *A* to *C* you pass *B* on the way, but if you move from *B* to *C* you do not pass *A* on the way, and if you move from *A* to *B* you do not pass *C* on

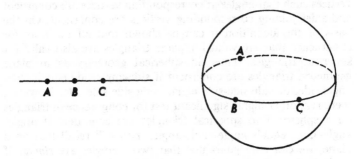

the way. However, on the great circle *ABC*, to pass from any one of the three points, *A*, *B*, and *C* to another, you may choose from two opposite directions of motion. One of these directions will always take you past the third point. Consequently it makes no sense to say one of the points is between the other two. Because of the *between* relation that exists on a line of a plane, two points on such a line determine *one* segment on the line.

On a line of a sphere, where the *between* relation does not exist, two points on a line determine *two* segments.

The absence of the *between* relation for points on a line of a sphere does not mean that they have no order relation at all. It merely means that they do not have an order relation that satisfies the Hilbert axioms II,1–4. These axioms characterize what is known as *linear order*. Actually the points on a line of a sphere have *circular order* which may be characterized by a different set of properties, related to the fact that there are two segments determined by any two points on a line of a sphere.

Congruence of segments. Hilbert's axiom IV,1 asserts for plane geometry that a given segment can be laid off on a line on either of two sides of any point on the line. This applies in modified form for spherical geometry: a given segment can be laid off on a segment of a line on either of two sides of any point of the *segment, provided that the given segment is not too long.* Except for this modification, the congruence properties of segments on a line of a sphere are the same as those on a line of a plane.

Congruence of angles. Angles on a sphere and on a plane have essentially the same congruence properties.

Congruence of triangles. Congruence of triangles can be defined in the same way on a sphere as on a plane: Two triangles are congruent if there is a one-to-one correspondence of their vertices such that angles at corresponding vertices are congruent and sides joining corresponding vertices are congruent. On the basis of this definition it can be shown that all the tests for congruence that are valid for plane triangles are also valid for spherical triangles. Thus, in spherical geometry as in plane geometry, triangles are congruent if side-angle-side equals side-angle-side, or side-side-side equals side-side-side, etc. However, there is another highly significant test for congruence of triangles on a sphere: Two spherical triangles are congruent if angle-angle-angle equals angle-angle-angle. You will recall that on a plane, we can only guarantee that two triangles are *similar* if angle-angle-angle equals angle-angle-angle. That is, on a plane, two such triangles have the same shape but may or may not have the same size. But on a sphere, if two triangles have the same shape, they also have the same size, and are congruent. That is, *there are no such things in spherical geometry as similar triangles (that are not congruent).*

The line. On both a sphere as well as on a plane, a line is *unbounded.* That is, no matter how far you move along a line in a particular direction there is always more of the line ahead of you. However, while a line on a plane is infinite in extent, a line on a sphere, since it is a great circle of the sphere, is finite in extent. Consequently there is an upper limit to the possible length of a line segment on a sphere, namely the length of the circumference of a great circle.

Units of measure. Since angles on a sphere are defined in terms of angles on a plane (see page 87), they have the same system of measure. We have already seen that on the Euclidean plane there is a natural unit of angle measure, namely the right angle. Consequently, the sphere also has a natural unit of angle measure. In practice, it is customary to use either of two other units of angle measure, the *degree,* or the *radian,* which are related to the natural unit by these equations:

$$180 \text{ degrees} = \pi \text{ radians} = 2 \text{ right angles.}$$

On a plane, there is no natural unit of measure of length. However, *there is a natural unit of measure of length on a sphere,* namely the length of the circumference of a great circle on the sphere. However, since the length of an arc of a circle in comparison to the circumference of the circle is measured by the angle that the arc subtends at the center of the circle, it is customary to measure lengths on a sphere in degrees or radians, where these units as units of measure of length are related to the natural unit by these equations:

$$180 \text{ degrees} = \pi \text{ radians} = 2 \text{ quadrants}$$
$$= 2 \text{ fourths of a great circle.}$$

Angles of a triangle. In a plane triangle, the sum of the angles is always *equal* to two right angles. In a spherical triangle, the

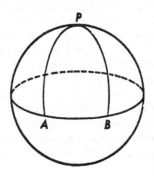

sum of the angles is always *more than* two right angles. For example, if *P* is the north pole on the earth, and *A* and *B* are two points on the equator that are 90° apart, then each of the angles of triangle *PAB* is a right angle, and the sum of the angles of triangle *PAB* is 3 right angles.

From the rules about the sum of the angles of a triangle it is easy to derive the rules for the sum of the angles of a quadrilateral. If a diagonal of the quadrilateral is drawn, as in the diagram below, it is seen that the sum of the angles of a quadrilateral is the sum of the angles of two triangles. Consequently, while the sum of the angles of a plane quadrilateral is 4 right angles, the sum of the angles of a spherical quadrilateral is more than 4 right angles. An interesting consequence of this rule is

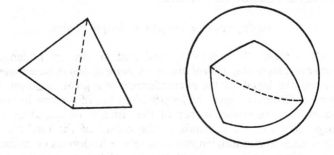

the fact that there is no such thing as a rectangle (a quadrilateral with four right angles) on a sphere.

Plane Trigonometry

Another by-product of the study of astronomy in ancient and medieval times was the development of the subject of *plane trigonometry*, in which formulas were discovered relating the measures of the sides and angles of a triangle. Plane trigonometry first appeared in embryonic form as part of astronomy in the work of the Greek astronomer Hipparchus (about 140 B.C.). It was developed further by Ptolemy (about A.D. 150) and the Hindu astronomer Aryabhata (about A.D. 510). It appeared as a separate subject for the first time in a book by the Persian astronomer Nasîr ed-dîn al-Tûsî (about A.D. 1250).

If *ABC* is a right triangle, the ratios of the sides of the triangle are found to be related to the measures of the acute angles in the triangle. Three of these ratios, which we shall have occasion to

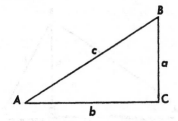

refer to later, are now called sine, cosine, and tangent, abbreviated as sin, cos, and tan, and are defined as follows:

$$\text{sine of an angle} \quad = \frac{\text{leg opposite the angle}}{\text{hypotenuse}}$$

$$\text{cosine of an angle} \quad = \frac{\text{leg adjacent to the angle}}{\text{hypotenuse}}$$

$$\text{tangent of an angle} = \frac{\text{leg opposite the angle}}{\text{leg adjacent to the angle}}.$$

Thus, in triangle *ABC* shown above,

$$\sin A = \frac{a}{c}; \quad \sin B = \frac{b}{c};$$

$$\cos A = \frac{b}{c}; \quad \cos B = \frac{a}{c};$$

$$\tan A = \frac{a}{b}; \quad \tan B = \frac{b}{a}.$$

The Law of Sines

With the help of these formulas that apply only to a right triangle, other formulas that apply to any triangle in a plane can be derived. As an example we derive the formula known as the *law of sines*.

Let *ABC* be any triangle, and denote the sides opposite *A*, *B* and *C* by *a*, *b* and *c* respectively. Draw *BD* perpendicular to *AC*, and denote it by *h*. Then, in right triangle *ABD* we have sin $A = h/c$, and in right triangle *CBD* we have sin $C = h/a$. Dividing these two equations, we obtain the law in equation (1).

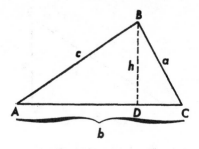

$$(1) \qquad \frac{\sin A}{\sin C} = \frac{a}{c}.$$

Generalized Trigonometric Functions

If x is the number of radians in an acute angle, it is shown in the calculus that $\sin x$ and $\cos x$ may be expressed as infinite series as follows:

$$(2) \quad \sin x = \frac{x}{1} - \frac{x^3}{1\cdot2\cdot3} + \frac{x^5}{1\cdot2\cdot3\cdot4\cdot5} - \frac{x^7}{1\cdot2\cdot3\cdot4\cdot5\cdot6\cdot7} + \cdots.$$

$$(3) \quad \cos x = 1 - \frac{x^2}{1\cdot2} + \frac{x^4}{1\cdot2\cdot3\cdot4} - \frac{x^6}{1\cdot2\cdot3\cdot4\cdot5\cdot6} + \cdots.$$

These formulas permit an extensive generalization of the trigonometric functions $\sin x$ and $\cos x$. First, they show how the values of $\sin x$ and $\cos x$ may be calculated from the value of the *number* x, without first drawing a right triangle that contains an angle with x radians. Secondly, it is found that the series have meaning, in the sense explained on page 54, for many more numbers besides those that are measures of acute angles. In fact, the series have meaning when x is any member of the real number system or the complex number system. Consequently we use the series to extend the definition of $\sin x$ and $\cos x$ to any real or complex number x. In the series for $\sin x$, if x is replaced by $-x$, the sign of every term of the series is changed. Consequently $\sin(-x) = -\sin x$. We shall use this fact on page 241 in Chapter 8.

In the series for $\sin x$ and $\cos x$, the signs of alternate terms are minus. If these minus signs are replaced by plus signs, two other series are obtained that define new functions known as the *hyperbolic sine of x* and the *hyperbolic cosine of x*, abbreviated as $\sinh x$ and $\cosh x$ and expressed in equation (4) and equation (5).

(4) $\sinh x = \dfrac{x}{1} + \dfrac{x^3}{1\cdot 2\cdot 3} + \dfrac{x^5}{1\cdot 2\cdot 3\cdot 4\cdot 5} + \dfrac{x^7}{1\cdot 2\cdot 3\cdot 4\cdot 5\cdot 6\cdot 7} + \cdots .$

(5) $\cosh x = 1 + \dfrac{x^2}{1\cdot 2} + \dfrac{x^4}{1\cdot 2\cdot 3\cdot 4} + \dfrac{x^6}{1\cdot 2\cdot 3\cdot 4\cdot 5\cdot 6} + \cdots .$

If we substitute iy for x in equation (2), and note that i is the complex number with the property that $i^2 = -1$, then it is easy to verify that

(6) $\sin iy = i \sinh y.$

We shall have occasion to refer to this formula in Chapter 8.

Spherical Trigonometry

Just as *plane trigonometry* provides us with formulas that relate the measures of the sides and angles of a *plane triangle*, *spherical trigonometry* provides us with formulas that relate the measures of the sides and angles of a *spherical triangle*. As an illustration, we derive the law of sines for a spherical triangle.

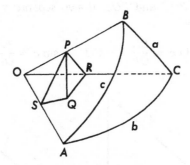

Let ABC be a spherical triangle on a sphere whose center is at O. Denote the sides opposite A, B, and C by a, b, and c respectively. Draw the radii OA, OB, and OC. Let P be any point on OB, and draw PQ perpendicular to plane OAC. Through PQ draw a plane perpendicular to OA, cutting planes OBA and OAC in PS and SQ respectively. Similarly, through PQ draw a plane perpendicular to OC, cutting planes OBC and OAC in PR and RQ respectively. Then angles PQS, PQR, PSO, and PRO are all right angles. Moreover, as indicated on page 87, angle PSQ = spherical angle A, and angle PRQ = spherical angle C.

Also angle $POR = a$, and angle $POS = c$, since an arc of a circle is measured by the angle it subtends at the center of the circle.

In triangles PSQ and PRQ, drawn separately below, we see immediately that

$$\sin A = \frac{PQ}{PS} \quad \text{and} \quad \sin C = \frac{PQ}{PR}.$$

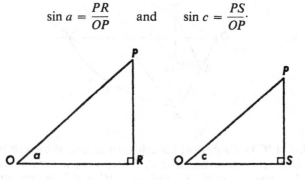

Dividing these two equations, we get

$$\frac{\sin A}{\sin C} = \frac{PR}{PS}.$$

In triangles POR and POS, drawn separately below, we see immediately that

$$\sin a = \frac{PR}{OP} \quad \text{and} \quad \sin c = \frac{PS}{OP}.$$

Dividing these two equations, we get

$$\frac{\sin a}{\sin c} = \frac{PR}{PS}.$$

Equating the two values of PR/PS that we have obtained, we get the statement of equality expressed by equation (7) below.

(7) $$\frac{\sin A}{\sin C} = \frac{\sin a}{\sin c}.$$

This is the law of *sines* in spherical trigonometry.

In the formula above, a and c represent the measures of arcs BC and BA expressed in spherical units of length, that is, radians. We shall find it useful to have another version of the formula in which the lengths of the arcs are measured in the same linear units as the radius r of the sphere. To do so, we must first show how measures expressed in terms of the linear units are related to measures expressed in terms of spherical units of length. Let x' be the length in linear units of an arc of x radians. We shall express in two forms the ratio

$$\frac{\text{length of } x}{\text{length of a great circle}}.$$

Using linear units, the ratio is $x'/2\pi r$. Using radians, the ratio is $x/2\pi$. By equating these two versions of the ratio, we find that $x = x'/r$. If, in the law of sines for a spherical triangle we replace a by a'/r, and replace c by c'/r, the law takes this form:

$$\frac{\sin A}{\sin C} = \frac{\sin \dfrac{a'}{r}}{\sin \dfrac{c'}{r}}.$$

To simplify the notation, we now write a instead of a' and c instead of c', thus getting this version of the law of sines, in which it is understood that a and c are measured in the same linear units as r:

(8) $$\frac{\sin A}{\sin C} = \frac{\sin \dfrac{a}{r}}{\sin \dfrac{c}{r}}.$$

Orientation in the Plane

If A and B are two points on a Euclidean line, we can specify two opposite *senses* or directions on the line. One sense is from A to B, and the other sense is from B to A, as indicated by the arrows in the diagram. Similarly, if A, B, and C are three points

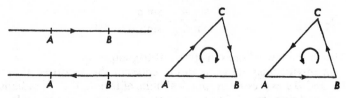

Two senses on a Euclidean line   Two senses on a Euclidean plane

in a Euclidean plane that are not on the same straight line, we can specify two opposite senses or directions around the perimeter of triangle *ABC*. One sense is from *A* to *C* to *B* to *A*, and the other sense is from *A* to *B* to *C* to *A*, as indicated by the arrows in the diagram. If we view these diagrams from above, the first sense may be described as the clockwise sense, and the second sense as the counter-clockwise sense. When a sense is chosen for a triangle, it automatically induces a particular sense on each of its sides. For example, the clockwise sense in triangle *ABC* induces the sense from *B* to *A* on side *AB*.

Suppose we take any triangle in a Euclidean plane and specify a particular sense around its perimeter. If we move the triangle by sliding it in the plane, it will carry this sense with it to its new position. If, after moving about along any path in the plane, the triangle returns to its original position, it will still have the same sense that it had originally. Because a sense of rotation moved about in the plane returns unchanged to its starting position, we say that the plane is an *orientable* surface.

It is possible to define the concept of orientability without invoking the intuitive notion of "moving" a triangle in the plane. To do so, notice first that if two triangles with a common side, such as triangles *ABC* and *BCD* in the diagram below are assigned the *same sense*, say the clockwise sense, the sense of each triangle induces a *different sense* on the common side. For example, in the diagram, the sense of triangle *ABC* induces the sense from *B* to *C* on side *BC*, but the sense of triangle *CBD*

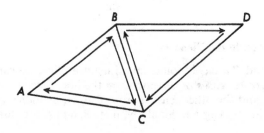

induces the sense from C to B on that side. Suppose the entire plane is covered by a network of triangles. If we want to assign a sense to each triangle in such a way that all the senses look the same (either clockwise or counter-clockwise) as seen from above the plane, then we must do so in such a way that the senses of adjacent triangles induce opposite senses on their common side. This fact is the basis of the following definition: A surface is *orientable* if, when it is covered by a network of triangles, it is possible to assign a sense to each triangle in such a way that each pair of adjacent triangles induces opposite senses on their common side. When a sense has been assigned in this way to all the triangles of a triangular network, we say that the surface is *oriented*. Because such an assignment of senses to the triangles of a triangular network on a Euclidean plane is possible, a Euclidean plane is orientable.

Suppose a Euclidean plane is covered by a network of triangles, and we orient the plane by assigning senses to the triangles so that they all are clockwise as seen from above the plane. Then they will all be counter-clockwise as seen from below the plane. This fact makes it possible for us to distinguish the topside from the underside of the plane, by the appearance of the assigned sense of the triangles as seen from each side. Consequently, the fact that the Euclidean plane is orientable is related to the fact that it has two sides. In general, any orientable surface is a two-sided surface, and vice versa.

A One-sided Surface

Not all surfaces are orientable and two-sided. To prove this fact we give directions for making a simple surface that is non-orientable and is one-sided. Cut a long narrow rectangular strip of paper or cellophane, and label the vertices at both ends A and B as shown in the diagram. Give one end a half-twist and bring the two ends together to make the A's coincide and the B's coincide, as shown in the diagram. Fasten the ends together in this position with masking tape. The loop with a half-twist that is formed in this way is called a *Möbius strip*. We shall verify in two ways that the Möbius strip is non-orientable.

Draw a circle on the Möbius strip and put an arrowhead on it to specify a particular sense of rotation on it. Cut a rectangular sheet of cellophane that is slightly more than double the width of the Möbius strip. Fold it in half, and hang it over the strip. Seal the ends with tape so that the folded sheet becomes a sleeve that you can slide along the Möbius strip. Slide it into position

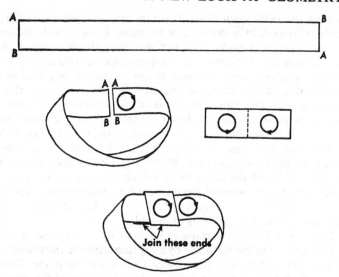

directly over the circle on the strip, and trace the circle and its arrowhead onto each half of the sleeve. (The drawing next to the Möbius strip shows how the traced circles would look if the sleeve were opened up again.) The circles on the sleeve now have the same sense of rotation as the circle on the strip. You can move this sense of rotation on the surface of the strip by sliding the sleeve along the strip. If you slide the sleeve around the strip until it returns to its original position over the circle, you will find that the sense of rotation indicated on the sleeve no longer matches the sense of rotation indicated on the circle drawn on the strip. Since the sense of rotation did not return unchanged, the Möbius strip is non-orientable.

To show in another way that a Möbius strip is non-orientable, take a small narrow rectangle suitable for making a small Möbius strip, and cover it with a network of triangles, as shown in the diagram below. Draw three arrows inside triangle XYC, as shown in the drawing, to give it a clockwise orientation. Then

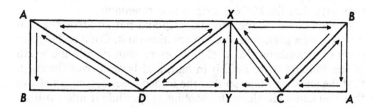

the sense on XY is from Y to X. We shall now try to give every triangle a sense of rotation so that adjacent triangles induce opposite senses on their common side. In triangle XYC, the sense on XC is from X to C. So in triangle XCB, the sense on XC has to be from C to X. Consequently, the sense of triangle XCB, as seen by the reader, will be clockwise, and the sense on BC is from B to C. Proceed in this way from each triangle to the next one. In triangle ACB, the sense on CB will be from C to B, and the sense on AB will be from B to A. Consequently, in triangle ABD, which will be adjacent to it on the Möbius strip, the sense on AB will be from A to B, and the sense on AD will be from D to A. In triangle ADX, the sense on AD will be from A to D, and the sense on DX will be from D to X. Then, in triangle DXY, the sense on DX will be from X to D, and the sense on XY will be from Y to X. Now notice that the two adjacent triangles that have XY as a common side induce on it the same sense, from Y to X. This shows that it is impossible to assign a sense to each triangle in such a way that each pair of adjacent triangles will induce opposite senses on their common side. Consequently, in view of the definition given on page 99, the Möbius strip is not orientable.

To verify directly that the Möbius strip is one-sided, draw a dot on the strip, and behind it, on what appears to be the "other side of the strip," put another dot. Now draw a line that starts at the first dot and runs around the strip as shown in the drawing. After one trip around, you will reach the other dot. The fact that you could join the two dots by a line that does not go over

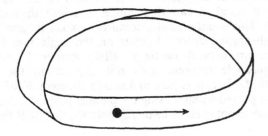

the edge of the strip shows that they are on the same side of the strip, and the strip has only one side. Incidentally, the strip also has only one edge, too. You can verify this fact by running your finger along it.

The Möbius strip is not a mere curiosity. We shall encounter it again in Chapters 8 and 11.

1. *Euclid's common notions.* Euclid's common notions are valid for line segments and angles if the word "equal" is interpreted to mean "congruent." However, they need not be valid if the word "equal" is given some other interpretation. Suppose, for example, we interpret the word "equal" to mean "equivalent as point sets." Suppose, too, that we say a set is greater than one of its subsets if it is not equivalent to this subset. Consider the sets A, B and C defined below:

$A = \{1, 2, 3, \ldots\}$, (the set of all positive integers)
$B = \{3, 4, 5, \ldots\}$, (the set of positive integers greater than 2)
$C = \{5, 6, 7, \ldots\}$, (the set of positive integers greater than 4)

a) Show that A, B and C are equivalent.

b) B is a part of A in the sense that B is a proper subset of A. Is A greater than B? Is Euclid's common notion 5 valid in this case?

c) Define $A - B$ as the set of elements left over when the elements of B are removed from A. Define $A - C$ in a similar manner. Are the sets $A - B$ and $A - C$ equivalent? Is Euclid's common notion 3 valid in this case?

2. *The side-angle-angle theorem.* In many high school geometry textbooks, the side-angle-angle congruence theorem is proved with the help of the theorem that the sum of the angles of a triangle is equal to two right angles (Euclid's theorem 32 of Book I). This is unfortunate, because it makes it seem that the side-angle-angle theorem depends on the angle sum theorem, which in turn depends on the parallel postulate. However, the side-angle-angle theorem does not depend on the parallel postulate. In fact, it is part of Euclid's theorem 26 of Book I, which he proved without using the parallel postulate. An outline of Euclid's proof is given below. Supply the details of the proof.

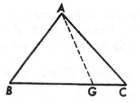 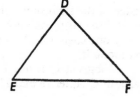

Given: angle ABC = angle DEF
angle ACB = angle DFE
$AB = DE$
Prove: triangle $ABC \cong$ triangle DEF

Proof: Either $BC = EF$, or BC is greater than EF, or EF is greater than BC. If BC is greater than EF, cut off BG on BC so that $BG = EF$. Triangle $ABG \cong$ triangle DEF. Then angle DFE equals angle AGB which is greater than angle ACB. But this is impossible. Therefore BC is not greater than EF. Similarly, EF cannot be greater than BC. Therefore $BC = EF$, and triangle $ABC \cong$ triangle DEF.

3. *Connection.* Using Hilbert's axioms of connection, prove that two straight lines intersect in at most one point.

4. *Order in the plane.* A point P is said to be *inside* triangle ABC if P is on the segment joining a vertex of the triangle to a point of the opposite side that is not a vertex of the triangle. Prove that, if P is inside a triangle, a line through P from any vertex of the triangle intersects the opposite side.

5. *Unit of length.* The surface of the earth is approximately a sphere. If it were exactly a sphere, it would have a natural unit of length, namely the length of a great circle. Unfortunately, great circles on the surface of the earth are not all the same length. The unit of length called the *meter* was originally defined as a sort of "semi-natural" unit by deriving it from an arbitrarily chosen great circle on the earth. Look up in an encyclopedia the original definition of a meter.

6. If YO is perpendicular to OX, and a line OP drawn inside angle YOX makes angles α and β respectively with OX and OY, prove that $\dfrac{\cos \beta}{\sin \alpha} = 1$.

7. Use the infinite series given on page 94 to prove that $\cos(-x) = \cos x$.

8. Prove that $\sin iy = i \sinh y$, and $\cos iy = \cosh y$.

9. Equation (2) on page 94 expresses $\sin x$ as an infinite series. When x is small, the first term of the series is approximately equal to the sum of the series, that is, $\sin x \approx x$, where the symbol $\approx$ means "is approximately equal to." Prove that if ABC is a spherical triangle on a sphere whose radius is r, then, if r is large, and a and c are small compared to r, equation (8) on

page 97 yields the following approximate equality:

$$\frac{\sin A}{\sin C} \approx \frac{a}{c}.$$

That is, the law of sines for a plane is approximately true for small spherical triangles on a large sphere.

10. Cut a rectangular strip of paper and cover it with a network of triangles, as shown in the diagram below. Join the ends marked *AB* so that the *A*'s coincide and the *B*'s coincide. The resulting surface is cylindrical. Verify by assigning senses to the triangles that the cylindrical surface is orientable.

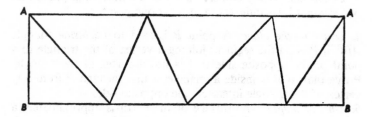

4

Geometry via Numbers

Space and Numbers Reunited

After the Pythagorean discovery that the ratio of some magnitudes of the same kind cannot be expressed as a rational number, arithmetic and geometry developed along separate paths. The two paths were united again in the seventeenth century when René Descartes (1596–1650) and Pierre Fermat (1601–1665) developed a way of expressing all geometric relationships as relationships among numbers, thus making it possible to solve geometric problems by the techniques of arithmetic and algebra. Geometry studied by these new algebraic techniques is called *analytic geometry*. In this chapter we develop some of the elementary ideas of analytic geometry for a threefold purpose: 1) to show how geometric problems can be solved by algebraic techniques; 2) to show how the new methods led to the development of the geometry of spaces of more than three dimensions; 3) to gather some facts that we need in later chapters.

Points in a Plane

The algebraic approach to plane geometry is based on the fact that we can associate with each point of a plane an ordered pair of numbers, just as, in a large city, we associate with a street intersection the numbers of the street and avenue that cross there. To do so, we proceed as follows: Draw two perpendicular lines OX and OY in the plane (shown in the diagram below as horizontal and vertical lines respectively). Choose a unit of length and place a number scale on each of these lines in the manner described on page 31, so that there is a one-to-one correspondence between the real number system and the points of each line, with zero assigned to the point O, and the positive numbers to the right of O on OX and above O on OY. Through any point P in the plane, draw a line perpendicular to OX and a line perpendicular to OY. Where these lines cross OX and OY there are scale numbers which we shall designate by x and y respectively. We associate

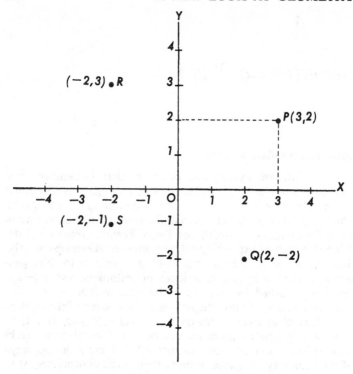

with the point P the ordered pair of numbers (x, y). (The pair is called *ordered* because the order in which the numbers x and y is written has been specified: the number x is written to the left of the comma, and the number y is written to the right of the comma.) The numbers x and y are called the *coordinates* of the point P. The number x is called the *abscissa* of the point, and the number y is called the *ordinate* of the point. In the preceding diagram, for the point P, $x = 3$, and $y = 2$; for the point R, $x = -2$, and $y = 3$; for the point S, $x = -2$, and $y = -1$; for the point Q, $x = 2$, and $y = -2$.

The lines OX and OY are called the *axes* of the coordinate system, and O is called the *origin* of the coordinate system. For any point on OX, $y = 0$. For any point on OY, $x = 0$. The coordinates of the origin are $(0, 0)$.

Directed Lines and Segments

When a point moves along a line, it may move in either of two possible directions. We shall find it useful to introduce a convention for distinguishing one of these directions from the other. A line is said to be *directed* if one of the two possible directions on the line is specified as the positive direction. A *directed segment* AB on the line is a segment with the specified direction *from A to B*. The *directed distance AB* is positive or negative according as the direction from A to B is the same as or opposite to the positive direction on the line. With this convention, $BA = -AB$. Note that we have already specified that on horizontal and vertical lines the positive directions are to the right and up respectively.

Vertical and Horizontal Lines

Since every point in the plane has associated with it an ordered pair of numbers (its coordinates), it is possible to identify a geometric configuration by identifying a numerical condition that is satisfied by the coordinates of its points. Where the configuration is a straight line we find that the condition satisfied by the coordinates of its points is an equation of a particularly simple type. We identify first the equation for a vertical or a horizontal line.

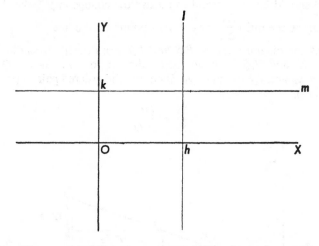

Let l be a vertical line that crosses OX at the scale number h. Then for every point on l, $x = h$. Conversely, any point for which $x = h$ lies on the line l. Consequently l may be described as the set of those points and only those points whose coordinates satisfy the equation $x = h$.

Let m be a horizontal line that crosses OY at the scale number

k. We see in the same way that m may be described as the et of those points and only those points whose coordinates satisfy the equation $y = k$.

Example: The vertical line that crosses OX at the scale number 2 is two units to the right of OY. Its equation is $x = 2$. The horizontal line that crosses OY at the scale number -3 is three units below OX. Its equation is $y = -3$.

Oblique Lines

We now identify the form taken by the equation of an oblique line.

Take any oblique line in the plane, and two points on the line, P_1 and P_2, with coordinates (x_1, y_1) and (x_2, y_2) respectively. If a point moves along the line from P_1 to P_2 its abscissa changes from x_1 to x_2. The amount of change is $x_2 - x_1$. Let us designate this change in the abscissa by the symbol Δx, which may be read as "delta x," and should be understood to mean "the change in x." Similarly, as a point moves from P_1 to P_2, its ordinate changes from y_1 to y_2 by an amount equal to $y_2 - y_1$. We denote this change in the ordinate by Δy, which may be read as "delta y," and should be understood to mean "the change in y." We shall examine the ratio $\dfrac{\Delta y}{\Delta x}$ for any two points on the line.

In the diagram below, P_1S and P_2T are drawn perpendicular to OX, and P_1Q is drawn perpendicular to P_2T. Then $P_1Q = x_2 - x_1$, and $QP_2 = y_2 - y_1$. Then, for the ordered pair of points (P_1, P_2),

$$\frac{\Delta y}{\Delta x} = \frac{QP_2}{P_1Q}.$$

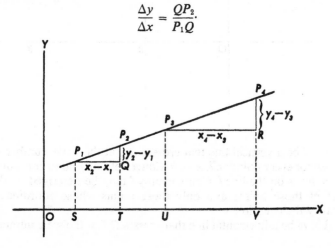

Let P_3 and P_4 be any other two points on the same line with coordinates (x_3, y_3) and (x_4, y_4) respectively. Draw P_3U and P_4V perpendicular to OX, and draw P_3R perpendicular to P_4V. We find in the same way that $P_3R = x_4 - x_3$, and $RP_4 = y_4 - y_3$. Consequently, for the ordered pair of points (P_3, P_4),

$$\frac{\Delta y}{\Delta x} = \frac{RP_4}{P_3R}.$$

However, triangles P_1P_2Q and P_3P_4R are similar, since their angles are respectively equal. Consequently,

$$\frac{QP_2}{P_1Q} = \frac{RP_4}{P_3R}.$$

That is, the ratio $\dfrac{\Delta y}{\Delta x}$ has the same value for any pair of points on the line. We call this value the *slope* of the line, and designate it by m. We have immediately the following formula for the slope of a straight line in terms of the coordinates of two points on the line:

(1) $$m = \frac{\Delta y}{\Delta x} = \frac{y_2 - y_1}{x_2 - x_1}.$$

Let P be an arbitrary point on the line P_1P_2 other than P_1 with coordinates (x, y). Then these coordinates must satisfy the equation

(2) $$\frac{y - y_1}{x - x_1} = m.$$

Multiplying by $x - x_1$, we find that

(3) $$y - y_1 = m(x - x_1).$$

Equation (3) is satisfied by the coordinates of all points on the line, including those of P_1. Conversely, it can be shown that any point whose coordinates satisfy equation (3) lies on the line P_1P_2. Consequently, the line P_1P_2 may be described as the set of those points and only those points whose coordinates satisfy equation (3).

Example: Let P_1 have coordinates $(2, 3)$ and let P_2 have coordinates $(5, 8)$. That is, $x_1 = 2$, $y_1 = 3$, $x_2 = 5$, and $y_2 = 8$. Then

$m = \dfrac{8 - 3}{5 - 2} = \dfrac{5}{3}$ is the slope of the line P_1P_2. The equation of the
line is obtained from (3) by substituting the values of x_1, y_1 and m:
$y - 3 = \dfrac{5}{3}(x - 2)$. To find out if a given point lies on the line
P_1P_2, all we need do is see if the coordinates of the point satisfy
the equation of the line. For example, if we substitute in the
equation the values $x = 8$ and $y = 13$, we obtain the true statement $13 - 3 = \dfrac{5}{3}(8 - 2)$. Therefore the point whose coordinates
are (8, 13) lies on the line P_1P_2. If we substitute in the equation
the values $x = 5$ and $y = 9$, we obtain the false statement
$9 - 3 = \dfrac{5}{3}(5 - 2)$. Therefore the point whose coordinates are
(5, 9) does not lie on the line P_1P_2.

If the line P_1P_2 is rotated around P_1 until it becomes horizontal,
y_2 approaches y_1 as a limit, Δy becomes zero, and m becomes zero.
Consequently, we define the slope of a horizontal line to be zero.
If we rotate the line into a vertical position, x_2 approaches x_1,
Δx becomes zero, and the ratio $\Delta y/\Delta x$ approaches no limit.
Consequently a vertical line has no slope. It is customary to say
that the slope of a vertical line is "infinity."

Any Line

Summing up, we see that for every line, there is an equation
that is satisfied by the coordinates of any point on the line, as
follows:

$$\begin{aligned}
&\text{vertical line:} &&x = h \\
&\text{horizontal line:} &&y = k \\
&\text{oblique line:} &&y - y_1 = m(x - x_1).
\end{aligned}$$

Each of these is an equation of the first degree in x and y. The
general equation of the first degree has the form

$$(4) \qquad\qquad\qquad ax + by = c.$$

(For a vertical line, $a \neq 0$, $b = 0$. For a horizontal line $a = 0$,
$b \neq 0$.) It can be shown that the set of points whose coordinates
satisfy an equation of form (4) is always a straight line. Consequently a straight line may be described as a set of all those

points and only those points whose coordinates satisfy some equation of the first degree.

It is possible to identify the slope of a line by means of the coefficients that appear in the equation of the line. Suppose the equation of the line is equation (4). If $b = 0$, the line is vertical and has no slope. Let us consider the case where $b \neq 0$. Let (x_1, y_1) be the coordinates of a point that lies on the line. Then these coordinates satisfy equation (4). That is,

$$(5) \qquad\qquad ax_1 + by_1 = c.$$

Subtracting equation (5) from equation (4), we find that

$$a(x - x_1) + b(y - y_1) = 0, \quad \text{or} \quad b(y - y_1) = -a(x - x_1).$$

Since b is not zero, we may divide by b and obtain

$$(6) \qquad\qquad y - y_1 = -\frac{a}{b}(x - x_1).$$

Comparing equation (6) with equation (3), we see that if b is not zero, the slope of the line is $m = -(a/b)$.

Example: In the equation $3x - 2y = 6$, $a = 3$ and $b = -2$. The slope of the line that has this equation is $-\left(\dfrac{a}{b}\right) = -\left(\dfrac{3}{-2}\right) = \dfrac{3}{2}.$

Parallel Lines

We now find a way of recognizing from their equations whether or not two lines are parallel.

Any two horizontal lines have the same slope, namely, zero. Any two vertical lines have the same slope, namely "infinity." We now show that any parallel oblique lines also have the same slope. In the diagram below, two parallel oblique lines are cut by the vertical lines PS and QT in the points P, S, Q, and T. PR and SW are drawn perpendicular to QT. It is easily seen that triangles QPR and TSW are congruent. Therefore $PR = SW$, and $RQ = WT$. Consequently the ratios RQ/PR and WT/SW are equal. That is, the two parallel lines PQ and ST have the same slope.

Thus, any two parallel lines have the same slope. It can also be shown, conversely, that if two distinct lines have the same slope,

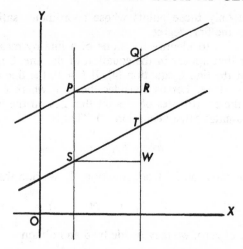

they are parallel. Therefore, if two distinct lines have slopes m and m' respectively, the lines are parallel if and only if $m = m'$. If the equations of the lines are $ax + by = c$, and $a'x + b'y = c'$, respectively, and the lines are not vertical, so that neither b nor b' is zero, then, since $m = -(a/b)$ and $m' = -(a'/b')$, we see that the lines are parallel if and only if

$$(7) \qquad\qquad \frac{a}{b} = \frac{a'}{b'}$$

Equation (7) implies

$$(8) \qquad\qquad ab' - a'b = 0.$$

Equation (8) is also true when $b = b' = 0$, that is, when the two lines are vertical, so it provides a test for parallelism without any exceptions: Two distinct lines whose equations are $ax + by = c$ and $a'x + b'y = c'$ are parallel if and only if equation (8) is satisfied.

Example: The lines whose equations are $2x + 3y = 5$ and $4x + 6y = 7$ respectively are parallel because $2 \cdot 6 - 4 \cdot 3 = 0$.

Perpendicular Lines

There is also a way of recognizing from their equations whether or not two lines are perpendicular.

Let r and s be perpendicular lines that are oblique and that intersect at B. Let C be any other point on s. Draw a line from C perpendicular to OX, and denote by A its intersection with r. Draw AD and BE parallel to OX. Triangles AEB and BEC are similar, and consequently

$$\frac{AE}{EB} = \frac{EB}{EC}.$$

Let m and m' be the slopes of r and s respectively.

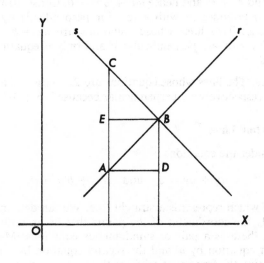

Then

$$m = \frac{DB}{AD} = \frac{AE}{EB} = \frac{EB}{EC}.$$

and

$$m' = \frac{EC}{BE} = -\frac{EC}{EB}.$$

Consequently $mm' = \left(\frac{EB}{EC}\right)\left(-\frac{EC}{EB}\right) = -1$. It is also possible to show that, conversely, if $mm' = -1$, two lines whose slopes are m and m' respectively are perpendicular. Consequently two oblique lines are perpendicular if and only if the product of their slopes is -1.

If the equations of the lines are $ax + by = c$ and $a'x + b'y = c'$, then since $m = -(a/b)$, and $m' = -(a'/b')$, the condition for perpendicularity takes the form of the following equation (9).

(9)
$$\frac{aa'}{bb'} = -1.$$

Equation (9) implies

(10) $aa' + bb' = 0.$

Equation (10) also holds for perpendicular lines that are not oblique. For example, if r is horizontal and s is vertical, then $a = 0$ and $b' = 0$, and hence $aa' + bb' = 0$. Consequently equation (10) provides us with a test for perpendicularity without exceptions: Two lines whose equations are $ax + by = c$ and $a'x + b'y = c'$ are perpendicular if and only if equation (10) is satisfied.

Example: The lines whose equations are $2x + 3y = 5$ and $6x - 4y = 3$ respectively are perpendicular because $2\cdot6 + 3\cdot(-4) = 0$.

Intersecting Lines

Consider the equations

$$ax + by = c, \quad \text{and} \quad a'x + b'y = c',$$

each of which represents a straight line. We can determine algebraically the coordinates of the points that lie on both lines by solving these as a pair of simultaneous equations. Multiplying the first equation by a' and the second equation by a and then subtracting the first equation from the second, we get,

(11) $(ab' - a'b)y = ac' - a'c.$

Also, multiplying the first equation by b' and the second equation by b and then subtracting the second equation from the first, we get

(12) $(ab' - a'b)x = b'c - bc'.$

If $ab' - a'b \neq 0$, equations (11) and (12) can be solved for x and y, giving

(13) $x = \dfrac{b'c - bc'}{ab' - a'b}, \qquad y = \dfrac{ac' - a'c}{ab' - a'b}.$

It is easy to verify by substitution that these values of x and y satisfy the equations of both lines. Consequently, if $ab' - a'b \neq$

0, the two straight lines have one and only one point in common, namely the point whose coordinates are given by (13).

The method used above for finding a unique point that lies on both lines breaks down when $ab' - a'b = 0$. The algebraic reason is that division by zero is not permitted. It is easy to see the geometric reason, too. If $ab' - a'b = 0$, there are two possibilities. Either the two lines represented by the original equations coincide, or they are distinct. If they coincide, they don't have a unique common point because they have all their points in common. If they are distinct, then since equation (8) is satisfied, the two lines are parallel. In this case, they don't have any common point at all.

Determinants

It is useful to express the result given in equation (13) with the notation of *determinants*. A determinant is a number represented by a square array of numbers displayed between two vertical bars, and computed from the numbers in the square array according to the following rules: For a two-row array,

$$\begin{vmatrix} a & b \\ a' & b' \end{vmatrix} = ab' - a'b.$$

For a three-row array,

$$\begin{vmatrix} a & b & c \\ a' & b' & c' \\ a'' & b'' & c'' \end{vmatrix} = aA - bB + cC,$$

where A is the determinant obtained by deleting the row and column containing a, B is the determinant obtained by deleting the row and column containing b, etc. That is,

$$A = \begin{vmatrix} b' & c' \\ b'' & c'' \end{vmatrix}, \qquad B = \begin{vmatrix} a' & c' \\ a'' & c'' \end{vmatrix}, \qquad C = \begin{vmatrix} a' & b' \\ a'' & b'' \end{vmatrix}.$$

For a four-row array,

$$\begin{vmatrix} a & b & c & d \\ a' & b' & c' & d' \\ a'' & b'' & c'' & d'' \\ a''' & b''' & c''' & d''' \end{vmatrix} = aA - bB + cC - dD,$$

where A is the determinant obtained by deleting the row and column containing a, etc. The definition may be extended in this way to any square array of numbers.

It is shown in college algebra textbooks that if two rows or two columns of a determinant are identical, then the value of the determinant is zero. The reader can easily verify this for a two- or three-row determinant.

Using the determinant notation, equations (13) may be written in this form:

$$(14) \qquad x = \frac{\begin{vmatrix} c & b \\ c' & b' \end{vmatrix}}{\begin{vmatrix} a & b \\ a' & b' \end{vmatrix}}, \qquad y = \frac{\begin{vmatrix} a & c \\ a' & c' \end{vmatrix}}{\begin{vmatrix} a & b \\ a' & b' \end{vmatrix}}.$$

The determinant notation provides an easy way of writing the equation of a straight line through two given points. Suppose the coordinates of the points are (x_1, y_1) and (x_2, y_2). Then the equation of the line through these points is

$$(15) \qquad \begin{vmatrix} x & y & 1 \\ x_1 & y_1 & 1 \\ x_2 & y_2 & 1 \end{vmatrix} = 0.$$

To prove this fact, we show first that equation (15) is indeed the equation of a straight line, and secondly that the two given points lie on this line. First, by the definition of the determinant of a three-row array of numbers, equation (15) can be rewritten as follows:

$$x \begin{vmatrix} y_1 & 1 \\ y_2 & 1 \end{vmatrix} - y \begin{vmatrix} x_1 & 1 \\ x_2 & 1 \end{vmatrix} + 1 \begin{vmatrix} x_1 & y_1 \\ x_2 & y_2 \end{vmatrix} = 0.$$

Since this is a first degree equation in x and y, it is the equation of a straight line. Secondly, if we substitute x_1 for x and y_1 for y in equation (15), the left-hand side of the equation becomes zero, since the value of a determinant is zero if two of its rows are identical. That is, (x_1, y_1) satisfies equation (15). Consequently the point with coordinates (x_1, y_1) lies on the line represented by equation (15). The point with coordinates (x_2, y_2) lies on the line for the same reason.

Example: The equation of the straight line through the points whose coordinates are $(2, 3)$ and $(8, 5)$ follows on page 117.

$$\begin{vmatrix} x & y & 1 \\ 2 & 3 & 1 \\ 8 & 5 & 1 \end{vmatrix} = 0,$$

or $\quad x \begin{vmatrix} 3 & 1 \\ 5 & 1 \end{vmatrix} - y \begin{vmatrix} 2 & 1 \\ 8 & 1 \end{vmatrix} + 1 \begin{vmatrix} 2 & 3 \\ 8 & 5 \end{vmatrix} = 0,$

or $-2x + 6y - 14 = 0$. The reader can easily verify by substituting 2 for x and 3 for y that the coordinates $(2, 3)$ satisfy this equation. Substitution of 8 for x and 5 for y will show that the coordinates $(8, 5)$ also satisfy this equation.

Distance Between Two Points

Let P_1 and P_2 be two distinct points with coordinates (x_1, y_1) and (x_2, y_2) respectively. To find the distance d from P_1 to P_2,

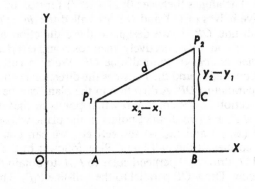

draw P_1A and P_2B perpendicular to OX, and draw P_1C perpendicular to P_2B. Then $P_1C = x_2 - x_1$, and $CP_2 = y_2 - y_1$. Then, by the Pythagorean Theorem, we have

(16) $\qquad d^2 = (x_2 - x_1)^2 + (y_2 - y_1)^2,$

so

(17) $\qquad d = \sqrt{(x_2 - x_1)^2 + (y_2 - y_1)^2}.$

Example: The distance between the points whose coordinates

are $(2, 3)$ and $(8, 5)$ is $d = \sqrt{(8 - 2)^2 + (5 - 3)^2} = \sqrt{40}.$

Directions

If P is any point other than the origin O, the half-line that starts at O and extends through P points out a particular direction in

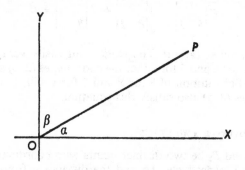

the plane. The angles (between 0° and 180°) formed by OP with the positive halves of OX and OY are called the *direction angles* of the half-line OP. If we designate these direction angles by Greek letters α and β respectively, then cos α and cos β are called the *direction cosines* of the half-line OP. We also call α and β the direction angles, and their cosines the direction cosines of any half-line parallel to OP. A direction in the plane can be specified by the direction cosines of any ray that points in that direction.

If P_1 and P_2 are two distinct points in the plane whose coordinates are (x_1, y_1) and (x_2, y_2) respectively, we can compute the direction cosines of the half-line P_1P_2 as follows: In the diagram on page 117, draw P_2D perpendicular to P_1A, to obtain the figure shown below. Draw OP parallel to the half-line P_1P_2. Then, since

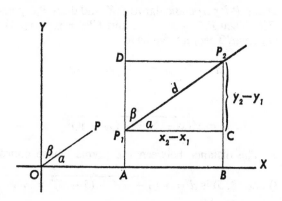

P_1C is parallel to OX, and P_1D is parallel to OY, angle $CP_1P_2 = \alpha$, and angle $DP_1P_2 = \beta$, the direction angles of OP and P_1P_2. Let us denote their direction cosines by l and m respectively. Then

$$(18) \qquad l = \cos \alpha = \frac{x_2 - x_1}{d}, \qquad m = \cos \beta = \frac{y_2 - y_1}{d}.$$

Although we have derived these formulas from a diagram in which α and β are acute angles, it can be shown that the formulas are valid even when α and β are not acute.

Example: If P_1 and P_2 have coordinates $(5, 7)$ and $(11, 15)$ respectively, we find by equation (17) that the distance between P_1 and P_2 is 10. Then the direction cosines of the half-line P_1P_2 are $l = \dfrac{11 - 5}{10} = .6$, and $m = \dfrac{15 - 7}{10} = .8$.

The Projection of a Segment on a Line

When an opaque object is held above the ground in sunlight, it casts a shadow on the ground. The concept of (orthogonal) projection is an abstraction from this phenomenon. Let h and k be directed lines, and let AB be a directed segment on h. If AC and BD are drawn perpendicular to k, the directed segment CD is called the *orthogonal projection* of AB on k. If AC and BD

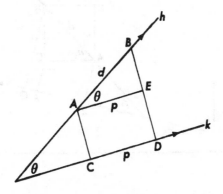

were rays of sunlight, and AB were a stick, CD would be the shadow of the stick on k. Let θ be the angle between the directions specified on h and k. Denote CD by p, and AB by d. Draw AE

perpendicular to BD. Then $AE = p$, and angle $EAB = \theta$. Consequently we have $\cos \theta = \dfrac{p}{d}$, and

(19) $$p = d \cos \theta.$$

Example: If a segment of length 8 makes an angle of 45° with a given line, its projection on the line has length $p = 8 \cos 45° = 8(\frac{1}{2}\sqrt{2}) = 4\sqrt{2}$.

The Angle Between Two Lines

Let h and k be two directed lines, and let θ be the angle between the specified directions on these lines. We shall find a formula expressing $\cos \theta$ in terms of the direction cosines of the specified directions on h and k.

Let the direction angles of the specified direction on h be α_1 and β_1, and let the direction angles of the specified direction on k be α_2 and β_2. Let $l_1 = \cos \alpha_1$, $m_1 = \cos \beta_1$, $l_2 = \cos \alpha_2$, and

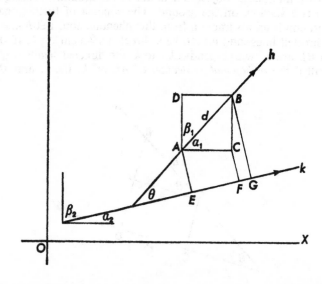

$m_2 = \cos \beta_2$. Let AB be a directed segment on h, and denote it by d. Draw AD and BC parallel to OY, draw BD perpendicular to AD, and draw AC perpendicular to BC. Draw AE, BG, and CF perpendicular to k. Then, we obtain the following equation (20).

(20) $$EG = EF + FG.$$

The direction angles of h are α_1 and β_1. Therefore, since $\cos \alpha_1 = \dfrac{AC}{d}$, $AC = d \cos \alpha_1 = dl_1$. Similarly $CB = AD = d \cos \beta_1 = dm_1$. The angle between h and k is θ, and EG is the projection of AB on m. Therefore, by equation (19), $EG = d \cos \theta$. The direction angles of k are α_2 and β_2. EF is the projection of the horizontal segment AC on k, and FG is the projection of the vertical segment CB on k. Therefore, by equation (19), $EF = AC \cos \alpha_2 = dl_1l_2$, and $FG = CB \cos \beta_2 = dm_1m_2$. Substituting into equation (20), we get

(21) $$d \cos \theta = dl_1l_2 + dm_1m_2.$$

Therefore

(22) $$\cos \theta = l_1l_2 + m_1m_2.$$

Example: If the direction cosines of a line are $l_1 = \dfrac{\sqrt{3}}{2}$, $m_1 = \frac{1}{2}$, and the direction cosines of another line are $l_2 = \frac{1}{2}$, $m_2 = \dfrac{\sqrt{3}}{2}$, and θ is the angle between the two lines, then $\cos \theta = \left(\dfrac{\sqrt{3}}{2}\right)\left(\dfrac{1}{2}\right) + \left(\dfrac{1}{2}\right)\left(\dfrac{\sqrt{3}}{2}\right) = \dfrac{\sqrt{3}}{2}$, and $\theta = 30°$.

Algebraic Curves

We have observed that points whose coordinates satisfy an equation of the first degree lie on a straight line. It is natural to ask what may be said about the set of points whose coordinates satisfy an algebraic equation of degree higher than one. The answer turns out to be that in general the set of points constitutes a curved line, and the higher the degree of the equation, the more complicated the curve looks. These curves are known as *algebraic curves*.

Second-Degree Curves

The simplest algebraic curves are those whose equations are of the second degree. Any second degree equation in x and y may be put in the form of equation (23) on page 122.

(23) $Ax^2 + Bxy + Cy^2 + Dx + Ey + F = 0,$

where A, B, C, D, E and F are real numbers. In certain special
"degenerate" cases, the points whose coordinates satisfy such an
equation lie on two straight lines or one straight line or they all
coincide. Except for these unusual "degenerate" cases, the curve
represented by such an equation is found to be either an ellipse,
a parabola or a hyperbola. It is an ellipse if $B^2 - 4AC$ is negative,
a parabola if $B^2 - 4AC$ is zero, and a hyperbola if $B^2 - 4AC$

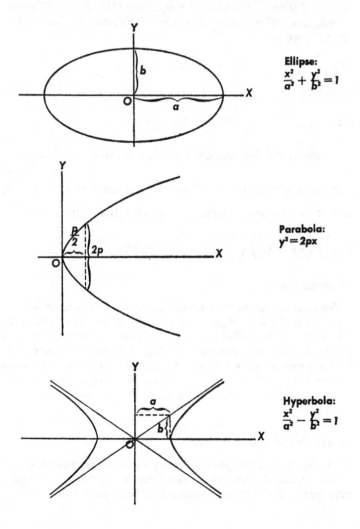

Ellipse:
$\dfrac{x^2}{a^2} + \dfrac{y^2}{b^2} = 1$

Parabola:
$y^2 = 2px$

Hyperbola:
$\dfrac{x^2}{a^2} - \dfrac{y^2}{b^2} = 1$

is positive. If the curve is symmetrically placed with respect to the coordinate axes, the equation in each of these cases takes a particularly simple form, as shown in the illustration on page 122.

The ellipse, the parabola, and the hyperbola are known as *conic sections* because they are the curves that are obtained when a plane intersects a conical surface. Their properties were studied in detail by Appolonius (about 260–170 B.C.).

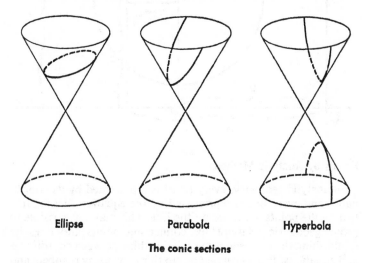

Ellipse   Parabola   Hyperbola

The conic sections

Example: In the equation $2x^2 + 3y^2 + 2x - 5 = 0$, $A = 2$, $B = 0$ and $C = 3$. $B^2 - 4AC = 0^2 - 4(2)(3) = -24$. Therefore the curve whose points satisfy this equation is an ellipse.

The Circle

The circle is a special case of the ellipse. Since we shall use it in the next paragraph, we now derive the equation of a circle whose center is at the point A with coordinates (a, b) and whose radius is r.

Let P be any point on the circle, with coordinates (x, y). Let C be the intersection of the vertical line through P and the horizontal line through A. Then $AC = x - a$, $CP = y - b$, and the length of AP is r. Since AC, CP and AP are the sides of a right triangle, we know from the Pythagorean theorem that the coordinates (x, y) satisfy the equation

(24) $$(x - a)^2 + (y - b)^2 = r^2.$$

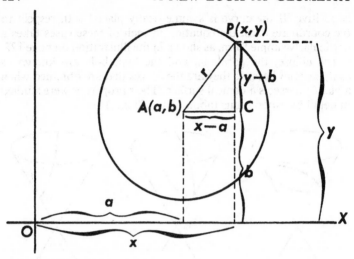

Proofs by Analytic Methods

In analytic geometry, every point is represented by its coordinates, and every line is represented by the equation that is satisfied by the points which lie on the line. This makes it possible to prove geometric relationships connecting points and lines by establishing the corresponding relationships between coordinates and equations. Proofs carried out in this way, using numbers and equations, are said to be done by *analytic* methods. As an example of a proof done by analytic methods we shall prove the following theorem:

An angle inscribed in a semicircle is a right angle.

Proof: Let the center of a circle be A, with coordinates (a, b), and let the radius of the circle be r. Draw diameter RS parallel to OX, and inscribe angle RPS in the semicircle that is above RS. Let the coordinates of P be (x, y). Since P is on the circle, x and y satisfy equation (24). From this equation we find that

$$(25) \qquad (x - a)^2 - r^2 = -(y - b)^2.$$

Let m and m' be the slopes of RP and SP respectively. Then, since the coordinates of R are $(a - r, b)$, and the coordinates of S are $(a + r, b)$, we find by equation (1) of page 109 that

$$m = \text{slope of } RP = \frac{y - b}{x - (a - r)} = \frac{y - b}{(x - a) + r},$$

$$m' = \text{slope of } SP = \frac{y-b}{x-(a+r)} = \frac{y-b}{(x-a)-r}.$$

$$\text{Then } mm' = \frac{y-b}{(x-a)+r} \cdot \frac{y-b}{(x-a)-r} = \frac{(y-b)^2}{(x-a)^2-r^2}$$

$$= \frac{(y-b)^2}{-(y-b)^2} = -1.$$

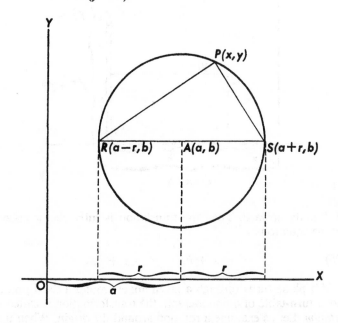

Then by the test for perpendicularity developed on page 113, RP is perpendicular to SP, and angle RPS is a right angle.

Isometry

The concept of congruent segments plays a central role in Euclidean geometry. For this reason those one-to-one onto transformations of a plane that "move" every segment into a congruent segment are of particular importance. Such transformations are called *isometries*, because the length of each segment is the same in its old and new positions. (Isometry means "same measure.")

Examples: 1. If a plane moves along itself for a given distance in a given direction, the transformation is called a *translation*. Let P, with coordinates (x, y), be any point in the plane, and let P', with coordinates (x', y'), be the position to which it is moved by a translation. The translation involves a horizontal displacement, say of h units to the right, and a vertical displacement, say

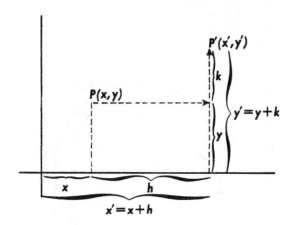

of k units upward. Then the translation is fully characterized by the equations

(26) $$x' = x + h, \qquad y' = y + k.$$

2. If a plane turns through a given angle around a fixed point, like a turn-table of a phonograph, the transformation is called a *rotation*. Let us examine a rotation around the origin. When the plane is turned through a given angle around the origin, the effect on the coordinates of each point is the same as would be obtained by turning the coordinate axes an equal amount in the opposite direction. Suppose that the turning of the axes brings OX into the position of OX', and brings OY into the position of OY'. Let the direction cosines of OX' with respect to OX and OY be l_1 and m_1. Let the direction cosines of OY' be l_2 and m_2. Let P be any point in the plane with coordinates (x, y) with respect to the axes OX and OY. Let (x', y') be its coordinates with respect to OX' and OY'. Let the direction cosines of OP with respect to OX and OY be l and m. Let θ be the angle between OP and OX', and let ϕ be the angle between OP and OY'. Denote by d the directed segment OP. Then, since x' is the projection of d on

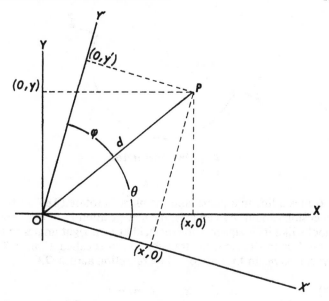

OX', and y' is the projection of d on OY', we have $x' = d \cos \theta$, and $y' = d \cos \phi$. Since x is the projection of d on OX, and y is the projection of d on OY, we have $x = dl$, and $y = dm$. From equation (22) we have

$$\cos \theta = ll_1 + mm_1, \qquad \cos \phi = ll_2 + mm_2.$$

Multiplying each of these equations by d, we get

$$d \cos \theta = (dl)l_1 + (dm)m_1,$$
$$d \cos \phi = (dl)l_2 + (dm)m_2,$$

or

(27)
$$x' = l_1 x + m_1 y,$$
$$y' = l_2 x + m_2 y.$$

In the special case of a rotation of 180° around a point A, each point P moves to a position P' on the line AP so that P and P' are on opposite sides of A and at equal distances from A. For this reason a rotation of 180° around A is called a *point reflection* across A. In the case of a point reflection across the origin it is obvious that $x' = -x$, and $y' = -y$.

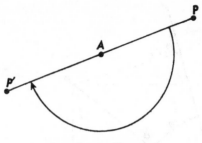

A point reflection across A

3. If *m* is a line in a plane, and the plane is rotated 180° in three dimensions about *m* as an axis, the new position *P′* of each point *P* looks like its image in a mirror held at *m* at right angles to the plane. For this reason, the transformation is called a *line reflection* across *m*. In the case of a line reflection across *OX*,

$$(28) \qquad\qquad x' = x, \qquad y' = -y.$$

A line reflection across m

Other isometries of a plane can be obtained by applying in succession any two of the examples given above. Thus, a translation followed by a rotation, a translation followed by a line reflection, etc., are also isometries.

The product of two isometries, in the sense defined on page 57, is also an isometry. The identity transformation *e* defined on page 57 is an isometry, and the inverse of an isometry is an isometry. Consequently the set of all isometries of a plane is a subgroup of the group of all one-to-one onto transformations of the plane.

Geometry Without Pictures

In our outline of the elements of analytic geometry we began with the geometric entities known as points and lines in a Euclidean plane. We pictured these in the usual way, visualizing a point as an idealized dot, and a line as an idealized taut string. Then, by introducing coordinate axes in the plane, we found that we could associate with each point an ordered pair of real numbers, and with each straight line an equation of the first degree. Then after that, all geometric reasoning could be done with the numbers and equations alone. This fact opens up the possibility of discarding as unnecessary the original "pictures" of points and lines, and reconstructing the geometry of the Euclidean plane through the use of numbers and equations. This can be done as follows: Define a point to be any ordered pair of real numbers (x, y). Define a straight line to be the set of points satisfying an equation of the form $ax + by = c$, where a and b are not both zero. Define order relations on a line with the help of the inequality relation among real numbers. Define the distance between two points by means of equation (17) on page 117. Define the direction cosines of a half-line by means of equation (18). Define the angle between two half-lines as the angle between $0°$ and $180°$ whose cosine is given by equation (22) on page 121. Define congruent segments as those whose endpoints are equal distances apart. Define congruent angles as those which have the same number of degrees. Then it can be shown that the points, lines and relations defined in this way satisfy the Hilbert axioms for a Euclidean plane. Consequently they provide a concrete representation of a Euclidean plane that is not necessarily tied to the usual visual picture of the plane.

In this approach to Euclidean plane geometry, an ordered pair of real numbers is not merely *associated* with a point. It *is* the point. The set of ordered pairs that satisfies an equation of the first degree is not merely associated with a straight line. It *is* the straight line. Relations among numbers and equations are not merely asscciated with geometric relations. They *are* the geometric relations. From this point of view, geometry is not an independent subject, but is merely a branch of algebra. The relationship between algebra and geometry will be discussed further in Chapters 5, 6, 10 and 13.

Analytic Geometry of Space

The methods of analytic geometry can also be used to study the geometry of Euclidean space of three dimensions. Three mutually perpendicular axes OX, OY and OZ are drawn in space, and a real number scale is placed on each of them. The positive sides of the three axes are shown in the diagram. Through

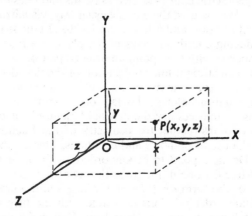

any point P in space, planes are drawn perpendicular to OX, OY and OZ respectively. The scale numbers x, y, and z where these planes intersect the axes are the coordinates of the point P. In this way we associate with each point in space the ordered triple (x, y, z). We state without proof some of the results that are obtained. They are completely analogous to those obtained for the analytic geometry of a Euclidean plane.

1. A plane in space is represented by an equation of the form $ax + by + cz = d$, where a, b, and c are not all zero. If (x_1, y_1, z_1), (x_2, y_2, z_2), and (x_3, y_3, z_3) are the coordinates of three points that are not on the same straight line, then the equation of the plane that is determined by them is

$$\begin{vmatrix} x & y & z & 1 \\ x_1 & y_1 & z_1 & 1 \\ x_2 & y_2 & z_2 & 1 \\ x_3 & y_3 & z_3 & 1 \end{vmatrix} = 0.$$

2. Since a straight line can be determined as the intersection of two planes, it can be represented by the set of coordinates that satisfy the equations of the two planes.

3. If d is the distance between two points P_1 and P_2 whose coordinates are (x_1, y_1, z_1) and (x_2, y_2, z_2), then

$$d = \sqrt{(x_2 - x_1)^2 + (y_2 - y_1)^2 + (z_2 - z_1)^2}.$$

4. The direction cosines of the directed line P_1P_2 are

$$l = \frac{x_2 - x_1}{d}, \qquad m = \frac{y_2 - y_1}{d}, \qquad n = \frac{z_2 - z_1}{d}.$$

5. If θ is the angle between two directed lines whose direction cosines are l_1, m_1, n_1 and l_2, m_2, n_2 respectively, then

$$\cos \theta = l_1l_2 + m_1m_2 + n_1n_2.$$

Geometry of n Dimensions

Our visual intuition allows us to picture a space of at most three dimensions. However, the abstract approach to analytic geometry, in which a point of a plane is simply an ordered pair of real numbers, makes our geometric concepts independent of visual intuition. This independence opens the way to an extension of these concepts to spaces with more than three dimensions. A Euclidean space of n dimensions, designated briefly as an *n-space*, can be constructed for any positive integer n in the following way. Define a point in the n-space to be an ordered n-tuple of real numbers $(x_1, x_2, \ldots, x_n)$. Then,

1. Define an $(n - 1)$-space in the n-space to be the set of n-tuples that satisfy an equation of the form

$$a_1x_1 + a_2x_2 + \ldots + a_nx_n = a_{n+1},$$

where not all the numbers $a_1, \ldots, a_n$ are zero.

2. A straight line (that is, a 1-space) can be determined as the intersection of $n - 1$ $(n - 1)$-spaces.

3. The distance d between two points P and P' whose coordinates are $(x_1, x_2, \ldots, x_n)$ and $(x'_1, x'_2, \ldots, x'_n)$ respectively is defined by the equation

$$d = \sqrt{(x'_1 - x_1)^2 + (x'_2 - x_2)^2 + \ldots + (x'_n - x_n)^2}.$$

4. The direction cosines of the directed line PP' are defined by

$$l_1 = \frac{x'_1 - x_1}{d}, \qquad l_2 = \frac{x'_2 - x_2}{d}, \ldots, \qquad l_n = \frac{x'_n - x_n}{d}.$$

5. The angle θ between two directed lines whose direction cosines are $l_1, l_2, \ldots, l_n$ and $l'_1, l'_2, \ldots, l'_n$ respectively is defined by

$$cos\ \theta = l_1l'_1 + l_2l'_2 + \ldots + l_nl'_n.$$

The first systematic study of the geometry of n dimensions was made by Hermann Grassmann (1809–1877).

Polytopes

A polyhedron is the three-dimensional analogue of a polygon. The corresponding analogue in a space with more than three dimensions is called a *polytope*. A simple polyhedron is one that can be continuously deformed into a sphere. Similarly, a simple polytope in a space of n dimensions is one that can be continuously deformed into the n-dimensional analogue of a sphere. A simple polytope in a space of n dimensions is called a *cell* of n dimensions.

The boundary of a simple polyhedron includes cells of 2 dimensions (faces), cells of 1 dimension (edges), and cells of 0 dimensions (vertices). Similarly, the boundary of a simple polytope of n dimensions includes cells of $n - 1$ dimensions, cells of $n - 2$ dimensions, etc., down to cells of 0 dimensions. We shall denote by N_r the number of cells of r dimensions in the boundary of a simple polytope and of a simple polyhedron.

Euler's formula, derived on page 26, asserts that $V - E + F = 2$, for every simple polyhedron. In our present notation, $V = N_0$, $E = N_1$, and $F = N_2$, and Euler's formula takes this form:

$$N_0 - N_1 + N_2 = 2.$$

A generalization of this formula for a simple polytope of n dimensions was established by Henri Poincaré (1854–1912):

(29) $\quad N_0 - N_1 + N_2 - \ldots + (-1)^{n-1}N_{n-1} = 1 - (-1)^n.$

The right-hand side of this equation is 0 or 2 depending on whether n is even or odd.

Regular Polytopes

The n-dimensional analogue of a regular polyhedron is a *regular polytope*. We shall name a regular polytope or a regular polyhedron of n dimensions by specifying how many cells of $n - 1$ dimensions are in its boundary: *a cell of n dimensions is called a k-cell if it is bounded by k cells of $n - 1$ dimensions.* For example, since a tetrahedron has four faces, it is a 4-cell; a cube is a 6-cell;

an octahedron is an 8-cell; a dodecahedron is a 12-cell; and an icosahedron is a 20-cell. Similarly, since a triangle has three sides, it is a 3-cell; a square is a 4-cell; and a pentagon is a 5-cell.

We saw in Chapter 2 that there are exactly five kinds of regular polyhedra. Using the new nomenclature for them, we can summarize part of the information given about them on page 27 in the following table:

REGULAR POLYHEDRA (3 DIMENSIONS)

| Type | Boundary cells of 2 dimensions | | Vertices | | Duality |
	Number (N_2)	Type	Number (N_0)		
4-cell	4	3-cell	4		self-dual
6-cell	6	4-cell	8		} mutually dual
8-cell	8	3-cell	6		
12-cell	12	5-cell	20		} mutually dual
20-cell	20	3-cell	12		

Just as the number of types of regular polyhedra is finite, the number of types of regular polytopes of more than 3 dimensions is also finite. There are exactly six kinds of regular polytopes of 4 dimensions. Some of their properties are recorded in the following table:

REGULAR POLYTOPES (4 DIMENSIONS)

| Type | Boundary cells of 3 dimensions | | Vertices | | Duality |
	Number (N_3)	Type	Number (N_0)		
5-cell	5	4-cell	5		self-dual
8-cell	8	6-cell	16		} mutually dual
16-cell	16	4-cell	8		
24-cell	24	8-cell	24		self-dual
120-cell	120	12-cell	600		} mutually dual
600-cell	600	4-cell	120		

The examples of 3 dimensions and 4 dimensions would seem to suggest that the number of regular polytopes increases with the number of dimensions. However, this suggestion turns out to be false. For every value of n above 4, the number of regular polytopes of n dimensions is exactly three. The corresponding table for these polytopes is shown below:

REGULAR POLYTOPES (n DIMENSIONS, $n \geqq 5$)

Type	Boundary cells of $n - 1$ dimensions		Vertices		Duality
	Number (N_{n-1})	Type	Number (N_0)		
$(n + 1)$-cell	$n + 1$	n-cell	$n + 1$		self-dual
$2n$-cell	$2n$	$(2n - 2)$-cell	2^n		mutually dual
2^n-cell	2^n	n-cell	$2n$		

EXERCISES FOR CHAPTER 4

1. *Coordinates.* Draw coordinate axes OX and OY on a sheet of graph paper and locate the points whose coordinates are $(3, 1)$, $(-3, 1)$, $(-3, -1)$, $(3, 0)$ and $(0, 3)$.

2. *Directed segments.* Let A, B and C be on the same straight line, with A between B and C. Using directed segments, show that $AC = AB + BC$.

3. *Horizontal and vertical lines.* a) What is the equation of a horizontal line that is 5 units above the x axis? b) What is the equation of a vertical line that is 4 units to the left of the y axis? c) What is the equation of the x axis? d) What is the equation of the y axis?

4. *Oblique lines.* a) What is the slope of the line determined by the two points whose coordinates are $(1, 1)$ and $(2, 4)$ respectively? b) Write the equation of this line in the form of equation (3) on page 109. c) Does the point whose coordinates are $(3, 7)$ lie on this line? d) Does the point whose coordinates are $(4, 8)$ lie on this line?

5. *Slopes.* Identify from the coefficients of x and y in its equation the slope of the line whose equation is a) $2y = 6$; b) $3x = 5$; c) $3x + 4y = 7$; d) $8x - 2y = 5$.

6. *Parallel lines.* a) Are the lines whose equations are $3x - 4y = 5$ and $6x - 8y = 4$ parallel? b) Are the lines whose equations are $3x - 4y = 5$ and $5x + 7y = 2$ parallel?

7. *Perpendicular lines.* a) Are the lines whose equations are $3x - 4y = 5$ and $2x + y = 2$ perpendicular? b) Are the lines whose equations are $3x - 4y = 5$ and $8x + 6y = 9$ perpendicular?

8. *Intersection of lines.* Determine whether the two lines whose equations are given coincide or are parallel or intersect. If they intersect, find the coordinates of the point of intersection.

a) $2x + 3y = 5.$ b) $2x + 3y = 5.$ c) $2x + 3y = 5.$
   $4x + 6y = 4.$ $4x + 6y = 10.$ $x - y = 10.$

9. *Determinants.* Find the value of each of these determinants:

a) $\begin{vmatrix} 2 & 1 \\ 1 & 4 \end{vmatrix}$ b) $\begin{vmatrix} 1 & 2 & 1 \\ 0 & 3 & 2 \\ 4 & 1 & 5 \end{vmatrix}$

c) Verify that the value of this determinant, which has two identical columns, is zero:

$$\begin{vmatrix} 2 & 2 & 1 \\ 1 & 1 & 3 \\ 5 & 5 & 4 \end{vmatrix}$$

10. a) Write in the form of equation (15) on page 116 the equation of the straight line that passes through the points whose coordinates are (3, 1) and (2, 2).

b) Write in the form of equation (3) on page 109 the equation of the same straight line. c) Solve each of these equations for y and compare the results.

11. *Distance between two points.* Find the distance between two points whose coordinates are (3, 1) and (8, 13) respectively.

12. *Direction cosines.* Find the direction cosines of the half-line from the point whose coordinates are (3, 1) through the point whose coordinates are (8, 13).

13. *Projection of a segment.* If a segment of length 8 units makes

an angle of 60° with a given line, what is the length of its orthogonal projection on the line?

14. *Angle between two lines.* If the direction cosines of a line are $l_1 = 3/5$, $m_1 = 4/5$, and the direction cosines of another line are $l_2 = \frac{1}{2}$, $m_2 = \frac{1}{2}\sqrt{3}$, find the cosine of the angle between the two lines.

15. *Conic sections.* Use the rule given on page 122 to determine whether the graph of each of these equations is an ellipse, a parabola, or a hyperbola:

a) $x^2 - 4x - 4y + 16 = 0$.
b) $xy = 12$.
c) $4x^2 + 9y^2 = 36$

16. *The circle.* Write the equation of the circle whose radius is 8 and whose center has coordinates $(-2, 3)$.

17. *The tesseract.* The analogue of a cube in four-dimensional Euclidean space is called a *tesseract*. It is the regular 8-cell listed in the table on page 133. As the table indicates, it is enclosed by eight three-dimensional 6-cells (cubes). The way in which these cubes are joined to each other is shown in the diagram below, which represents a three-dimensional projection of the tesseract. The large cube and the small cube inside it represent two of the boundary cubes of the tesseract. Each of the other six cubes of the boundary joins a face of the inner cube to the adjacent face of the outer cube. Determine from the diagram the values of N_0, N_1, N_2 and N_3. Then, keeping in mind that $n = 4$, verify that they satisfy equation (29) on page 132.

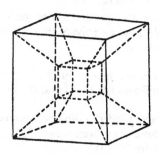

5

Geometry via Arrows

Vectors

In the preceding chapter we made frequent use of directed segments of a straight line. There is another way of studying plane geometry in which the concept of a directed segment plays a central role, and is the basis of the concept of a *vector*. For any two points A and B we denote by AB the directed segment from A to B, and we may picture it as an arrow with tail at A and head at B. We derive the concept of the *vector AB* by detaching this arrow from its tail point A and allowing it to be placed anywhere in the plane without changing its length and direction. Thus the *vector AB is the directed segment AB or any other directed segment that is parallel and equal to AB*. We shall use the symbol $\overrightarrow{AB}$ to represent the vector AB. If the directed segment DC is parallel and equal to the directed segment AB, then $\overrightarrow{AB}$ and $\overrightarrow{DC}$ both stand for the same vector. We shall find it convenient, too, to designate vectors by bold-face lower-case letters, such as **u, v, w,** etc. For example, we may represent the vector $\overrightarrow{AB}$ by the symbol **v**. Then, if the directed segments AB and DC are parallel and equal, $\overrightarrow{AB} = \overrightarrow{DC} = \mathbf{v}$.

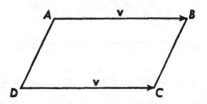

For any point A, the vector $\overrightarrow{AA}$ whose tail coincides with its head has length 0 and no specified direction. We call it the *zero* vector, and represent it by **0** (a bold-face zero).

For any given vector **v**, we shall designate by −**v** the vector with the same length as **v** that points in the opposite direction. For example, if $\overrightarrow{AB}$ = **v**, then $\overrightarrow{BA}$ = −**v**.

Vector Addition

An operation called addition of vectors, and represented by a plus sign, is defined in the following way. To find the sum **u** + **v**, first draw **u**; then draw **v** with its tail at the head of **u**. Then **u** + **v** is the vector from the tail of **u** to the head of **v**. For example, if **u** = $\overrightarrow{AB}$, and **v** = $\overrightarrow{BC}$, then **u** + **v** = $\overrightarrow{AC}$.

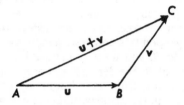

Addition of vectors has the following properties:

I.   (**u** + **v**) + **w** = **u** + (**v** + **w**).
II. **u** + **v** = **v** + **u**.
III. **u** + **0** = **0** + **u** = **u**.
IV. **u** + (−**u**) = (−**u**) + **u** = **0**.

The proof of property I is established by the diagram below.

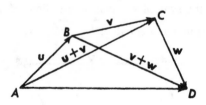

Since (**u** + **v**) + **w** = $\overrightarrow{AC}$ + $\overrightarrow{CD}$ = $\overrightarrow{AD}$, and **u** + (**v** + **w**) = $\overrightarrow{AB}$ + $\overrightarrow{BD}$ = $\overrightarrow{AD}$, it follows that (**u** + **v**) + **w** = **u** + (**v** + **w**).
The proof of property II is seen in the next diagram, where ABCD is a parallelogram. Since **u** + **v** = $\overrightarrow{AB}$ + $\overrightarrow{BC}$ = $\overrightarrow{AC}$, and **v** + **u** = $\overrightarrow{AD}$ + $\overrightarrow{DC}$ = $\overrightarrow{AC}$, it follows that **u** + **v** = **v** + **u**.

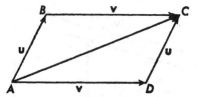

Properties III and IV are obvious.

Because vector addition has the four properties listed above, the vectors in a plane form a commutative group with respect to the operation addition. (See the definition of a group on page 39.)

Subtraction of vectors is defined in terms of addition: $\mathbf{u} - \mathbf{v}$ means $\mathbf{u} + (-\mathbf{v})$. That is, to subtract a vector, add its negative.

A Number Times a Vector

A multiplication operation in which any vector $\mathbf{u}$ may be multiplied by any real number n is defined as follows: If n is positive, $n\mathbf{u}$ or $\mathbf{u}n$ is the vector that is n times as long as $\mathbf{u}$, with the same direction as $\mathbf{u}$; $(-n)\mathbf{u}$ or $\mathbf{u}(-n)$ is the vector n times as long as $\mathbf{u}$ and with the opposite direction. If $n = 0$ or $\mathbf{u} = \mathbf{0}$, the product $n\mathbf{u} = \mathbf{0}$. For example, if $\mathbf{u} = \overrightarrow{AB}$, and M is the midpoint of AB, then $\frac{1}{2}\mathbf{u} = \overrightarrow{AM} = \overrightarrow{MB}$, and $-\frac{1}{2}\mathbf{u} = \overrightarrow{MA} = \overrightarrow{BM}$.

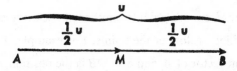

In this context we call every real number a *scalar*, and we call the multiplication of a vector by a scalar *scalar multiplication*. Notice that the product of a scalar and a vector is a vector.

The scalar multiplication defined above has the following properties:

V. $n(\mathbf{u} + \mathbf{v}) = n\mathbf{u} + n\mathbf{v}$.
VI. $(m + n)\mathbf{u} = m\mathbf{u} + n\mathbf{u}$.
VII. $m(n\mathbf{u}) = (mn)\mathbf{u}$.
VIII. $1\mathbf{u} = \mathbf{u}$.

Property V is a consequence of the fact that the corresponding sides of similar triangles are proportional. Properties VI, VII and VIII are obvious.

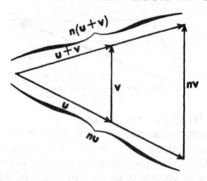

Properties I to VIII imply that addition and scalar multiplication of vectors may be carried out with the help of the formal rules ordinarily used in high school algebra.

Position Vector

Let O be the origin of a system of coordinates in a plane, and let P be any point in the plane. The vector $\overrightarrow{OP}$, with its tail fixed at O, is called the *position vector* of the point P. When a point is represented by a particular capital letter of the alphabet, we shall represent the position vector of that point by the same letter printed in bold-face lower-case type. Thus, **p** is the position vector of P.

The vector from one point to another can be expressed in terms of the position vectors of the points. For example, if $\mathbf{a} = \overrightarrow{OA}$ is the position vector of A, and $\mathbf{b} = \overrightarrow{OB}$ is the position vector of B,

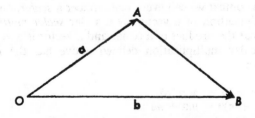

then, by the definition of vector addition, $\overrightarrow{OA} + \overrightarrow{AB} = \overrightarrow{OB}$. That is, $\mathbf{a} + \overrightarrow{AB} = \mathbf{b}$. Consequently, $\overrightarrow{AB} = \mathbf{b} - \mathbf{a}$. *Any vector equals the position vector of its head point minus the position vector of its tail point.*

Since every point in the plane has one and only one position vector, relationships among points may be expressed as relationships among their position vectors. This fact opens the door to another way of studying plane geometry by algebraic methods. In the preceding chapter we studied geometry by means of the algebra of the coordinates of points. In this chapter we study geometry by means of the algebra of vectors. In particular, we shall see how some theorems of Euclidean geometry can be proved by vector methods.

The Points on a Line

Let U be any point on a given line, and let $\mathbf{v}$ be any non-zero vector on the line. Then any vector on the line is a multiple of $\mathbf{v}$. Consequently, if P is any point on the line, $\overrightarrow{UP} = r\mathbf{v}$, where r is some real number. Let O be the origin. The diagram shows that

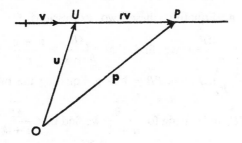

$\overrightarrow{OP} = \overrightarrow{OU} + \overrightarrow{UP}$. That is, $\mathbf{p} = \mathbf{u} + r\mathbf{v}$. Conversely, if $\mathbf{p} = \mathbf{u} + r\mathbf{v}$, then P is on the line through U in the direction of $\mathbf{v}$.

Dividing a Segment in a Given Ratio

Let A and B be distinct points, and let P be a point on AB that divides it in the ratio $h:k$ so that $\dfrac{AP}{PB} = \dfrac{h}{k}$, where h and k are real numbers such that neither h nor k nor $h + k$ is zero. We shall derive a formula that expresses the position vector of P in terms of the position vectors of A and B.

Divide AB into $h + k$ equal segments. Then AP contains h of them, and PB contains k of them. $\overrightarrow{AP}$ has the same direction as $\overrightarrow{AB}$. Since AP contains h out of the $h + k$ equal segments in AB, $\overrightarrow{AP}$ is $\dfrac{h}{h+k}$ times as long as $\overrightarrow{AB}$. Consequently $\overrightarrow{AP} = \left(\dfrac{h}{h+k}\right)\overrightarrow{AB}$. (For example, if $h = 2$ and $k = 3$, as in the diagram below, $\overrightarrow{AP} = \tfrac{2}{5}\overrightarrow{AB}$.) However, by the rule expressing

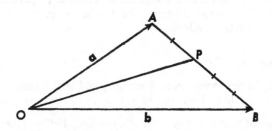

a vector in terms of the position vectors of its endpoints, $\overrightarrow{AP} = \mathbf{p} - \mathbf{a}$. Equating these two expressions for $\overrightarrow{AP}$, and solving for $\dfrac{\overrightarrow{AB}}{h+k}$, we find that $\dfrac{\overrightarrow{AB}}{h+k} = \dfrac{\mathbf{p} - \mathbf{a}}{h}$. Similarly, $\overrightarrow{PB} = \left(\dfrac{k}{h+k}\right)\overrightarrow{AB}$, and $\overrightarrow{PB} = \mathbf{b} - \mathbf{p}$. Equating the two expressions for $\overrightarrow{PB}$, and solving for $\dfrac{\overrightarrow{AB}}{h+k}$, we find that $\dfrac{\overrightarrow{AB}}{h+k} = \dfrac{\mathbf{b} - \mathbf{p}}{k}$.

Equating the two expressions for $\dfrac{\overrightarrow{AB}}{h+k}$ obtained in this way, we have $\dfrac{\mathbf{p} - \mathbf{a}}{h} = \dfrac{\mathbf{b} - \mathbf{p}}{k}$. If we solve this equation for $\mathbf{p}$, we obtain the formula

(1) $$\mathbf{p} = \frac{k\mathbf{a} + h\mathbf{b}}{k + h}.$$

Conversely, we can show that if $\mathbf{p}$ is related to $\mathbf{a}$ and $\mathbf{b}$ by equation (1), where neither h, k, nor $h + k$ is zero, then P is on AB, and $AP/PB = h/k$. We note first that $\mathbf{p} - \mathbf{a} \neq 0$, because if $\mathbf{a} = \mathbf{p}$, then, by equation (1), $\mathbf{p} = \dfrac{k\mathbf{p} + h\mathbf{b}}{k + h}$, and if we solve for $\mathbf{p}$, we get $\mathbf{p} = \mathbf{b}$, which implies that $\mathbf{a} = \mathbf{b}$, and the points A and B coincide, contrary to our assumption that they are distinct points. Equation (1) implies that $\dfrac{\mathbf{p} - \mathbf{a}}{h} = \dfrac{\mathbf{b} - \mathbf{p}}{k}$, that is, that $\dfrac{\overrightarrow{AP}}{h} = \dfrac{\overrightarrow{PB}}{k}.$ This last equation implies that AP and PB have the same direction. Moreover, they have a common point P. Therefore A, P, and B must be collinear, that is, the point P must lie on the line AB. The equation $\dfrac{\overrightarrow{AP}}{h} = \dfrac{\overrightarrow{PB}}{k}$ also implies that $AP/PB = h/k.$

In the special case where P is the midpoint of AB, $h = k$, and equation (1) reduces to

$$(2) \qquad\qquad \mathbf{p} = \tfrac{1}{2}(\mathbf{a} + \mathbf{b}).$$

Proofs by Vector Methods

With equations (1) and (2) as tools, we are ready to prove some theorems of plane geometry.

Theorem. The diagonals of a parallelogram bisect each other.

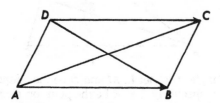

Proof. In parallelogram $ABCD$, $\overrightarrow{AB} = \overrightarrow{DC}$. Expressing both of these vectors in terms of position vectors, we have $\mathbf{b} - \mathbf{a} = \mathbf{c} - \mathbf{d}$. Consequently $\mathbf{b} + \mathbf{d} = \mathbf{a} + \mathbf{c}$, and therefore $\tfrac{1}{2}(\mathbf{b} + \mathbf{d}) = \tfrac{1}{2}(\mathbf{a} + \mathbf{c})$. By equation (2), $\tfrac{1}{2}(\mathbf{b} + \mathbf{d})$ is the position vector of the midpoint of BD, and $\tfrac{1}{2}(\mathbf{a} + \mathbf{c})$ is the position vector of the midpoint of AC. Since we have found that these two position vectors are equal, it means that the midpoint of BD coincides with the midpoint of AC. That is, AC and BD bisect each other.

Theorem. The line segment joining the midpoints of two sides of a triangle is parallel to the third side and equal to half of it.

Proof. In triangle ABC, let M and L be the midpoints of AB and AC respectively. Equation (2) gives us their position vectors.

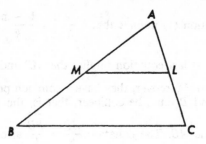

Thus, $\mathbf{l} = \frac{1}{2}(\mathbf{a} + \mathbf{c})$, and $\mathbf{m} = \frac{1}{2}(\mathbf{a} + \mathbf{b})$.

Subtracting the second of these equations from the first, we find that $\mathbf{l} - \mathbf{m} = \frac{1}{2}(\mathbf{c} - \mathbf{b})$. That is, $\overrightarrow{ML} = \frac{1}{2}\overrightarrow{BC}$.

Theorem. The medians of a triangle are concurrent.

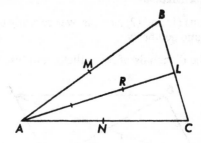

Proof. In triangle ABC, let L, M and N be the midpoints of BC, AB, and AC respectively. Let R be the point on median AL that divides it so that $AR/RL = 2/1$. Then, by equation (1),

$$\mathbf{r} = \frac{1\mathbf{a} + 2\mathbf{l}}{1 + 2} = \frac{\mathbf{a} + 2\mathbf{l}}{3}.$$

By equation (2), $\mathbf{l} = \frac{1}{2}(\mathbf{b} + \mathbf{c})$, so $2\mathbf{l} = \mathbf{b} + \mathbf{c}$. Substituting this value for $2\mathbf{l}$ in the formula above for $\mathbf{r}$, we get

(3) $$\mathbf{r} = \frac{\mathbf{a} + \mathbf{b} + \mathbf{c}}{3}.$$

Similarly, if R' is the point on CM such that $CR'/R'M = 2/1$, and if R'' is the point on BN such that $BR''/R''N = 2/1$, then $\mathbf{r}' = \dfrac{\mathbf{a} + \mathbf{b} + \mathbf{c}}{3}$, and $\mathbf{r}'' = \dfrac{\mathbf{a} + \mathbf{b} + \mathbf{c}}{3}$. That is, the three position vectors **r**, **r'**, and **r''** are all equal. Therefore the three points R, R', and R'' coincide. The point R where the medians of triangle ABC intersect is called the *centroid* of the triangle. Its position vector is given by equation (3). If a finite mass were uniformly distributed over a triangle, as in a triangular piece of sheet metal, the centroid of the triangle would be the center of gravity of the mass.

Components of a Vector

Let OX and OY be the horizontal and vertical axes respectively of a coordinate system in the plane. Let **i** be a vector of unit length in the positive direction of OX and let **j** be a vector of unit length in the positive direction of OY. Every horizontal vector is a multiple of **i**. That is, every horizontal vector has the form $x\mathbf{i}$, where x is a real number. Similarly, every vertical vector is a multiple of **j**; that is, it has the form $y\mathbf{j}$, where y is a real number.

Let **u** be any vector in the plane. If we draw a horizontal line through the tail of **u** and a vertical line through the head of **u**, we obtain a right triangle of which **u** is the hypotenuse. We see in this right triangle that **u** is the sum of a horizontal vector and a

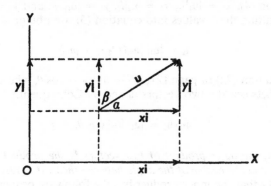

vertical vector. If the horizontal vector is $x\mathbf{i}$, and the vertical vector is $y\mathbf{j}$, then $\mathbf{u} = x\mathbf{i} + y\mathbf{j}$. The numbers x and y are called the horizontal and vertical *components* respectively of the vector **u**. They are uniquely determined for any given vector.

The vector $x\mathbf{i}$ is the orthogonal projection of **u** on OX, and the vector $y\mathbf{j}$ is the orthogonal projection of **u** on OY. If α and β are the direction angles of a half line pointing in the direction of **u**, if l and m are the corresponding direction cosines, and if $|\mathbf{u}|$

denotes the length of **u**, then we find by applying equation (19) of page 120 that

(4) $\quad x = |\mathbf{u}| \cos \alpha = |\mathbf{u}|l,$ and $\quad y = |\mathbf{u}| \cos \beta = |\mathbf{u}|m.$

It follows from equations (4) that if a vector **u** has components x and y, then a vector $r\mathbf{u}$ has components rx and ry.

Inner Product

Let the vector $\mathbf{u}_1$ have components x_1 and y_1, and let the vector $\mathbf{u}_2$ have components x_2 and y_2. The *inner product* of $\mathbf{u}_1$ and $\mathbf{u}_2$, denoted by $\mathbf{u}_1 \cdot \mathbf{u}_2$, is defined as follows:

(5) $\qquad\qquad \mathbf{u}_1 \cdot \mathbf{u}_2 = x_1 x_2 + y_1 y_2.$

Notice that the inner product of two vectors is a scalar, that is, a real number. Example: If $\mathbf{u}_1 = 4\mathbf{i} - 2\mathbf{j}$, and $\mathbf{u}_2 = 3\mathbf{i} + 5\mathbf{j}$, $\mathbf{u}_1 \cdot \mathbf{u}_2 = (4)(3) + (-2)(5) = 2$.

The geometric significance of the inner product is easily derived from equation (5). Let l_1 and m_1 be the direction cosines of $\mathbf{u}_1$, and let l_2 and m_2 be the direction cosines of $\mathbf{u}_2$. Then, by equation (4), $x_1 = |\mathbf{u}_1|l_1$, $x_2 = |\mathbf{u}_2|l_2$, $y_1 = |\mathbf{u}_1|m_1$, and $y_2 = |\mathbf{u}_2|m_2$. Substituting these values into equation (5), we obtain

(6) $\qquad\qquad \mathbf{u}_1 \cdot \mathbf{u}_2 = |\mathbf{u}_1|\,|\mathbf{u}_2|(l_1 l_2 + m_1 m_2).$

By equation (22) of page 121, $l_1 l_2 + m_1 m_2 = \cos \theta$, where θ is the angle between the two vectors $\mathbf{u}_1$ and $\mathbf{u}_2$. Consequently

(7) $\qquad\qquad \mathbf{u}_1 \cdot \mathbf{u}_2 = |\mathbf{u}_1|\,|\mathbf{u}_2| \cos \theta.$

That is, *the inner product of two vectors is the product of their lengths and the cosine of the angle between them.* It is not difficult to verify that the inner product has the following properties:

IX. $\mathbf{u} \cdot \mathbf{v} = \mathbf{v} \cdot \mathbf{u}.$
X. $\mathbf{u} \cdot (\mathbf{v} + \mathbf{w}) = \mathbf{u} \cdot \mathbf{v} + \mathbf{u} \cdot \mathbf{w}.$
XI. $(r\mathbf{u}) \cdot \mathbf{v} = r(\mathbf{u} \cdot \mathbf{v}).$

If a vector **u** has components x and y, then

(8) $\qquad\qquad \mathbf{u} \cdot \mathbf{u} = xx + yy = x^2 + y^2.$

If $\mathbf{u} = \mathbf{0}$, then $x = 0$, and $y = 0$, and $x^2 + y^2 = 0$. In all other cases, $x^2 + y^2$ is positive. Consequently the inner product has this additional property:

XII. $\mathbf{u} \cdot \mathbf{u}$ is positive, unless $\mathbf{u} = \mathbf{0}$.

The Length of a Vector

We see in the diagram on page 145 that if $\mathbf{u}$ is a vector whose components are x and y, then $\mathbf{u}$ is the hypotenuse of a right triangle whose legs have directed lengths x and y respectively. Then, by the Pythagorean Theorem, the length $|\mathbf{u}|$ of the vector $\mathbf{u}$ is related to x and y by the equation $|\mathbf{u}|^2 = x^2 + y^2$. Combining this result with equation (8), we get

$$(9) \qquad \mathbf{u} \cdot \mathbf{u} = |\mathbf{u}|^2, \quad \text{or} \quad |\mathbf{u}| = \sqrt{\mathbf{u} \cdot \mathbf{u}}.$$

Equation (9) is the vector equivalent of equation (17) on page 117.

Example. If $\mathbf{u} = 6\mathbf{i} + 8\mathbf{j}$ its length is $|\mathbf{u}| = \sqrt{\mathbf{u} \cdot \mathbf{u}} = \sqrt{6 \cdot 6 + 8 \cdot 8} = 10$.

The Angle Between Two Vectors

Let θ be the angle between two vectors $\mathbf{u}_1$ and $\mathbf{u}_2$. If we solve equation (7) for $\cos \theta$, we get

$$(10) \qquad \cos \theta = \frac{\mathbf{u}_1 \cdot \mathbf{u}_2}{|\mathbf{u}_1| \, |\mathbf{u}_2|} = \frac{\mathbf{u}_1 \cdot \mathbf{u}_2}{\sqrt{\mathbf{u}_1 \cdot \mathbf{u}_1} \sqrt{\mathbf{u}_2 \cdot \mathbf{u}_2}}.$$

Equation (10) is the vector equivalent of equation (22) on page 121.

Example. If $\mathbf{u}_1 = \mathbf{i} - 2\mathbf{j}$, and $\mathbf{u}_2 = 2\mathbf{i} + 2\mathbf{j}$,

$$\mathbf{u}_1 \cdot \mathbf{u}_2 = (1)(2) + (-2)(2) = -2;$$

$$|\mathbf{u}_1| = \sqrt{(1)(1) + (-2)(-2)} = \sqrt{5}; \quad |\mathbf{u}_2| = \sqrt{2 \cdot 2 + 2 \cdot 2} = \sqrt{8}.$$

Then the cosine of the angle between $\mathbf{u}_1$ and $\mathbf{u}_2$ is

$$\frac{-2}{\sqrt{5}\sqrt{8}} = \frac{-2}{\sqrt{40}} = \frac{-1}{\sqrt{10}}.$$

A Test for Perpendicularity

Suppose neither **u** nor **v** has length 0. If θ is the angle between **u** and **v**, then by equation (7), $\mathbf{u} \cdot \mathbf{v} = |\mathbf{u}|\,|\mathbf{v}| \cos \theta$. Since neither $|\mathbf{u}|$ nor $|\mathbf{v}|$ is zero, the product $\mathbf{u} \cdot \mathbf{v}$ is zero if and only if $\cos \theta = 0$. But $\cos \theta = 0$ if and only if $\theta = 90°$. Consequently

(11)   **u** is perpendicular to **v** if and only if $\mathbf{u} \cdot \mathbf{v} = 0$,

$$(\mathbf{u} \neq \mathbf{0},\, \mathbf{v} \neq \mathbf{0}).$$

Two non-zero vectors are perpendicular if and only if their inner product is 0. This test for perpendicularity is the vector equivalent of the test given by equation (10) on page 114.

More Proofs by Vector Methods

By using the inner product of vectors, and the test for perpendicularity expressed in equation (11) above, we can employ vector methods to prove some more theorems of plane geometry. We give three examples of such proofs below.

Theorem. The altitudes of a triangle are concurrent.

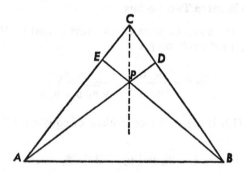

Proof. In triangle ABC, let AD be perpendicular to BC, and let BE be perpendicular to AC. AD and BE intersect at P. We shall show that CP is perpendicular to AB.

Since AP is perpendicular to BC, the inner product of the vectors $\overrightarrow{AP}$ and $\overrightarrow{BC}$ is 0. Moreover, $\overrightarrow{AP} = \mathbf{p} - \mathbf{a}$, and $\overrightarrow{BC} = \mathbf{c} - \mathbf{b}$. Consequently,

(12) $(\mathbf{p} - \mathbf{a}) \cdot (\mathbf{c} - \mathbf{b}) = 0.$

Similarly, since *BP* is perpendicular to *CA*,

(13) $(p - b) \cdot (a - c) = 0.$

Properties IX, X, and XI of inner products imply that we may expand these products by using the ordinary rules of elementary algebra. Expanding, we get

(14) $p \cdot c - p \cdot b - a \cdot c + a \cdot b = 0,$
(15) $-p \cdot c + p \cdot a + b \cdot c - a \cdot b = 0.$

Adding equations (14) and (15), we get

(16) $p \cdot (a - b) - c \cdot (a - b) = 0,$

or

(17) $(p - c) \cdot (a - b) = 0.$

Equation (17) asserts that the inner product of $\overrightarrow{CP}$ and $\overrightarrow{BA}$ is zero. Since neither of these vectors is zero, it follows from rule (11) that *CP* is perpendicular to *BA*. Hence all the altitudes of the triangle pass through *P*. The point *P* through which all the altitudes of a triangle pass is called its *orthocenter*.

Theorem. The perpendicular bisectors of the sides of a triangle are concurrent.

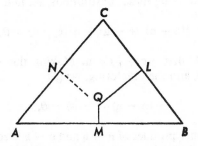

Proof. In triangle *ABC*, let *M* be the midpoint of *AB*, let *L* be the midpoint of *BC*, and let *N* be the midpoint of *AC*. Let *Q* be the intersection of the perpendicular bisectors of *AB* and *BC*. We shall show that the perpendicular bisector of *AC* passes through *Q*.

Since QM is perpendicular to AB,

$$(18) \qquad (\mathbf{m} - \mathbf{q}) \cdot (\mathbf{b} - \mathbf{a}) = 0.$$

Since QL is perpendicular to BC,

$$(19) \qquad (\mathbf{l} - \mathbf{q}) \cdot (\mathbf{c} - \mathbf{b}) = 0.$$

From equation (18), we get $\mathbf{m} \cdot (\mathbf{b} - \mathbf{a}) - \mathbf{q} \cdot (\mathbf{b} - \mathbf{a}) = 0$, or

$$(20) \qquad \mathbf{m} \cdot (\mathbf{b} - \mathbf{a}) - \mathbf{q} \cdot \mathbf{b} + \mathbf{q} \cdot \mathbf{a} = 0.$$

Similarly, from equation (19) we get

$$(21) \qquad \mathbf{l} \cdot (\mathbf{c} - \mathbf{b}) - \mathbf{q} \cdot \mathbf{c} + \mathbf{q} \cdot \mathbf{b} = 0.$$

However, $\mathbf{m} = \frac{1}{2}(\mathbf{a} + \mathbf{b})$, and $\mathbf{l} = \frac{1}{2}(\mathbf{b} + \mathbf{c})$. Substituting these values of $\mathbf{m}$ and $\mathbf{l}$ into equations (20) and (21) respectively and expanding, we get

$$(22) \qquad \tfrac{1}{2}\mathbf{b} \cdot \mathbf{b} - \tfrac{1}{2}\mathbf{a} \cdot \mathbf{a} - \mathbf{q} \cdot \mathbf{b} + \mathbf{q} \cdot \mathbf{a} = 0,$$
$$(23) \qquad \tfrac{1}{2}\mathbf{c} \cdot \mathbf{c} - \tfrac{1}{2}\mathbf{b} \cdot \mathbf{b} - \mathbf{q} \cdot \mathbf{c} + \mathbf{q} \cdot \mathbf{b} = 0.$$

Adding equations (22) and (23), we find that

$$(24) \qquad \tfrac{1}{2}(\mathbf{c} \cdot \mathbf{c} - \mathbf{a} \cdot \mathbf{a}) + \mathbf{q} \cdot \mathbf{a} - \mathbf{q} \cdot \mathbf{c} = 0.$$

We observe that $\mathbf{c} \cdot \mathbf{c} - \mathbf{a} \cdot \mathbf{a} = (\mathbf{c} + \mathbf{a}) \cdot (\mathbf{c} - \mathbf{a})$, and $\mathbf{q} \cdot \mathbf{a} - \mathbf{q} \cdot \mathbf{c} = -\mathbf{q} \cdot (\mathbf{c} - \mathbf{a})$. Making these substitutions, we find that

$$(25) \qquad \tfrac{1}{2}(\mathbf{c} + \mathbf{a}) \cdot (\mathbf{c} - \mathbf{a}) - \mathbf{q} \cdot (\mathbf{c} - \mathbf{a}) = 0.$$

We note now that $\frac{1}{2}(\mathbf{c} + \mathbf{a}) = \mathbf{n}$. Making this substitution in equation (25), and then factoring, we get

$$(26) \qquad (\mathbf{n} - \mathbf{q}) \cdot (\mathbf{c} - \mathbf{a}) = 0,$$

that is, the inner product of $\mathbf{n} - \mathbf{q}$ and $\mathbf{c} - \mathbf{a}$ is zero. The vector $\mathbf{c} - \mathbf{a}$ is not zero, because the points C and A are distinct. However, the vector $\mathbf{n} - \mathbf{q}$ may or may not be zero. If $\mathbf{n} - \mathbf{q} = \mathbf{0}$, then $\mathbf{n} = \mathbf{q}$, which implies that the point N coincides with the point Q, since the points have the same position vector. If $\mathbf{n} - \mathbf{q}$ is not zero, then by rule (11) of page 148, $\mathbf{n} - \mathbf{q}$ is perpendicular to $\mathbf{c} - \mathbf{a}$, that is, QN is perpendicular to AC. In either case, the perpendicular bisector of *AC* passes through *Q*. The point *Q* that lies on all three perpendicular bisectors of the sides of a triangle is called its *circumcenter*. It is the center of the circumscribed circle of the triangle.

Theorem. The orthocenter, circumcenter and centroid of a triangle lie on a straight line.

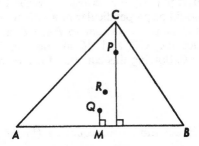

Proof. In triangle *ABC*, let *P* be the orthocenter of the triangle, let *Q* be the circumcenter of the triangle, let *R* be the centroid of the triangle, and let *M* be the midpoint of *AB*. Since *CP* is perpendicular to *BA*,

(27) $(\mathbf{p} - \mathbf{c}) \cdot (\mathbf{a} - \mathbf{b}) = 0.$

Since *MQ* is perpendicular to *BA*,

(28) $(\mathbf{q} - \mathbf{m}) \cdot (\mathbf{a} - \mathbf{b}) = 0.$

We know that $\mathbf{m} = \frac{1}{2}(\mathbf{a} + \mathbf{b})$. Making this substitution into equation (28) and multiplying by *2* we get

(29) $(2\mathbf{q} - \mathbf{a} - \mathbf{b}) \cdot (\mathbf{a} - \mathbf{b}) = 0.$

Adding equations (27) and (29), we get

(30) $(\mathbf{p} + 2\mathbf{q} - [\mathbf{a} + \mathbf{b} + \mathbf{c}]) \cdot (\mathbf{a} - \mathbf{b}) = 0.$

From equation (3) on page 144 we see that $\mathbf{a} + \mathbf{b} + \mathbf{c} = 3\mathbf{r}$. Making this substitution into equation (30), we get

(31) $(\mathbf{p} + 2\mathbf{q} - 3\mathbf{r}) \cdot (\mathbf{a} - \mathbf{b}) = 0.$

By similar reasoning we can show that

(32) $(\mathbf{p} + 2\mathbf{q} - 3\mathbf{r})\cdot(\mathbf{b} - \mathbf{c}) = 0,$

(33) $(\mathbf{p} + 2\mathbf{q} - 3\mathbf{r})\cdot(\mathbf{c} - \mathbf{a}) = 0.$

Since the points A, B and C are distinct, $\mathbf{a} - \mathbf{b} \neq \mathbf{0}$, $\mathbf{b} - \mathbf{c} \neq \mathbf{0}$, and $\mathbf{c} - \mathbf{a} \neq \mathbf{0}$. If $\mathbf{p} + 2\mathbf{q} - 3\mathbf{r}$ is also not zero, then by rule (11) $\mathbf{p} + 2\mathbf{q} - 3\mathbf{r}$ would be perpendicular to $\mathbf{a} - \mathbf{b}$, $\mathbf{b} - \mathbf{c}$, and $\mathbf{c} - \mathbf{a}$. That is, it would be perpendicular to BA, CB, and AC. This is impossible, since BA, CB, and AC are not parallel. Therefore $\mathbf{p} + 2\mathbf{q} - 3\mathbf{r} = \mathbf{0}$. Solving this equation for $\mathbf{r}$ we obtain

(34) $$\mathbf{r} = \frac{I\mathbf{p} + 2\mathbf{q}}{3}.$$

Comparing this equation with equation (1) on page 142, we see that R is the point on PQ that divides it so that $\dfrac{PR}{RQ} = \dfrac{2}{1}$. Hence P, Q, and R lie on a straight line.

Linear Independence

Let $\mathbf{u}$ and $\mathbf{v}$ be any two non-zero vectors in the plane that are not parallel. Then $r\mathbf{u}$ is a vector r times as long as $\mathbf{u}$ and parallel to $\mathbf{u}$, and $s\mathbf{v}$ is a vector s times as long as $\mathbf{v}$ and parallel to $\mathbf{v}$. The vector $r\mathbf{u} + s\mathbf{v}$ can be obtained by the triangle construction shown in the diagram. It is obvious from this construction that $r\mathbf{u} + s\mathbf{v} = \mathbf{0}$ only if the coefficients of $\mathbf{u}$ and $\mathbf{v}$ are zero, that is, only if $r = 0$ and $s = 0$. On the other hand, if

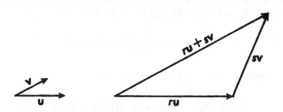

$\mathbf{w}$ is any third vector $\overrightarrow{AB}$ in the plane, we can draw a triangle ABC with $\overrightarrow{AC}$ parallel to $\mathbf{u}$ and $\overrightarrow{CB}$ parallel to $\mathbf{v}$. Then $\overrightarrow{AC}$ is some multiple $a\mathbf{u}$ of $\mathbf{u}$, $\overrightarrow{CB}$ is some multiple $b\mathbf{v}$ of $\mathbf{v}$, and $\overrightarrow{AB}$ is their sum. That is, $\mathbf{w} = a\mathbf{u} + b\mathbf{v}$. From this equation we see that $a\mathbf{u} + b\mathbf{v} - 1\mathbf{w} = \mathbf{0}$, and the coefficients of $\mathbf{u}$, $\mathbf{v}$, and $\mathbf{w}$ are not all zero, since the coefficient of $\mathbf{w}$ is -1. This difference in properties of a set of two vectors in the plane as compared to a set of

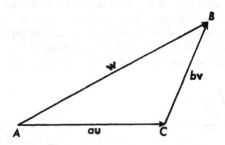

three vectors in the plane leads to the following definition: Let $\mathbf{u}_1, \ldots, \mathbf{u}_n$ be any n vectors. Then if $a_1\mathbf{u}_1 + \ldots + a_n\mathbf{u}_n = \mathbf{0}$ only if the coefficients of all the n vectors are zero, we say that the n vectors are linearly independent. If we apply this terminology to the examples just given, we may say that the set of vectors $\{\mathbf{u}, \mathbf{v}\}$ is linearly independent, but the set of vectors $\{\mathbf{u}, \mathbf{v}, \mathbf{w}\}$ is not linearly independent. Indeed, although there are many sets of two vectors in the plane that are linearly independent, namely any two non-zero vectors that are not parallel, any set of three vectors in the plane is not linearly independent. It can be shown from this fact that the largest number of vectors there may be in a set of linearly independent vectors in the plane is 2. Before we consider the significance of this fact, let us examine two more examples.

Let us choose any straight line, and consider only vectors that may be drawn on that line. Let $\mathbf{u}$ be any non-zero vector on the line. Then $a\mathbf{u} = \mathbf{0}$ only if $a = 0$. Therefore the set of vectors whose only member is $\mathbf{u}$ is linearly independent. Let $\mathbf{v}$ be any other vector on the line. The vector $\mathbf{v}$, since it has the same direction as $\mathbf{u}$ or the opposite direction, is some multiple $a\mathbf{u}$ of $\mathbf{u}$. That is $\mathbf{v} = a\mathbf{u}$, or $a\mathbf{u} - 1\mathbf{v} = \mathbf{0}$, and in this equation the coefficients of $\mathbf{u}$ and $\mathbf{v}$ are not all zero. Consequently the set of

vectors $\{\mathbf{u}, \mathbf{v}\}$ is not linearly independent. Indeed, while a set consisting of a single vector on the line may be linearly independent, any set of two vectors on the line is not linearly

independent. It can be shown from this fact that the largest
number of vectors there may be in a set of linearly independent
vectors on the line is 1.

Let us consider now vectors in three-dimensional space. Let
u, **v**, and **w** be three non-zero vectors that are not parallel to
the same plane. We may draw any multiples r**u**, s**v**, and t**w** of
these vectors as the adjacent edges of a parallelepiped, as shown
in the diagram. Then r**u** $+ s$**v** $+ t$**w** is a diagonal of the parallelepiped. Consequently r**u** $+ s$**v** $+ t$**w** $= \mathbf{0}$ only if $r = 0$, $s = 0$,

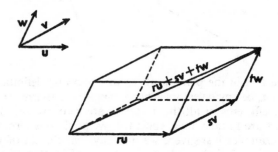

and $t = 0$. This means that the set of three vectors $\{\mathbf{u}, \mathbf{v}, \mathbf{w}\}$ is
linearly independent. However, if **z** is any fourth vector in space,
we can make **z** the diagonal AD of a parallelepiped whose edges
AB, BC and CD are parallel to **u**, **v**, and **w** respectively. Then $\overrightarrow{AB}$
is some multiple a**u** of **u**, $\overrightarrow{BC}$ is some multiple b**v** of **v**, $\overrightarrow{CD}$ is some
multiple c**w** of **w**, and $\overrightarrow{AD}$ is their sum. That is, $\mathbf{z} = a\mathbf{u} + b\mathbf{v} + c\mathbf{w}$.
From this equation we get $a\mathbf{u} + b\mathbf{v} + c\mathbf{w} - 1\mathbf{z} = \mathbf{0}$, in which the

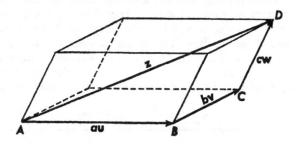

coefficients of the vectors are not all zero. Consequently the set
of vectors $\{\mathbf{u}, \mathbf{v}, \mathbf{w}, \mathbf{z}\}$ is not linearly independent. Indeed,
although there are many sets of three vectors in three-dimensional

space that are linearly independent, namely any three non-zero vectors that are not parallel to the same plane, any set of four vectors in three-dimensional space is not linearly independent. It can be shown from this fact that the largest number of vectors there may be in a set of linearly independent vectors in three-dimensional space is 3.

We may sum up our observations as follows: On a line (which is a one-dimensional space), the largest number of vectors there may be in a set of linearly independent vectors is 1. On a plane (which is a two-dimensional space), the largest number of vectors there may be in a set of linearly independent vectors is 2. In three-dimensional space, the largest number of vectors there may be in a set of linearly independent vectors is 3. We can now see the geometric significance of these observed facts: *The largest number of vectors there may be in a set of linearly independent vectors in a given space is the number of dimensions of the space.*

A Vector Approach to Geometry

In the preceding paragraphs, we began with a Euclidean plane and then introduced vectors as arrows joining points of the plane. We found that we could associate with each point of the plane one and only one position vector. This fact opens up the possibility of reversing the procedure. We can begin with suitably chosen objects that behave like vectors, and then use these vectors, construed as position vectors, to define the points of a plane. The procedure for constructing Euclidean plane geometry in this manner may be outlined as follows: Take any set of objects in which an addition operation is defined and for which multiplication by a real number (called scalar multiplication) is defined such that addition and scalar multiplication have properties I to VIII listed on pages 138 and 139. Such a set of objects is an algebraic structure called a *vector space*, and its members are called *vectors*. Assume that the largest number of vectors in any set of linearly independent vectors of this vector space is 2. Then the vector space is two-dimensional. We define a point to be any vector in the vector space. We call the set of all these points a plane. We define a line to be a set of points p satisfying the condition $p = u + rv$, where u is any fixed vector and v is any fixed non-zero vector. (See page 141.) Assume further that there is defined in this vector space an inner product that has properties IX to XII. (See pages 146–147.) Define the length of a vector by means of equation (9), and define the angle between two vectors by means of equation (10). With the concepts of point, line, length, and angle introduced in this way, it can be shown that the plane has all the properties of a Euclidean plane. From this point of view, a Euclidean plane can be described as a two-dimensional vector space in which an inner product is defined.

In general, let n be the largest number of vectors in any set of linearly independent vectors in a vector space in which an inner product is defined. Then, if point, line, length and angle are defined as above, the vector space is an n-dimensional Euclidean space. Since a vector space is an algebraic structure, this procedure gives us another way of obtaining geometry as a branch of algebra.

EXERCISES FOR CHAPTER 5

1. *Position vectors.* Express in terms of the position vectors of A, B and C the vectors $\overrightarrow{AB}$, $\overrightarrow{BC}$ and $\overrightarrow{CA}$.

2. *Vector equation of a line.* Let C and D be two distinct points, and let P be an arbitrary point of the straight line determined by C and D. a) What is the symbol for the position vector of C? b) Express the vector $\overrightarrow{CD}$ in terms of the position vectors of C and D. c) Express the position vector of P in terms of your answer to a) and b), assuming that CP is r times as long as CD. (See page 141.)

3. *Dividing a segment in a given ratio.* If P is on the segment AB and divides it so that $AP:PB = 3:4$, express the position vector of P in terms of the position vectors of A and B. (See equation (1) on page 142.)

4. *Midpoints.* If L is the midpoint of AB, M is the midpoint of CD, and P is the midpoint of LM, express the position vector of P in terms of the position vectors of A, B, C and D.

5. If ABC is any triangle in a plane with origin O, and L, M and N are the midpoints of AB, BC and CA respectively, prove that $\overrightarrow{OL} + \overrightarrow{OM} + \overrightarrow{ON} = \overrightarrow{OA} + \overrightarrow{OB} + \overrightarrow{OC}$.

6. *Inner product.* Find the inner product of each pair of vectors:

a) $2\mathbf{i} + 3\mathbf{j}$ and $4\mathbf{i} - \mathbf{j}$;   b) $\mathbf{i} + \mathbf{j}$ and $\mathbf{i} - \mathbf{j}$;
c) $3\mathbf{i} + 4\mathbf{j}$ and $3\mathbf{i} + 4\mathbf{j}$.

7. Using $\mathbf{u} = a\mathbf{i} + b\mathbf{j}$, $\mathbf{v} = c\mathbf{i} + d\mathbf{j}$, and $\mathbf{w} = e\mathbf{i} + f\mathbf{j}$, verify properties IX, X and XI of the inner product. (See page 146.)

8. *Length of a vector.* Find the length of the vector $5\mathbf{i} + 12\mathbf{j}$.

9. *The angle between two vectors.* If θ is the angle between $\mathbf{i} - 2\mathbf{j}$ and $3\mathbf{i} + 4\mathbf{j}$, find $\cos \theta$.

10. *Perpendicular vectors.* Which pair of vectors are not perpendicular?

a) $\mathbf{i} + \mathbf{j}$ and $\mathbf{i} - \mathbf{j}$;   b) $3\mathbf{i} + 2\mathbf{j}$ and $2\mathbf{i} - 3\mathbf{j}$;
c) $\mathbf{i} + \mathbf{j}$ and $3\mathbf{i} + 2\mathbf{j}$.

11. *Isosceles triangle.* Let ABC be an isosceles triangle in the plane, with $AB = BC$. Let M be the midpoint of AC. a) Express $\overrightarrow{AB}$ and $\overrightarrow{BC}$ in terms of the position vectors $\mathbf{a}$, $\mathbf{b}$ and $\mathbf{c}$.
b) Express the lengths $|\overrightarrow{AB}|$ and $|\overrightarrow{BC}|$ in terms of $\mathbf{a}$, $\mathbf{b}$ and $\mathbf{c}$.
c) Using the fact that $|\overrightarrow{AB}| = |\overrightarrow{BC}|$, write an equation that is satisfied by $\mathbf{a}$, $\mathbf{b}$ and $\mathbf{c}$.
d) Express $\mathbf{m}$, the position vector of M, in terms of $\mathbf{a}$ and $\mathbf{c}$.
e) Express $\overrightarrow{BM}$ in terms of $\mathbf{a}$, $\mathbf{b}$ and $\mathbf{c}$.
f) Express $\overrightarrow{AC}$ in terms of $\mathbf{a}$ and $\mathbf{c}$.
g) Using the answers to c), e) and f), prove that BM is perpendicular to AC.

6

Geometry via Reflections

The Algebra of Isometries

On pages 125 to 129 we singled out for special mention those transformations of a plane, called isometries, which move any given line segment into a congruent segment. We observed that the set of all isometries of a plane constitutes a group. Since a group is an algebraic structure, the use of isometries provides us with another way of studying plane geometry by algebraic methods. In Chapter 4 we studied geometry by means of the algebra of the coordinates of points. In Chapter 5 we studied geometry by means of the algebra of vectors. In this chapter we study geometry by means of the algebra of isometries. It will provide us with a third algebraic technique by means of which geometric theorems can be proved.

To prepare the ground for this new approach to geometry we examine first some elementary properties of any group of transformations and some special properties of the isometries of a plane. Then we show how every statement about points and lines can be translated into a corresponding statement about isometries. By using such translations, we shall convert geometric problems about points and lines into algebraic problems about isometries, and we shall solve the problems algebraically with the help of the observed properties of groups and of isometries.

Some Properties of a Group of Transformations

In this chapter we shall use Roman capital letters to stand for isometries. The set of isometries of a plane is a group because it has these three properties which we transcribe from page 58:

A. If R, S and T are isometries, R(ST) = (RS)T. (The associative law.)
C. There exists an isometry 1, the identity transformation, such that for every isometry T, T1 = 1T = T.

D. For every isometry T there exists an isometry T^{-1}, the inverse of T, that has the property $TT^{-1} = T^{-1}T = 1$.

Notice that condition B of page 39 is not listed, because the group of isometries is not a commutative group. That is, if R and S are isometries, the products RS and SR need not be equal. For example, if R is the translation that associates with each point of the plane the point that is one inch to the right of it, and S is a clockwise rotation of 90° about a point P, which associates with each point of the plane the point to which it would be moved by such a rotation, then RS is the transformation that results

R moves P to Q   and S moves Q to T....

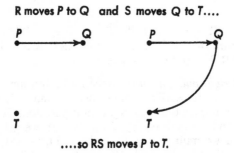

....so RS moves P to T.

from first moving the plane one inch to the right and then rotating the plane 90° clockwise about P, while SR is the transformation that results from first rotating the plane 90° clockwise about P and then moving the plane one inch to the right. If Q is one inch to the right of P, and T is one inch below P, the transformation RS moves P to T, but the transformation SR moves P to Q. (See the diagram below.) Consequently, in this case, RS and SR are two different transformations, that is, RS $\neq$ SR.

S leaves P fixed   and   R moves P to Q....

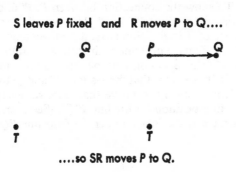

....so SR moves P to Q.

Use of parentheses. It can be shown that the associative law implies that, in any product of several transformations, parentheses may be inserted or removed arbitrarily. We shall use this rule frequently to simplify some products. For example, the product $TSS^{-1}R$ can be simplified if we insert and remove parentheses as shown below, and take note of the fact that $SS^{-1} = 1$, and $T1 = T$:

$$TSS^{-1}R = T(SS^{-1})R = T1R = (T1)R = TR.$$

Inverse of a product. To obtain the inverse of a product of several transformations, multiply in inverse order the inverses of the factors. For example, the inverse of RST is $T^{-1}S^{-1}R^{-1}$, because

$$(RST)(T^{-1}S^{-1}R^{-1}) = RSTT^{-1}S^{-1}R^{-1} = RS(TT^{-1})S^{-1}R^{-1}$$
$$= RS1S^{-1}R^{-1} = R(S1)S^{-1}R^{-1} = RSS^{-1}R^{-1} = R(SS^{-1})R^{-1}$$
$$= R1R^{-1} = (R1)R^{-1} = RR^{-1} = 1.$$

Left and right multiplication. Since multiplication of transformations is not necessarily commutative, we must distinguish between left multiplication and right multiplication. For example, if $A = B$, and we multiply by T on the left, we get $TA = TB$. However, if we multiply by T on the right, we get $AT = BT$. As we have already observed, AT is not necessarily equal to TA, and BT is not necessarily equal to TB.

Cancellation. If $AT = BT$, we may multiply by T^{-1} on the right to obtain $ATT^{-1} = BTT^{-1}$. By simplifying this equation we find that $A = B$. Similarly, if $TA = TB$, we find by multiplying by T^{-1} on the left that $A = B$. That is, in any equation of the form $AT = BT$ or $TA = TB$ the T may be canceled.

The image of a point or a line. If a transformation T moves a point P to P', we say that P' is the image of P under the transformation T. To show the connection between P, P' and T we use this abbreviated notation: $PT = P'$. This equation may be read as "P is moved by T to P'." Similarly, if T moves a line a to a', we say that a' is the image of a under the transformation T, and we write $aT = a'$. This equation may be read as "a is moved by T to a'." If T is an isometry that moves P to P' and Q to Q', where $P \neq Q$, it is not difficult to prove that it moves every point on the line PQ to a position on the line $P'Q'$. (See exercise 6, page 184.) Consequently the image under T of the line PQ is the line $P'Q'$.

Involutions. A transformation T is called an involution if
T $\neq$ 1, and TT = 1. If we multiply the equation TT = 1 by T^{-1}
on the left we get T^{-1}TT = T^{-1}1, which simplifies to T = T^{-1}.
That is, *an involution is equal to its inverse.* Let P be any point in
the plane, and let P' be its image under an involution T. Then T
moves P to P'. By definition of the inverse T^{-1}, T^{-1} moves P' to
P. But we have observed that T^{-1} = T. Consequently T moves
P' to P. That is, T interchanges the pair of points P and P'.

The Principle of Rigidity

Let P be any point in the plane, and let h be any half-line whose
vertex is P. The line that contains h divides the plane into two
sides. Let S be one of these two sides. The configuration (P, h, S)
is represented in the diagram by a flag-shaped figure, in which
the "flagpole" shows the position of h, and the "flag" shows the
position of S. Let P' be any other point in the plane, let h' be

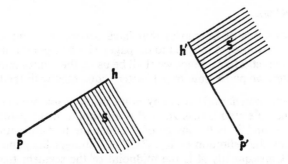

any half-line whose vertex is P', and let S' be one of the two sides
into which the plane is divided by the line that contains h'. The
configuration (P', h', S') is represented by another flag-shaped
figure in the diagram. It can be proved *that there is one and only
one isometry that moves P to P', h to h', and S to S'.* (See exercise 8,
page 184.) This theorem is known as the *principle of the rigidity
of the plane.* We shall use this principle repeatedly in the paragraphs that follow. Any time that we encounter two isometries
A and B each of which moves a flag-shaped configuration
(P, h, S) into coincidence with a flag-shaped configuration
(P', h', S'), the principle of rigidity allows us to conclude that
A = B.

Example: Let k be a half-line whose vertex is Q, and let S_1 and S_2 be the two sides into which the plane is divided by the line that contains k. The identity transformation 1 "moves" Q to Q,

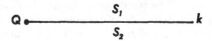

k to k, and S_1 to S_1. Consequently if an isometry T moves Q to Q, k to k, and S_1 to S_1, we conclude by the principle of rigidity that T = 1. Let a be the line that contains k, and let R_a be the reflection of the plane across a. R_a "moves" Q to Q, k to k, and S_1 to S_2. Consequently, if an isometry T moves Q to Q, k to k, and S_1 to S_2, we conclude that T = R_a. The transformations 1 and R_a are the only isometries that move Q to Q and k to k. *If an isometry moves a given half-line onto itself the isometry is either 1 or the reflection across the line that contains the half-line.*

Reflections

Two kinds of isometries that have particularly simple properties are those we referred to on pages 127–8 as point reflections and line reflections. Since we shall be using them throughout this chapter, we pause now to get better acquainted with them.

Point reflections. If A is any point in the plane, we denote by R_A the reflection across A. If P is any point in the plane other than A, and P' is the image of P under the transformation R_A, then by the definition of R_A, PAP' is a straight line, and $PA = AP'$. Consequently A is the midpoint of the segment that joins any point to its image under R_A. (This property may, in fact, be taken as a definition of R_A.) It is obvious that while R_A moves P to P', it also moves P' to P. Consequently the product $R_A R_A$ moves P to P, that is, it doesn't move it at all. In other words $R_A R_A = 1$, the identity transformation. Moreover, R_A itself is not the identity transformation, since it actually moves to a different position every point except A. Therefore, since $R_A \neq 1$, and $R_A R_A = 1$, R_A is an involution. *Every point reflection is an involution.*

The image of A under the transformation R_A is A. If a point is its own image under a transformation, we call it a *fixed point* of the transformation. So we may say that A is a fixed point of R_A. Moreover, it is obvious that A is the only fixed point of R_A.

Consider any line PA through A. The image under R_A of every point on that line is also on the line. Consequently the image under R_A of the line PA is itself. That is, while R_A moves most of the points on PA, it does not move the line through these points. If a line is its own image under a transformation, so that the line is not moved by the transformation (although individual points on it may be moved to new positions on the line) we call it a *fixed line* of the transformation. Every line through A is a fixed line under R_A.

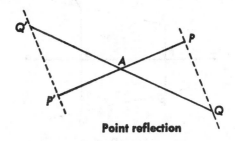

Point reflection

Let P and Q be two points each distinct from A, and let P' and Q' be their images respectively under R_A. The image under R_A of the line PQ is the line $P'Q'$. It can be shown that if A is not on PQ, then $P'Q'$ and PQ are two distinct lines. (See exercise 10, page 185.) Consequently the only fixed lines under R_A are the lines through A.

We may sum up the observed properties of R_A as follows: 1) R_A is an involution. 2) If P and P' are distinct points that are interchanged by the involution R_A, then A is the midpoint of PP'. 3) R_A has one and only one fixed point, namely A. 4) A line is a fixed line under R_A if and only if it passes through A.

Line reflections. If a is any line in the plane, we denote by R_a the reflection across a. If P is any point in the plane that is not on a, and P' is the image of P under the transformation R_a, then PP' is perpendicular to a. If R is the intersection of a and PP', then $PR = RP'$. Consequently a is the perpendicular bisector of the segment that joins any point to its image. (This property may, in fact, be taken as a definition of R_a.) It is obvious that while R_a moves P to P', it also moves P' to P. Consequently $R_aR_a = 1$. Moreover, R_a itself is not the identity transformation. Therefore R_a is an involution. *Every line reflection is an involution.*

The image under R_a of any point R on a is R itself. Consequently every point of a is a fixed point of R_a. Moreover, it is obvious that the points on a are the only fixed points of R_a.

Since every point of a is a fixed point of R_a, it follows that a is a fixed line of R_a. If a is the perpendicular bisector of a segment PP', R_a moves P to P' and it moves P' to P. Consequently it moves the line PP' to the line $P'P$. It follows that every line perpendicular to a is a fixed line of R_a. It is not difficult to see that any line PQ that is neither a nor a perpendicular to a is not a fixed line under R_a. (See exercise 11, page 185.)

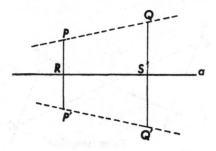

Line reflection

We may sum up the observed properties of R_a as follows: 1) R_a is an involution. 2) If P and P' are distinct points that are interchanged by the involution R_a, then a is the perpendicular bisector of PP'. 3) A point is a fixed point of R_a if and only if it lies on a. 4) A line is a fixed line of R_a if and only if it is a itself or a line perpendicular to a.

Since both point reflections and line reflections are involutions, it is natural to ask what other isometries are involutions. The answer is that there are no other involutions, because we can prove the following theorem: If an isometry is an involution, it is either a point reflection or a line reflection.

Proof: Let T be an isometry that is an involution. Since T is an involution, it is not the identity transformation. Therefore there are points P that are distinct from their images P' under T. T moves each such point P to its image P' and moves P' to P. Therefore it moves the line PP' into itself. Let M be the midpoint of the segment PP'. Let M' be the point on PP' that is the image of M under T. $PM = MP'$. The image of PM is $P'M'$, and the image of MP' is $M'P$. Since an isometry preserves lengths, these images, too, are equal. That is, $P'M' = M'P$. So M' is the midpoint of $P'P$, or $M' = M$. In other words, for every point pair (P, P') interchanged by the involution T, the midpoint M of the segment PP' is a fixed point of T. Then the segment MP is moved by T onto the segment MP'. Consequently the half-line MP whose vertex is M is moved onto the half-line MP' whose vertex is M. There are two possibilities that may arise: either 1) all the point pairs interchanged by T have the same midpoint M, or 2) at least two point pairs interchanged by T have distinct midpoints M_1 and M_2. In case 1), T is obviously the point reflection R_M. In case 2), since the image of M_1 under T is M_1, and the image of M_2 under T is M_2, then T moves the half-line M_1M_2 whose vertex is M_1 onto itself. Consequently, as we noted in the example on page 162, T is either the identity transformation or the reflection across the line M_1M_2. Since T is not the identity, it must be the reflection across M_1M_2.

We see then that the set of all point reflections and line reflections is the set of all isometries of the plane that are involutions. This is one reason why the reflections are being given special attention.

Product of Two Line Reflections

It will be useful to know what kind of transformation results from multiplying two reflections. We examine first the product of two line reflections R_a and R_b, where a and b are any two distinct lines in the plane. (We know already that if $a = b$, $R_aR_b = R_aR_a = 1$.) There are two possibilities that arise: I. a and b intersect; II. a and b are parallel. We examine these possibilities separately.

I. a and b intersect at P. Choose on a and b the half-lines h and k respectively with vertices at P so that the angle between them is not obtuse. Denote by θ = angle (a, b) = angle (h, k), the directed angle *from h to k*. (The directed angle from h to k is different from the directed angle from k to h, just as a directed segment AB is different from the directed segment BA.) Draw a half-line h' with vertex P so that angle $(k, h') = \theta$. The line a divides the plane into two sides S and $\bar{S}$. Let S be the side which contains k. (In the diagram S is the side that h moves into when h is rotated clockwise around P.) Let S' be the side of the line containing h' that h' moves into if h and h' turn in the same direction as h moves into S. (In the diagram, S' is the side that h' moves into when h' is rotated clockwise around P.) We observe now that P is a fixed point of R_a and also of R_b. Consequently P is a fixed point of R_aR_b. Notice, too, that R_a leaves h fixed, and R_b moves h to h'. Consequently R_aR_b moves h to h'. Notice next that R_a moves S into $\bar{S}$, and R_b moves $\bar{S}$ into S'. Consequently R_aR_b

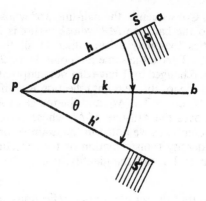

moves S into S'. Therefore R_aR_b moves the flag-shaped configuration (P, h, S) into the flag-shaped configuration (P, h', S'). But so does a rotation of 2θ, or a rotation of 2 angle (a, b). Therefore, by the principle of rigidity,

(1) $R_aR_b =$ a rotation through 2 angle (a, b).

If a and b are interchanged in equation (1), we see that $R_bR_a =$ rotation through 2 angle (b, a). In general, a rotation through 2 angle (a, b) and a rotation through 2 angle (b, a) are not the same because they are in opposite directions. Therefore, in general, $R_aR_b \neq R_bR_a$.

In the special case where a and b are perpendicular to each other, 2 angle $(a, b) = 180°$. But a rotation of $180°$ is a point reflection across the center of rotation. Therefore,

(2)   if $a \perp b$ at P, $R_aR_b = R_P$.

Under the same conditions, $R_bR_a = R_P$. Therefore,

(3)   if $a \perp b$, $R_aR_b = R_bR_a$.

Suppose a, b, c and d are concurrent at P, and angle $(a, b) =$ angle (d, c). Since $R_aR_b =$ a rotation around P through 2 angle (a, b), and $R_dR_c =$ a rotation around P through 2 angle (d, c), it follows that $R_aR_b = R_dR_c$.

II. a and b are parallel. Draw g perpendicular to a and b at A and B respectively. R_g is an involution. Therefore $R_gR_g = 1$. If we insert 1 between the two factors in the product R_aR_b we get

$$R_aR_b = R_a1R_b = R_a(R_gR_g)R_b = (R_aR_g)(R_gR_b) = R_AR_B.$$

The last step in this sequence of equalities follows from equation (2). Now all we have to do is identify the product R_AR_B.

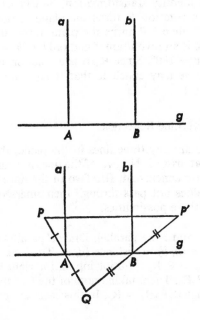

Let P be any point in the plane that is not on g. (See the second part of the diagram above.) Let Q be the image of P under R_A, and let P' be the image of Q under R_B. Then $PA = AQ$, and $QB = BP'$. Using the theorem proved on page 143 we conclude that PP' is parallel to AB, and $PP' = 2AB$. It is easily verified that if P is on g, P' is also on g, and $PP' = 2AB$. (See exercise 12, page 185.) If we denote the distance from a to b by distance (a, b), it follows that

(4) if $a \parallel b$, R_aR_b = translation through twice distance (a, b),

in the direction from a to b along a common perpendicular.

Suppose a, b, c, and d are parallel, and distance (a, b) = distance (d, c). Then a, b, c and d have a common perpendicular. Since R_aR_b = translation through twice distance (a, b), and R_dR_c = translation through twice distance (d, c), it follows that $R_aR_b = R_dR_c$.

Combining the results of cases I and II, we see that *every product of two line reflections is either a rotation or a translation.*

If a translation T moves the plane through a positive distance d, then TT moves the plane through a distance $2d$. Consequently TT is not the identity transformation, and hence T is not an involution. If a rotation R turns the plane through a positive angle $\theta \leq 180°$, then RR turns the plane through an angle 2θ. RR = 1 and R is an involution if and only if $2\theta = 360°$, that is, if and only if $\theta = 180°$. Since R_aR_b is a rotation of 180° if and only if $a \perp b$, we may conclude that R_aR_b is an involution if and only if $a \perp b$.

Product of Three Line Reflections

If a, b and c are any three lines in the plane, there are three possibilities that arise: I. No two of the lines a, b and c intersect. II. a, b and c are concurrent. III. Two of the lines intersect and the third line does not pass through their intersection. We examine each of these possibilities.

I. Assume a, b and c are parallel. Draw d parallel to a, b and c so that distance (a, b) = distance (d, c). Then, as we found on page 167, $R_aR_b = R_dR_c$. Multiplying on the right by R_c, we get $R_aR_bR_c = R_dR_cR_c$. Then, taking note of the fact that $R_cR_c = 1$, we conclude that $R_aR_bR_c = R_d$. In this case, the product of the

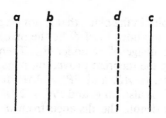

three line reflections R_a, R_b, and R_c is itself a line reflection. It is easy to see that the same conclusion follows if any two of the lines a, b and c coincide. (See exercise 13, page 185.)

II. a, b and c are concurrent at P. Draw d through P so that angle (a, b) = angle (d, c). Then, as we found on page 166, $R_aR_b = R_dR_c$. Consequently, as in case I, $R_aR_bR_c = R_d$. In this case too, the product of the three line reflections R_a, R_b, and R_c is itself a line reflection.

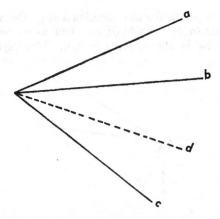

III. We consider first the special case where $a \perp c$, and $a \parallel b$. Denote by A and B the points where c intersects a and b respectively. $R_a R_b R_c = (R_a R_b)R_c$. By (4) on page 167, $R_a R_b$ is a translation along c. So in this case the product $R_a R_b R_c$ is a translation along c followed by a reflection across c. Such a product is called a *glide-reflection* in c. By equation (3) on page 166, $R_b R_c = R_c R_b$. Consequently $R_a R_b R_c = R_a R_c R_b$. By equation (2) on page 166, $R_b R_c = R_B$, and $R_a R_c = R_A$. Consequently the glide-reflection

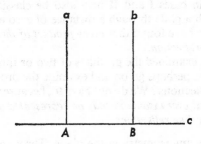

$R_a R_b R_c$ can also be represented by $R_a R_B$ or $R_A R_b$. Conversely, every product of the form $R_a R_B$ is a glide-reflection. In fact, let c be the perpendicular from B to a, meeting a at A, and let b be the perpendicular to c at B. Then we have the situation described above, with $R_a R_B = R_a R_b R_c$. Similarly, every product of the form $R_A R_b$ is a glide-reflection. (See exercise 14, page 185.) We now consider the general case where two of the lines intersect and the third line does not pass through their intersection. Suppose a and b intersect at P, and c does not pass through P. Draw a line from P perpendicular to c, meeting it at Q. Draw d through P so that angle (a, b) = angle (d, e). Then, as we have already observed in case II above, $R_a R_b = R_d R_e$. Multiplying on the

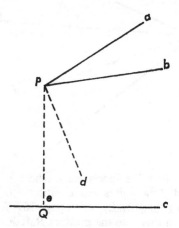

right by R_c, we get $R_a R_b R_c = R_d R_e R_c = R_d R_Q$, since $R_e R_c = R_Q$, by (2) on page 166. Consequently $R_a R_b R_c$ is a glide-reflection. Similarly, if b and c intersect, and a does not pass through their intersection, $R_a R_b R_c$ is a glide-reflection.

The results in cases I and II may also be classified as glide-reflections, with a glide through a distance of zero along the line d. Therefore we have found that *every product of three line reflections is a glide-reflection.*

After having examined the products of two or three line reflections, should we perhaps go on and examine the products of four or more line reflections? We do not have to, because we can prove the theorem that *every isometry may be represented as the product of at most three line reflections.*

Proof: Let T be any isometry in the plane. Through any point P in the plane draw any line a, choose a half-line h on a, and choose a side S of the plane to form the flag-shaped configuration (P, h, S). Either P is fixed under T, or P is moved by T to some point P'. We consider each case separately.

Case I. P is fixed under T. Let the image of a be a', and let the image of h be h' under T. The line a' divides the plane into two sides S_1 and S_2. T moves S to either S_1 or S_2. These cases are illustrated in the diagram below, as cases IA and IB. Let b be the bisector of angle(h, h'). In case IA, T moves the flag-shaped

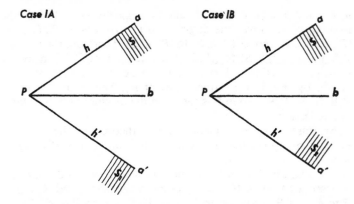

configuration (P, h, S) to (P, h', S_1). But so does R_aR_b, which is a rotation through twice angle (a, b). Then, by the principle of rigidity, $T = R_aR_b$, a product of two line reflections.

In case IB, T moves the flag-shaped configuration (P, h, S) to (P, h', S_2). But so does R_b, the reflection across b. In this case, by the principle of rigidity, $T = R_b$, a single line reflection.

Case II. P is moved by T to some point $P' \neq P$. Let c be the perpendicular bisector of PP'. R_c moves P to P'. Let the images of a, h and S under R_c be $\bar{a}$, $\bar{h}$ and $\bar{S}$ respectively. Let the images of

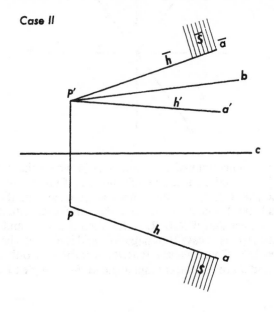

a, h and S under T be a', h' and S' respectively. Let V be the isometry that moves the flag-shaped configuration $(P', \bar{h}, \bar{S})$ to (P', h', S'). R_c moves (P, h, S) to $(P', \bar{h}, \bar{S})$. Therefore the product R_cV moves (P, h, S) to (P', h', S'). But so does T. Then by the principle of rigidity, $T = R_cV$. Let b be the bisector of angle $(\bar{a}, a')$. By case I, $V =$ either $R_{\bar{a}}R_b$ or R_b. Therefore $T =$ either $R_cR_{\bar{a}}R_b$ or R_cR_b. So in this case T is the product of two or three line reflections.

An immediate consequence of this theorem is the fact that every isometry is either a *translation*, a *rotation*, or a *glide-reflection*.

Let b be any line in the plane. Let a be a perpendicular to b intersecting it at B. If P is any point in the plane, and R_a moves P to $\bar{P}$, and R_B moves $\bar{P}$ to P', then R_aR_B moves P to P'. But so does R_b, as can be seen readily from the diagram. Therefore $R_b = R_aR_B$. We have already seen that every product of three line reflections also has the form R_aR_B. Consequently every isometry has the form R_aR_b or R_aR_B. The fact that every isometry is easily expressed as a product of reflections is the second reason why we give reflections particular attention.

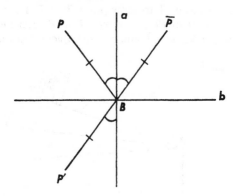

We have encountered two situations in which the product of three line reflections is a line reflection: 1) if no two of the lines intersect, and 2) if the three lines are concurrent. (See pages 168 and 169.) These are in fact the only such situations. It can be shown that if $R_aR_bR_c = R_d$, then if $a \neq b$, and a and b meet at P, we have case II of page 168, and if $a \parallel b$ we have case I of page 168. Consequently $R_aR_bR_c = R_d$ if and only if either a, b, c and d are concurrent and angle $(a, b) =$ angle (d, c), or no two of the lines a, b, c and d intersect and distance $(a, b) =$ distance (d, c).

Products of Point Reflections

If A and B are any two points in the plane, we have already proved that $R_A R_B$ is a translation of twice the distance between A and B in the direction from A to B. (See page 167.)

If A, B and C are any three points in the plane, the product $R_A R_B R_C$ is a point reflection.

Proof: Denote by v the line through A and B. Denote by w the line through C parallel to v. Draw a, b and c through A, B and C respectively perpendicular to v. Then a, b and c are parallel, and we have the conditions of case I on page 168. Consequently, if we choose d perpendicular to v and w and intersecting w at D so that distance (a, b) = distance (d, c), we have $R_a R_b R_c = R_d$.

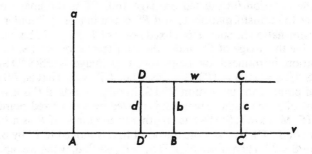

Multiplying this equation by R_w on the right we obtain $R_a R_b R_c R_w = R_d R_w$. Since $R_v R_v = 1$, we may insert it between the factors R_a and R_b in the left-hand member of this equation, to obtain $R_a R_v R_v R_b R_c R_w = R_d R_w$. By equation (2) of page 166, $R_a R_v = R_A$, $R_v R_b = R_B$, $R_c R_w = R_C$, and $R_d R_w = R_D$. Making these substitutions, we get $R_A R_B R_C = R_D$.

Since a point reflection is an involution, it is equal to its own inverse. If we take this fact into account while we apply to the product $R_A R_B R_C$ the rule for obtaining the inverse of a product (see page 160), we obtain the equation $R_C R_B R_A = R_D$. Consequently, if A, B and C are any three points in the plane, $R_A R_B R_C = R_C R_B R_A$. We shall make use of this result in a later paragraph of this chapter.

Conjugate Isometries

We introduce here one more property of the algebra of the group of isometries so that we may be free to use it in the paragraphs that follow.

Let T and S be any two isometries of the plane. We call the product $S^{-1}TS$ the *conjugate* of T with respect to S. We shall examine the properties of the conjugate of T with respect to S in the special case where T is an involution, that is, T is either a line reflection or a point reflection, so that $TT = 1$. If we multiply $S^{-1}TS$ by itself, we observe that $(S^{-1}TS)(S^{-1}TS) = S^{-1}T(SS^{-1})TS = S^{-1}T1TS = S^{-1}(TT)S = S^{-1}1S = 1$. Moreover, $S^{-1}TS \neq 1$, because if $S^{-1}TS = 1$, it is easy to see that $T = 1$, which is contrary to the hypothesis that T is an involution. Consequently, if T is an involution, $S^{-1}TS$ is also an involution.

We can determine the nature of the involution $S^{-1}TS$ by identifying its fixed points. Let P be a fixed point of the involution T. In the notation introduced on page 160, PT is the image of P under the transformation T, and PS is the image of P under the transformation S. Since P is a fixed point of T, $PT = P$. Let us determine the image of PS under the transformation $S^{-1}TS$. In the notation introduced on page 160, this image is $PS(S^{-1}TS) = P(SS^{-1})TS = PTS = (PT)S = PS$, since $PT = P$. That is, PS is a fixed point of the involution $S^{-1}TS$. In other words, if P is a fixed point of T, the point to which it is moved by S is a fixed point of $S^{-1}TS$. Moreover, $S^{-1}TS$ has no other fixed points: If P' is a fixed point of $S^{-1}TS$, let P denote the point that is moved to P' by S, so that $PS = P'$. Then, since $P'(S^{-1}TS) = P'$, we find, when we replace P' by PS, that $PSS^{-1}TS = PS$, or $PTS = PS$. Multiplying on the right by S^{-1}, we see that $PT = P$. That is, P is a fixed point of T. Consequently, P is a fixed point of T if and only if the image of P under S, namely PS, is a fixed point of $S^{-1}TS$. This fact implies that T and its conjugate $S^{-1}TS$ have the same number of fixed points. (See exercise 15, page 185.) We observed on pages 162 and 163 that a point reflection has exactly one fixed point, while a line reflection has an infinite number of fixed points. Consequently, if T is a point reflection R_A, its conjugate $S^{-1}R_AS$ is also a point reflection. Moreover, the fixed point of $S^{-1}R_AS$ is the image of A under S, that is, it is AS. Consequently

(5) $$S^{-1}R_AS = R_{AS}.$$

If T is a line reflection R_a, its conjugate $S^{-1}R_aS$ is also a line reflection. Moreover, the fixed points of $S^{-1}R_aS$ are the images under S of the points that lie on a. But these are the points that lie on aS. Consequently,

$$(6) \qquad\qquad S^{-1}R_aS = R_{aS}.$$

In particular, if S is itself an involution, then $S^{-1} = S$, and equations (5) and (6) take this form:

$$(7) \qquad\qquad SR_AS = R_{AS}, \qquad SR_aS = R_{aS}.$$

If we denote SR_AS by R_B, and denote SR_AS by R_b, a comparison with equation (7) shows that

$$(8) \qquad\qquad SR_AS = R_B \text{ if and only if } AS = B,$$

and

$$(9) \qquad\qquad SR_aS = R_b \text{ if and only if } aS = b,$$

where S is assumed to be an involution.

Statement (8) asserts that the equation $AS = B$ is equivalent to the equation $SR_AS = R_B$. To facilitate our use of this property, we give a simple rule in the form of a mnemonic device for obtaining the latter equation from the former: in the equation $AS = B$, replace A by R_A, replace B by R_B, and then multiply the left-hand member on the left by S. The result is the desired equation $SR_AS = R_B$.

Similarly, statement (9) asserts that the equation $aS = b$ is equivalent to the equation $SR_aS = R_b$. To obtain the latter equation from the former, use this mnemonic device: in the equation $aS = b$, replace a by R_a, replace b by R_b, and then multiply the left-hand member on the left by S. The result is the desired equation $SR_aS = R_b$.

A Geometry-Algebra Dictionary

There is associated with every point A an isometry R_A, the reflection across A. There is associated with every line a an isometry R_a, the reflection across a. By using this one-to-one correspondence between points and lines on the one hand and reflections on the other hand, we can translate every statement that expresses a geometric relationship among points and lines into a statement that expresses an algebraic relationship among reflections. This is the third reason why we have been paying particular attention to reflections.

To translate geometric statements about points and lines into equivalent statements about reflections we use a dictionary, part of which is given below:

	Geometric statement about points and lines	Equivalent algebraic statement about reflections
1.	P is on a.	$R_P R_a = R_a R_P$.
2.	$a \perp b$.	$a \neq b$, and $R_a R_b = R_b R_a$.
3.	a, b, c and d are concurrent and angle (a, b) = angle (d, c), or, no two of a, b, c and d intersect, and distance (a, b) = distance (d, c).	$R_a R_b = R_d R_c$.
4.	a, b and c are concurrent and b bisects angle (a, c), or, no two of a, b and c intersect, and b is half-way between a and c.	$R_a R_b = R_b R_c$.
5.	b and d are $\perp$ to PQ, and distance (P, b) = distance (d, Q).	$R_P R_b = R_d R_Q$.
6.	b is $\perp$ bisector of PQ.	$R_P R_b = R_b R_Q$.
7.	$b \parallel d$, and P is equidistant from b and d.	$b \neq d$, and $R_d R_P = R_P R_b$.
8.	M is the midpoint of AC.	$R_A R_M = R_M R_C$.
9.	$AB = DC$, and $AB \parallel DC$.	$R_A R_B = R_D R_C$.

Justification of the Dictionary

To justify the use of this dictionary, we prove the correctness of each of its entries.

1. The statement that P is on a is equivalent to the statement that P is a fixed point of R_a. In the notation introduced on page 160, the latter statement may be written in this form: $PR_a = P$. This equation has the form $AS = B$, with P in the place of A and B, and R_a in the place of S. We obtain from it an equivalent equation by applying the mnemonic device given on page 175. In this case, we transform the equation $PR_a = P$ into an equivalent equation by replacing P by R_P and then multiplying the left-hand member on the left by R_a. The resulting equation is $R_a R_P R_a = R_P$. If we multiply the latter equation on the left by R_a, and make use of the fact that $R_a R_a = 1$, we obtain the equivalent equation $R_P R_a = R_a R_P$.

2. We found on page 164 that the fixed lines of R_a are a itself and the lines that are perpendicular to a. Consequently the statement that $a \perp b$ is equivalent to the statement that a is not b, and b is a fixed line of R_a. In symbols, this statement takes the form: $a \neq b$, and $bR_a = b$. Using the mnemonic device given on page 175, we transform the equation $bR_a = b$ into an equivalent equation by replacing b by R_b and multiplying the left-hand member on the left by R_a. In this way we find that the statement that $a \neq b$ and $bR_a = b$ is equivalent to the statement that $a \neq b$ and $R_a R_b R_a = R_b$. Finally, by multiplying the equation $R_a R_b R_a = R_b$ by R_a on the right, we obtain the equivalent equation $R_a R_b = R_b R_a$.

3. On pages 168, 169 and 172 we showed that the statement in the left-hand column is equivalent to the equation $R_a R_b R_c = R_d$. By multiplying this equation by R_c on the right, we obtain the equivalent equation $R_a R_b = R_d R_c$.

4. This is obtained as a corollary of 3 when $d = b$.

5. Assume that b and d are perpendicular to PQ and that distance (P, b) = distance (d, Q). Let g be the line through P and Q. Draw h and k perpendicular to g at P and Q respectively. Then h, b, d and k are parallel, and distance (h, b) = distance (d, k). Consequently, by entry 3 of the dictionary, $R_h R_b = R_d R_k$. Since $h \perp g$, $R_P = R_g R_h$, and since $k \perp g$, $R_g R_k = R_Q$. Since $d \perp g$, $R_g R_d = R_d R_g$. Therefore $R_P R_b = (R_g R_h) R_b = R_g (R_h R_b) = R_g (R_d R_k) = (R_g R_d) R_k = (R_d R_g) R_k = R_d (R_g R_k) = R_d R_Q$. Conversely, suppose $R_P R_b = R_d R_Q$. $R_P R_b$ is a glide reflection in the line through P perpendicular to b, and $R_d R_Q$ is a glide reflection in the line through Q perpendicular to d. Since these two glide reflections

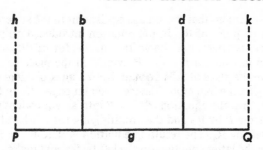

are equal, it follows that the line through P perpendicular to b coincides with the line through Q perpendicular to d. That is, b and d are perpendicular to PQ. Define g, h, and k as we did before. Then, from $R_P R_b = R_d R_Q$, we obtain the equation $R_g R_h R_b = R_d R_g R_k$ by substituting $R_g R_h$ for R_P and $R_g R_k$ for R_Q. Then, substituting $R_g R_d$ for $R_d R_g$, we get $R_g R_h R_b = R_g R_d R_k$. Canceling the R_g we find that $R_h R_b = R_d R_k$. Then, by 3, distance (h, b) = distance (d, k). That is, distance (P, b) = distance (d, Q).

6. The statement that b is the perpendicular bisector of PQ is equivalent to the statement that Q is the image of P under the line reflection R_b. That is, $P R_b = Q$. By the rule derived from (8) on page 175, this is equivalent to the equation $R_b R_P R_b = R_Q$. Multiplying on the left by R_b we obtain the equivalent equation $R_P R_b = R_b R_Q$.

7. This is obtained as a corollary of 5 when $Q = P$.

8. The statement that M is the midpoint of AC is equivalent to the statement that C is the image of A under the point reflection R_M. That is, $A R_M = C$. By the rule derived from (8) on page 175, this is equivalent to the equation $R_M R_A R_M = R_C$. Multiplying on the left by R_M, we obtain the equivalent equation $R_A R_M = R_M R_C$.

9. $R_A R_B$ is a translation of twice distance (A, B) in the direction from A to B, and $R_D R_C$ is a translation of twice distance (D, C) in the direction from D to C. These translations are equal if and only if $AB = DC$ and $AB \parallel DC$.

Proofs Using Reflections

The dictionary on page 176 gives us a third way of carrying out geometric proofs by algebraic methods. To show how this is done, we carry out two proofs in detail.

Example: Given: $AB = DE$, and $AB \parallel DE$.
$BC = EF$, and $BC \parallel EF$.
Prove: $AC = DF$, and $AC \parallel DF$.

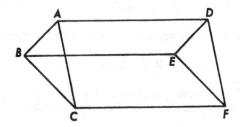

Proof: By entry 9 of the dictionary, the statements in the hypothesis are equivalent to the equations $R_A R_B = R_D R_E$, and $R_B R_C = R_E R_F$. Multiplying these equations, we obtain $R_A R_B R_B R_C = R_D R_E R_E R_F$, which simplifies to $R_A R_C = R_D R_F$, since $R_B R_B = 1$, and $R_E R_E = 1$. Using entry 9 of the dictionary again, we see that $AC = DF$ and $AC \parallel DF$.

Example: Prove that the perpendicular bisectors of the sides of a triangle are concurrent.

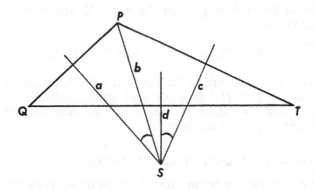

Proof: In triangle PQT, let a be the perpendicular bisector of PQ. Let c be the perpendicular bisector of PT. Let a and c intersect at S. Denote PS by b. Draw d through S so that angle $(a, b) =$ angle (d, c). We shall show that d is the perpendicular bisector of QT.

Since a is the perpendicular bisector of PQ, and c is the perpendicular bisector of PT, we know from entry 6 of the dictionary that

(10) $R_P R_a = R_a R_Q,$

and

(11) $R_P R_c = R_c R_T.$

Since P is on b, we know from entry 1 of the dictionary that

(12) $R_P R_b = R_b R_P.$

Since a, b, c and d are concurrent at S and angle (a, b) = angle (d, c), we know from entry 3 of the dictionary that

(13) $R_a R_b = R_d R_c.$

By multiplying equations (10) and (13), we find that

$$R_a R_Q R_d R_c = R_P R_a R_a R_b = R_P R_b = R_b R_P,$$

because of equation (12). Hence,

(14) $R_a R_Q R_d R_c = R_b R_P.$

By multiplying equation (14) on the left by R_a and on the right by R_c, we get

(15) $R_Q R_d = R_a R_b R_P R_c = R_d R_c R_c R_T = R_d R_T,$

by making the substitutions for $R_a R_b$ and $R_P R_c$ that are given by equations (13) and (11). Now, since we see in equation (15) that $R_Q R_d = R_d R_T$, we know from entry 6 of the dictionary that d is the perpendicular bisector of QT.

An Approach to Geometry Through Reflections

In this chapter we began with the geometric structure known as a Euclidean plane, and then introduced the isometries of the plane. Thus we were led to a study of an algebraic structure, the group of isometries of the plane. Within this group we found that we can associate with every point of the plane one and only one point reflection and we can associate with every line of the plane one and only one line reflection. This fact opens up the possibility of reversing the procedure. We can begin with a suitably chosen group that behaves like a group of isometries and that includes among its members things that behave like point reflections and line reflections. Then we can use the latter to define the points and lines of a plane. To show how this may be done, we first summarize some relevant properties of the Euclidean plane and its group of isometries.

Some properties of the Euclidean plane: The Euclidean plane has these seven familiar properties:

1. Given two points P and Q, there exists a line g with P and Q on g.

2. If points P and Q are on both lines g and h, either $P = Q$ or $g = h$.

3. If three lines a, b and c meet at a point P, there is a line d through P such that angle (a, b) = angle (d, c).

4. If the lines a, b and c are perpendicular to the line g, there is a line d perpendicular to g such that distance (a, b) = distance (d, c).

5. There exist three lines g, h and j such that g is perpendicular to h, j is not perpendicular to g, j is not perpendicular to h, and j does not pass through the intersection of g and h. (Existence of a right triangle.)

6. There exist four lines a, b, c and d such that $a \neq b$, $c \neq d$, and a and b are perpendicular to c and d. (Existence of a rectangle.)

7. Given any two lines a and b, then either there is a point C on a and b or there is a line c perpendicular to a and b.

Some properties of the group of isometries: Let $\mathbf{G}$ be the group of isometries of the Euclidean plane. Let $\mathbf{S}$ be the set of all line reflections in the plane. Let $\mathbf{T}$ be the set of all point reflections in the plane. We have already observed that $\mathbf{G}$, $\mathbf{S}$ and $\mathbf{T}$ have the following properties:

I. Every member of $\mathbf{S}$ is an involution.

II. $\mathbf{T}$ is the set of all products $R_a R_b$ that are involutions, where R_a and R_b are members of $\mathbf{S}$. (This is a restatement of two facts noted on pages 166 and 168: If $a \perp b$ at P, $R_a R_b = R_P$; and $R_a R_b$ is an involution if and only if $a \perp b$.)

III. Every member of $\mathbf{G}$ may be expressed as a product of at most three members of $\mathbf{S}$.

IV. Every conjugate of a member of **S** with respect to a member of **G** is a member of **S**. (That is, every conjugate of a line reflection is a line reflection.)

Expressing geometric properties in algebraic language: The dictionary on page 176 shows how certain geometric relations of points and lines may be expressed through algebraic relations of reflections. We call attention to two other possible entries in the dictionary:

10.	$a \perp b$	R_aR_b is an involution.
11.	P is on a.	R_PR_a is an involution.

Entry 10 is justified by the fact already noted that R_aR_b is an involution if and only if $a \perp b$. Entry 11 is a reformulation of entry 1. It is easy to verify in the diagram below that whether P is on a or not, R_PR_a moves at least one point Q to a new position

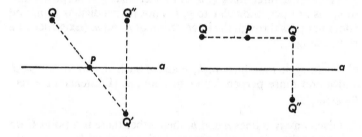

Q''. Hence $R_PR_a \neq 1$. If we multiply the equation $R_PR_a = R_aR_P$ by R_P on the left and by R_a on the right, we obtain $R_PR_PR_aR_a = R_PR_aR_PR_a$, which simplifies to $1 = (R_PR_a)(R_PR_a)$. Consequently, in view of the definition of an involution, if P is on a, R_PR_a is an involution. The converse is easily proved, so entry 1 is equivalent to entry 11.

Using this extended dictionary, we can translate the seven properties of the Euclidean plane listed on page 181 into algebraic language as follows:

1′. If R_P and R_Q are members of **T**, there is an R_g in **S** such that R_PR_g and R_QR_g are involutions.

2'. If R_P and R_Q are in T, R_g and R_h are in S, and if $R_P R_g$, $R_P R_h$, $R_Q R_g$, and $R_Q R_h$ are involutions, then either $R_P = R_Q$ or $R_g = R_h$.

3'. If R_a, R_b and R_c are members of S, and R_P is a member of T such that $R_P R_a$, $R_P R_b$ and $R_P R_c$ are involutions, then there is a member R_d of S such that $R_a R_b R_c = R_d$.

4'. If R_a, R_b, R_c and R_g are members of S such that $R_a R_g$, $R_b R_g$ and $R_c R_g$ are involutions, then there is a member R_d of S such that $R_a R_b R_c = R_d$.

5'. There exist in S three members R_g, R_h and R_j such that $R_g R_h$ is an involution, and $R_j R_g$, $R_j R_h$ and $R_g R_h R_j$ are not involutions.

6'. There exist in S four members R_a, R_b, R_c, and R_d such that $R_a \neq R_b$, $R_c \neq R_d$, and $R_a R_c$, $R_a R_d$, $R_b R_c$, and $R_b R_d$ are involutions.

7'. If R_a and R_b are in S, then either there is a member R_C in T such that $R_C R_a$ and $R_C R_b$ are involutions, or there is a member R_c in S such that $R_c R_a$ and $R_c R_b$ are involutions.

Constructing Euclidean plane geometry via reflections: Let G be a group and let S and T be sets within G such that G, S and T have properties I, II, III and IV listed on page 181, and properties 1', 2', 3', 4', 5', 6' and 7' listed above. Let S' be a duplicate copy of S, and let T' be a duplicate copy of T. If a is a member of S', denote by R_a the corresponding member of S. If P is a member of T', denote by R_P the corresponding member of T. Call the members of S' *lines*, call the members of T' *points*, and call the set of all these points and lines a *plane*. Taking our cue from entries 10 and 11 in the dictionary on page 182, we introduce the following definitions: Two lines a and b are perpendicular if $R_a R_b$ is an involution. A point P is on a line a if $R_P R_a$ is an involution. Taking our cue from entry 3 of the dictionary, we introduce the following definitions: Under the conditions stated in 3', we say that angle (a, b) = angle (d, c). Under the conditions stated in 4', we say that distance (a, b) = distance (d, c). Then the plane has properties 1 to 7 listed on page 181. It can be shown that a set of points and lines that has these properties has all the properties of a Euclidean plane. Since the group

G from which the plane is derived is an algebraic structure, this procedure gives us a third way of obtaining geometry as a branch of algebra.

EXERCISES FOR CHAPTER 6

1. *Identity transformation.* A clockwise rotation of the plane through x degrees about any point in the plane is the identity transformation if $x =$ _____ degrees.

2. *Inverse.* The inverse of a clockwise rotation of 100° about a point is a clockwise rotation of _____ degrees about the same point.

3. If A and B are isometries of a plane, what is the inverse of AB?

4. What is the inverse of R_A? of R_a?

5. *The image of a point.* Write in symbolic form

a) that A is moved by the transformation T to B;
b) that P is moved by R_a to Q.

6. *The image of a line.* If $B \neq C$, and A is on BC, and an isometry T moves A, B and C to A', B' and C' respectively, prove that A' is on $B'C'$.

7. *Involution.* If T is an isometry that is an involution, and AT = B, what is BT equal to?

8. Prove the principle of the rigidity of the plane. That is, if (P, h, S) and (P', h', S') are any two flag-shaped configurations in the plane, prove

a) that there is an isometry that moves P to P', h to h', and S to S';
b) if U and V are two such isometries, and A is any point of the plane, then the image of A under U is the same as the image of A under V.

9. *Reflections.* If triangle ABC is isosceles, with $AB = BC$, and h is the line containing the perpendicular bisector of AC, find
a) BR_h (the image of B under R_h); b) AR_h (the image of A under R_h); c) CR_h (the image of C under R_h).
   If M is the intersection of the diagonals of parallelogram

$ABCD$, find d) AR_M (the image of A under R_M); e) BR_M the image of B under R_M).

10. *Point reflection.* If $P \neq Q$, and A is not on PQ, and R_A moves P and Q to P' and Q' respectively, prove that $P'Q'$ and PQ are distinct lines.

11. *Line reflection.* If $b \neq a$, and b is not perpendicular to a, prove that bR_a (the image of b under R_a) is not b. Hint: Take a point P on b that is not on a and prove that P', the image of P under R_a, is not on b.

12. *Products of reflections.* If A and B are distinct points on g, P is any point on g, $PR_A = Q$, and $QR_B = P'$, prove that P' is on g, and $PP' = 2AB$.

13. a) Prove that $R_a R_b R_c$ is a line reflection if $a = b = c$.
   b) Prove that $R_a R_b R_c$ is a line reflection if $a = b$, and $a \neq c$.
   c) Prove that $R_a R_b R_c$ is a line reflection if $a = c$, and $a \neq b$.

14. Prove that every product of the form $R_A R_b$ is a glide-reflection.

15. *Conjugate of an involution.* Suppose S and T are isometries and T is an involution. On page 174 it was shown that P is a fixed point of T if and only if PS is a fixed point of $S^{-1}TS$. Prove that this implies that if P is the only fixed point of T, then PS is the only fixed point of $S^{-1}TS$, and vice versa.

16. Complete these statements, using as models equations (8) and (9) on page 175:

a) If S is an isometry that is an involution, and P and Q are points, $PS = Q$ if and only if _____.
b) If S is an isometry that is an involution, and C and D are points, $SR_C S = R_D$ if and only if _____.
c) If S is an isometry that is an involution, and h and k are lines, $hS = k$ if and only if _____.
d) If S is an isometry that is an involution, and f and g are lines, $SR_f S = R_g$ if and only if _____.
e) If Q is a point and b is a line $QR_b = Q$ if and only if
_____.

17. Using the dictionary on page 176, prove by the algebraic methods of this chapter that if the diagonals of a quadrilateral

bisect each other, a pair of opposite sides of the quadrilateral are equal and parallel. Hint: Use entries 8 and 9 and the fact that for any three points A, B and C, $R_A R_B R_C = R_C R_B R_A$, as shown on page 173.

18. a) What is the inverse of $R_P R_a$?
    b) Prove $R_P R_a \neq 1$.
    c) Derive from your answer to a) a proof of entry 11 in the dictionary on page 182.

7

Geometry in Newtonian Physics

Geometry and Physics

Physical events take place in physical space. Consequently physical concepts are necessarily intertwined with geometric concepts. The relationship between physics and geometry arises in two ways. First, since mathematical space is an abstraction from physical space, our geometric ideas are essentially distillations from our physical experience. Secondly, after we have formed definite geometric ideas, we interpret physical events in terms of these ideas. We project our concepts onto the physical world, so that to a certain extent we "see" in the physical world properties and relationships that really originated in our heads.

Geometry in the Physics of Aristotle

Physical thought in Europe during the first sixteen centuries of the Christian era was dominated by the ideas of Aristotle. These ideas were derived initially from the physical ideas of Pythagoras and Plato. We have already noted that Aristotle, like Plato, believed that physical space was finite and was a plenum. The physics of Aristotle was not an independent scientific discipline based on the systematic gathering of data through experiment. It was part of a system of philosophical-religious speculation, and as such was incorporated into medieval Christian theology. Consequently the Aristotelian interpretation of physical events was strongly influenced by pre-conceived and essentially extraneous ideas about divinity and perfection and their relation to man. Because of their symmetry, circles and spheres were considered to be perfect. The heavenly bodies were associated with divinity, and since God was assumed to be perfect, they could have only perfect shapes and perfect motions. Hence it was taken for granted that they were spheres moving uniformly in circular orbits. In the Ptolemaic system of astronomy, in which these pre-conceptions

187

were used with some attempt to make the theory fit the facts of observation, the planets were assumed to move in circles carried by circles around the earth. Even Copernicus (1473–1543), when he proposed his theory that the planets and the earth revolve around the sun, assumed that they move in circular orbits.

Motion itself was considered to be tainted with imperfection. Hence Aristotle assumed that the natural state of a body that is undisturbed by any outside force is a state of rest.

The Scientific Revolution

The fifteenth, sixteenth and seventeenth centuries were periods of great progress in the practical arts of warfare, agriculture and commerce. The requirements of these practical arts led to increasing reliance on observation and experiment for determining the truth about the physical world. The discoveries made during this period led to a revolution in scientific thought. The speculative physics of Aristotle was overthrown and replaced by the physics of Newton in which theory was solidly based on experimental fact. Even the conservative force of religion paradoxically played a part in stimulating the discoveries that brought about this scientific revolution.

Warfare made its contribution to the scientific revolution by creating an interest in the use of firearms. To be able to hit a target with a projectile, you had to be able to anticipate the path that the projectile would follow. Tartaglia (1500–1557) discovered that the path of a projectile is curved, that the range of a cannon is related to its angle of inclination, and that the range is a maximum when the angle is 45°. Simon Stevin (1548–1620) discovered that, contrary to Aristotle's belief, heavy bodies do not fall more quickly than light ones. Galileo Galilei (1564–1642) discovered that Aristotle was wrong, too, about the relationship between force and motion. He found that force is responsible not for motion itself, but for *acceleration*. A body not acted on by a force tends to maintain a state of uniform motion. Using these new ideas he was able to explain theoretically what Tartaglia had discovered empirically about how the range of a gun varies with its angle of elevation. These discoveries of Galileo were incorporated into the physics of Isaac Newton (1642–1727) as the first and second laws of motion.

Agriculture, commerce and religion contributed to the scientific revolution by stimulating observational astronomy. Agriculture, because of its dependence on the seasons, requires an accurate calendar. The Christian religion does, too, for the fixing of the date of Easter Sunday. The year in the Julian calendar was $11\frac{1}{4}$ minutes too long. The accumulated error in 16 centuries was more than ten days. The calendar reform introduced by Pope Gregory XIII in 1582 had to be based on accurate observations of the positions of the sun, moon and planets in the sky. Navigation, stimulated by growing commerce, was also dependent on accurate knowledge of the apparent motion of these heavenly bodies. The accurate astronomical observations of Tycho Brahe (1546–1601) were the first link in another chain of discoveries that culminated in the physics of Newton. Johannes Kepler (1571–1630), using Brahe's data, discovered that the orbit of each planet as it revolves around the sun is an ellipse, not a circle. Each planet moves so that the line drawn from the planet to the

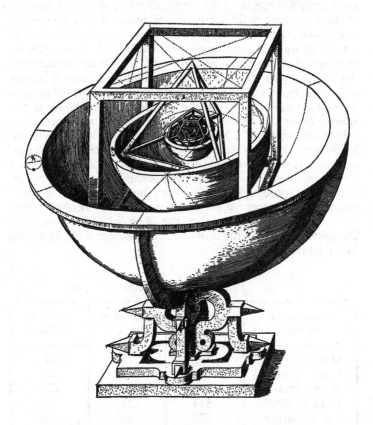

Kepler's model of the solar system

sun sweeps over equal areas in equal times. The period of each plane is related to the size of its orbit by the rule that the square of the period is proportional to the cube of the average distance from the planet to the sun. From these three discoveries, now known as *Kepler's Laws*, Newton derived his law of gravitation.

It is interesting to note that not all of Kepler's discoveries have stood the test of time. A fourth "law," which he "discovered" before the three laws for which he is justly famous, has turned out to be nonsense. There were six known planets in Kepler's time. In order of increasing distance from the sun they were Mercury, Venus, Earth, Mars, Jupiter and Saturn. While pondering over the spacing of the planets, it suddenly occurred to him that there are five spaces between consecutive planets, and there are exactly five regular solids. It struck him that the coincidence of these numbers could not be an accident, but must be the key to the mystery of the spacing of the planets. Each of the planets may be thought of as being on a sphere around the sun. Kepler proposed the theory that the spacing of the spheres was such that one of the regular solids could be inserted between each pair of adjacent spheres so that it was inscribed in the larger sphere and was circumscribed around the smaller sphere. Between the spheres of Saturn and Jupiter he inserted a cube. Between the spheres of Jupiter and Mars he inserted a regular tetrahedron. In the next three spaces he inserted a dodecahedron, an icosahedron, and an octahedron respectively. The model of the solar system that he obtained in this way is shown in the drawing on page 189, reproduced from Kepler's book *Mysterium Cosmographicum*. The spacing of the planets that is calculated from this model does not fit either the observed Copernican data known to Kepler, or the more accurate modern data, as shown in the table below.

Ratio of Orbit Widths

Consecutive Planets	Kepler's Model	Copernican Values	Modern Values
Mercury : Venus	.707	.723	.650
Venus : Earth	.795	.794	.741
Earth : Mars	.795	.757	.735
Mars : Jupiter	.333	.333	.337
Jupiter : Saturn	.577	.635	.604

But what is worse, we know now that the solar system includes other planets, Uranus, Neptune and Pluto, discovered after Kepler's time, and multitudes of smaller bodies called planetoids. So there are more spaces to be accounted for, but no more regular solids to perform that service.

Geometry in the Physics of Newton

The science of physics, as developed by Kepler, Galileo and Newton, was freed of the irrelevant pre-conceived ideas injected by Aristotelian philosophy. However, this does not mean that Newtonian physics is free of all geometric assumptions. Every model of the physical world includes a model of physical space, and therefore necessarily makes some assumptions about the nature of space. Newton, in his model, explicitly rejected the Aristotelian assumption that space is finite and is a plenum, and returned to the older idea of Democritus that space is an infinite void. Specifically, he assumed that there is an "absolute space" which "remains always similar and immovable," in which all things are placed, that this space is three-dimensional, and that it is correctly described by the geometry of Euclid. As Newtonian physics became firmly established, this view of space became generally accepted. There was some opposition to it, however, especially by the philosopher Gottfried Wilhelm von Leibniz (1646–1716), who insisted that space is an "order of coexistences, as time is an order of successions." From the Leibniz point of view, space is not a void, because space does not exist apart from the coexistence of the objects in space. This viewpoint, while it is a return to the position of Aristotle, also foreshadows the twentieth century conception of space that came in with Einstein's theory of relativity. We shall take up this discussion again in Chapter 11, where we discuss the geometry of relativity physics.

Statics and Geometry

Some geometrical assumptions about physical space are also introduced into physics by the usual treatment of the theory of statics based on the *Theory of the Lever* initiated by Archimedes. Archimedes, in his study of the lever, tacitly made the following assumption: A. When a lever is suspended from its middle point, it is in equilibrium if a weight 2W is suspended at one end, and at the other end there is a lever supported at its middle point with a weight W suspended at each of its ends. We shall show

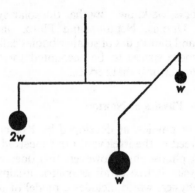

that, if we make the six basic assumptions of the theory of statics listed on page 254, this physical assumption of Archimedes is equivalent to the following geometric assumption:

A'. The line joining the midpoints of the legs of an isosceles triangle passes through the midpoint of the median to the base.

To show that A implies A' we first set up an appropriate geometric model: let PQR be an isosceles triangle with PQ equal to PR. Suppose that a weight W is suspended from each of the points Q and R, and a weight 2W is suspended from P. Let M and N be the midpoints of PQ and PR respectively. Let PS be the median to QR, and let T be the intersection of MN and PS. We shall show that if assumption A is made, then T is the midpoint of PS. This will establish proposition A'. If we support the triangle at the line MN, it will be like supporting each of the levers PR and PQ at its midpoint while a weight W is suspended at each of the four ends of the two levers. (The weight 2W at P is assumed to be made up of two separate weights W, one of which balances the weight at Q, while the other one balances

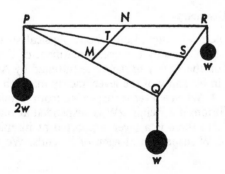

the weight at R.) Hence the system will be in equilibrium. The system is obviously also in equilibrium if the triangle is supported at the line PS, because then the lever QR, with a weight W at each end, is supported at its midpoint, while the point P, from which the weight 2W is suspended, is itself a point of support. Since the triangle is in equilibrium when it is supported at MN, and it is also in equilibrium when it is supported at PS, then it is in equilibrium when it is supported at their intersection T. However, PS is a lever with a weight 2W at one end and a lever QR at the other end, supported at its middle point S and with a weight W at each end. According to assumption A, under these conditions the system is in equilibrium when supported at the middle point of PS. Since, moreover, there is only one point of support on a lever with given weights at its ends for which the lever is in equilibrium, then T is the middle point of PS. Thus assumption A implies A'.

To show that A' implies A, let PS be a lever with a weight 2W at one end and a lever QR supported at S with S the middlepoint of QR, and with a weight W at Q and at R. Make QR perpendicular to PS. Then PQ = PR. As before, let M and N be the midpoints of PQ and PR respectively, and let T be the intersection of MN and PS. As we have already seen, under these conditions the system is in equilibrium if it is supported at T. According to assumption A', T is the midpoint of PS. Then it follows at once that the system of levers PS and QR is in equilibrium when it is supported at the midpoint of PS. That is, statement A is true. Hence A' implies A.

Consequently assumption A, which is expressed in terms of physical concepts, actually has a geometric content, as expressed in A'. It can be shown that proposition A' is equivalent to Euclid's postulate 5. After exploring in the next chapter the role in geometry of Euclid's postulate 5, we shall be better able to appraise the significance of this fact. So we postpone further discussion of it until Chapter 8 and Chapter 11, where we examine the geometry of relativity physics.

The Significance of Geometric Assumptions

Ever since the time of Plato it has been understood that a deductive science must have a set of axioms as its foundation. At first a distinction was made between two kinds of axioms, such as the common notions of Euclid on the one hand, and the postulates of Euclid on the other. The common notions were supposed to be axioms that are common to all sciences, while the postulates were supposed to be axioms that are peculiar to one particular science. This distinction is no longer made, because, as Aristotle pointed out, for the use of the common notions in any particular science it suffices to assume only that they are valid in that particular science. Thus, in so far as their use in geometry is concerned, the common notions are on a par with the postulates. So we shall class both common notions and postulates together as axioms or assumptions.

What then is an axiom? Aristotle answered by saying it is "that which is *per se* necessarily true." Proclus said, "They are treated as self-evident." This is the source of the traditional conception that an axiom is a self-evident truth. To give a precise meaning to this conception it is necessary to answer the question, "What is a self-evident truth?" We know already that it is not a "truth" that is deduced by reasoning from other "truths," for if it were, it would be a theorem. The use of the adjective "self-evident" also implies that it is not established empirically, that is, by reference to our experience. For if it were, we might say it is "evident," but not that it is "self-evident." The meaning of the concept of self-evident truth as applied to geometry was developed in detail by the philosopher Immanuel Kant (1724–1804) in his theory of knowledge. According to Kant, a sense-perception is compounded out of two parts, an *a posteriori* content impressed on our perceptions by the things-in-themselves that exist outside the perceiving mind, and an *a priori* form that arises from the act of perception itself. The *a priori* forms are supposed to be prior to all experience. They are imposed on experience by the nature of the mind itself. According to Kant, space is one of these *a priori* forms. In this view, space is not an object of perception, but a mode of perceiving objects, and the properties of space are imposed on our perceptions by our minds. From this point of view, the axioms of geometry are self-evident truths in the sense that they are properties of thought itself. According to the Kantian philosophy, as long as we think about objects at all, we must think about them as existing in the kind of space that was described by Euclid. From this point of view, the geometry of Euclid is the only possible geometry.

We shall see in the next chapter that Kant was wrong. Euclid's geometry is not the only geometry. There are other geometries that are no less valid than Euclid's. After we have established this fact we shall be able to complete our discussion of the significance of assumptions in a system of geometry.

8

Non-Euclidean Geometry

Euclid's Judgment Challenged

When Euclid organized plane geometry as a deductive system, he listed ten axioms as the foundation of the system. These axioms were the five postulates and five common notions listed on page 66. Almost immediately the fifth postulate became a subject of controversy. Other mathematicians thought that its inclusion among the axioms was an error in judgment. Because the statement of the postulate is not as simple as the statements of the other axioms, they thought that it should be proved as a theorem, rather than assumed as an axiom. They argued that the fifth postulate is far from being self-evident, as an axiom was supposed to be. Moreover, the converse of the fifth postulate is a theorem that can be proved, so it seemed reasonable to them that the so-called fifth postulate should also be a theorem. Consequently, for two thousand years, ranging from the first century B.C. to the nineteenth century A.D., mathematicians tried to prove that the fifth postulate was a consequence of Euclid's nine other axioms.

Euclid Vindicated and Dethroned

The outstanding fact about all attempts to prove the fifth postulate is that they failed. Some mathematicians thought that they had succeeded in proving the postulate, but in every case it turned out that the proof was fallacious. In most cases they had proved the fifth postulate by first assuming, consciously or unconsciously, some other proposition, such as the Playfair axiom, which is equivalent to it. But by doing so they had in effect assumed what they were trying to prove.

A significant technique, used by Gerolamo Saccheri (1667–1733), Johann Heinrich Lambert (1728–1777), and Adrien Marie Legendre (1752–1833), was the *reductio ad absurdum*. Each of them tried to show that the fifth postulate was a consequence of the other axioms of Euclid by showing that any assumption that is contrary to the fifth postulate is inconsistent with the other axioms. To show this inconsistency, they tried to prove that if the fifth postulate or its equivalent is replaced by a contrary assumption, the amended set of axioms has implications that are absurd in that they are self-contradictory. Each of them found that there are two contrary assumptions that can be substituted for the fifth postulate. Each of them thought he had succeeded in showing that both amended sets of axioms had absurd implications. It turned out later, however, that only one of the amended sets of axioms was absurd in the sense of being self-contradictory. The implications of the other amended set of axioms were not *absurd*. They were merely *different* from those derived from the original ten axioms of Euclid.

The failure of all attempts to prove the fifth postulate gave birth to a new conviction in the minds of Carl Friedrich Gauss (1777–1855), Nicolai Ivanovitsch Lobatschewsky (1793–1856) and Johann Bolyai (1802–1860). They decided that the reason why no one had succeeded in proving the fifth postulate as a consequence of the other axioms was because it was really independent of the other axioms. This conviction had two important implications. First, it implied that Euclid was right in taking the fifth postulate as an axiom. Secondly, it implied that some contrary postulate can be substituted for it. If this is done, we obtain another kind of geometry, different from Euclid's and hence called a non-Euclidean geometry. It has since been shown, as we shall see, that this new kind of geometry is as free of contradiction as Euclid's geometry, and hence is just as valid as a mathematical system. Thus the final outcome of two thousand years of investigation of the fifth postulate was that Euclid was both vindicated and dethroned. He was vindicated when it was shown that he was right in considering the fifth postulate to be an axiom. He was dethroned when it was shown that a contrary axiom could take its place. Euclidean geometry is no longer dominant. It is only one of several possible geometries.

To state explicitly what the alternative geometries are, it is best to do so in terms of Hilbert's axioms rather than Euclid's axioms, since they do not have the latter's imperfections. In the Hilbert axioms, the role of the fifth postulate is played by the Playfair axiom, which is equivalent to the fifth postulate: Through a point not on a given straight line there is one and only one straight line that does not intersect the given straight line. Gauss, Lobatschewsky and Bolyai developed another geometry by replacing this axiom by the contrary assumption: Through a point not on a given straight line there is more than one straight line that does not intersect the given line. The geometry obtained in this way is now known as *hyperbolic geometry*. A third kind of geometry can be obtained in which there are no parallel lines at all. It is the *elliptic geometry* of Riemann, referred to on page 83. However, it cannot be obtained by merely replacing the Playfair axiom by the assumption that through a point not on a given line there are no lines that do not intersect the given line (the Riemann axiom). As we noted on page 83, a line in elliptic geometry is finite in length, and the order axioms do not hold there. Consequently, to obtain the axioms of elliptic geometry it is necessary to modify more of the Hilbert axioms besides the Playfair axiom.

Absolute Geometry

All of the Hilbert axioms except the Playfair axiom are common to both Euclidean and hyperbolic geometry. Consequently all theorems derived from these axioms are theorems in both systems of geometry. Bolyai called this set of theorems that is common to both geometries *absolute geometry*. Since the first 28 theorems of Book I of Euclid's *Elements* are proved without the use of the fifth postulate, they belong to absolute geometry. Some of the theorems of absolute geometry, such as the theorem that the base angles of an isosceles triangle are equal, can be proved without using the axioms of order. These theorems are common to all three geometries, Euclidean, hyperbolic and elliptic.

The following important theorem of absolute geometry was proved by Legendre:

The sum of the angles of a triangle cannot be greater than two right angles.

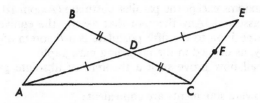

Proof: Let *ABC* be any triangle. Suppose that the sum of the angles of triangle *ABC* is *180° + d*, where *d > 0°*. We shall show that this is impossible. Let *D* be the midpoint of *BC*. Draw *AD* and extend it its own length to *E*, and draw *EC*. Triangles *ABD* and *CDE* are congruent, and hence angle *CED* = angle *BAD*, and angle ABD = angle DCE. $180° + d$ = angle BAC + angle ABC + angle BCA = (angle BAD + angle DAC) + angle ABD + angle BCA = (angle CED + angle DAC) + angle DCE + angle BCA = angle CED + angle DAC + (angle DCE + angle DCA) = the sum of the angles of triangle AEC. That is, the sum of the angles of triangle AEC is $180° + d$. One of the angles BAD and DAC is less than or equal to half of angle BAC. Consequently one of the angles CED and DAC is less than or equal to half of angle BAC. That is, one of the angles of triangle AEC is less than or equal to half of angle BAC. By using the same procedure, but starting with triangle AEC, letting F be the midpoint of EC, drawing AF, etc., we obtain a third triangle whose angle-sum is $180° + d$, and one of whose angles is less than or equal to $\frac{1}{2}(\frac{1}{2}$ angle $BAC)$. By using this procedure n times, we obtain a triangle whose angle sum is $180° + d$, and one of whose angles is less than or equal to $\frac{1}{2^n}$ angle BAC. If n is chosen large enough, this angle is smaller than d. Then the other two angles of the nth triangle must have a sum that is greater than two right angles. But this is impossible, by Euclid's proposition 17 of Book I.

Note that the proof of Legendre's theorem, like the proof of Euclid's proposition 16 of Book I, assumes that a line is infinite in length (see page 83). Therefore, it does not apply to elliptic geometry, where a line is finite.

Substitutes for the Fifth Postulate

An important incidental result of the attempts to prove Euclid's fifth postulate was the discovery of many propositions that are equivalent to it. A proposition is said to be equivalent to Euclid's fifth postulate if either can be derived from the other when all the Hilbert axioms except the parallel postulate (Axiom III on page 76) are assumed. Any theorem that asserts the equivalence of some proposition to the fifth postulate is a theorem of absolute geometry, as defined in the preceding paragraph.

We shall now prove such a theorem of absolute geometry:

The following statements are equivalent:

1. Through a point not on a given straight line there is only one straight line that does not intersect the given line. (Playfair axiom)

2. If a straight line falling on two straight lines makes the interior angles on the same side less than two right angles, then the two

straight lines meet on that side on which are the angles less than the two right angles. (Euclid's fifth postulate)

3. A straight line falling on lines that do not intersect makes the alternate interior angles equal. (Euclid's proposition 29 of Book I)

4. Two non-intersecting lines are everywhere equidistant. (A corollary of Euclid's proposition 33 of Book I)

5. There exists a triangle whose angle sum is two right angles.

6. The sum of the angles of any triangle is two right angles. (Euclid's proposition 32 of Book I)

To show that these statements are equivalent, we show that 1 implies 2, 2 implies 3, 3 implies 4, 4 implies 5, 5 implies 6, and 6 implies 1. Then it will follow that each of the six statements implies every other one.

Proof that 1 implies 2. Given that lines AB and CD are cut by EF so that angle BEF + angle DFE is less than two right angles. Assuming statement 1, we shall prove that AB and CD meet in the direction of B and D. Angle AEF + angle CFE is clearly more than two right angles. Draw GH through E so that angle HEF + angle DFE = 2 right angles. Then, by Euclid's proposition 28

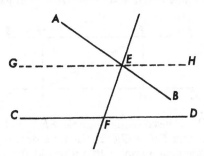

of Book I, GH and CD do not intersect. Then statement 1 implies that AB and CD do intersect. They cannot meet in the direction of A and C, for then a triangle would be formed containing two angles, AEF and CFE, whose sum is more than two right angles, which is impossible according to Euclid's proposition 17 of Book I. Consequently AB and CD meet in the direction of B and D.

Proof that 2 implies 3. Given that lines AB and CD do not intersect, and are cut by EF. Assuming statement 2, we shall prove that angle *AEF* = angle *EFD*. If angles *AEF* and *EFD* are not equal, one of them, say angle *EFD*, is the smaller of the two. Then

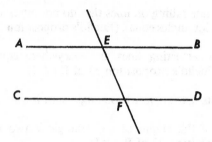

angle *EFD* < angle *AEF*. Adding angle *BEF* to both angles, we find that angle *EFD* + angle *BEF* < angle *AEF* + angle *BEF*, or angle *EFD* + angle *BEF* is less than two right angles. Then by statement 2, *AB* and *CD* would intersect, which is contrary to the hypothesis. Therefore angle *AEF* = angle *EFD*.

Proof that 3 implies 4. Given that lines *AB* and *CD* do not intersect. Let *F* and *G* be any two points on *AB*. Draw *FH* and *GK* perpendicular to *CD*. Assuming statement 3, we shall prove that *FH* = *GK*. By Euclid's proposition 28 of Book I, the lines *FH* and *GK* do not intersect. Draw *FK*. Statement 3 implies that angle *GFK* = angle *FKH*, and angle *HFK* = angle *FKG*. More-

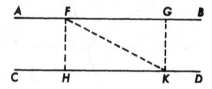

over *FK* = *FK*. Consequently triangle *FHK* and triangle *FGK* are congruent, and *FH* = *GK*. If *GK* were drawn perpendicular to *AB*, then statement 3 implies that it would also be perpendicular to *CD*. Then, in this case too, *FH* = *GK*.

Proof that 4 implies 5. Assuming statement 4, we show that there exists a triangle whose angle sum is two right angles. By Euclid's proposition 27 of Book I, there exist non-intersecting lines. Let *AB* and *CD* be non-intersecting lines. Let *F* and *K* be any two points on *CD*, and draw *FE* and *KJ* perpendicular to *AB*. Let *G* be any point on *EJ*, and draw *GH* perpendicular to *CD*. Statement 4 implies that *EF* = *GH* = *JK*. Then right triangles *GEF* and *GHF* are congruent, and right triangles *GHK* and *GJK* are congruent, since the hypotenuse and a leg are equal to the hypote-

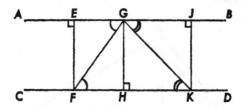

nuse and a leg. Now angle *EGF* + angle *FGK* + angle *JGK* = 2 right angles. But in the congruent triangles just noted, angle *EGF* = angle *GFH*, and angle *JGK* = angle *GKH*. Making these substitutions, we find that angle *GFH* + angle *FGK* + angle *GKH* = 2 right angles, that is, the sum of the angles of triangle *FGK* is 2 right angles.

Proof that 5 implies 6. To carry out this proof we first establish four preliminary propositions, L1, L2, L3, and L4, each of which is used to establish the next. Then we use the last of these four preliminary propositions to prove that 5 implies 6. The chain of argument is essentially that of Legendre.

L1. If there exists a triangle whose angle sum is two right angles, then the angle sum is two right angles for each triangle obtained from this one by drawing a line from a vertex to a point on the opposite side.

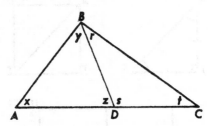

Proof. Given triangle *ABC* whose angle sum *S* is two right angles. Let *D* be any point on *AC*, and draw *BD*. Let S_1 and S_2 be the angle sums for triangles *ABD* and *BDC* respectively. Then $S_1 + S_2 = x + y + z + r + s + t = S + z + s = 4$ *right angles.* If S_1 and S_2 are not equal, then one of them is less than two right angles, and the other is more than two right angles. But this is impossible, in view of Legendre's theorem proved on page 197. Therefore $S_1 = S_2 = 2$ right angles.

L2. If there exists a triangle whose angle sum is two right angles, there exists an isosceles right triangle whose angle sum is two right angles.

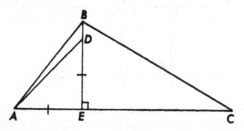

Proof. Given triangle *ABC* whose angle sum is two right angles. If *ABC* is not itself an isosceles right triangle, draw altitude *BE*. If triangle *ABE* is not isosceles, one of its legs, say *BE*, is longer than the other. Then measure off on *EB* a length *ED* equal to *AE*, and draw *AD*. Then triangle *ADE* is an isosceles right triangle. Moreover, by L1, since the angle sum for triangle *ABC* is two right angles so is the angle sum for triangle *ABE*, and hence also for triangle *ADE*.

L3. If there exists a triangle whose angle sum is two right angles, there exists an isosceles right triangle with legs greater than any given line segment and with angle sum equal to two right angles.

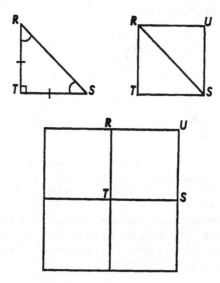

Proof. Given triangle *ABC* whose angle sum is two right angles. Then by L2 there exists an isosceles right triangle *RST* with angle *R* = angle *S* = 45°, and angle *T* = 90°. If we place two triangles congruent to triangle *RST* side by side so that their hypotenuses coincide, we obtain a quadrilateral with four right angles and all sides equal to *RT*. By using four quadrilaterals congruent to this one we can make a quadrilateral with four right angles and all sides equal to 2*RT*. By using four quadrilaterals congruent to this larger one, we can make a quadrilateral with four right angles and all sides equal to 4*RT*. By repeating *n* times the procedure of putting together four quadrilaterals congruent to the last already obtained, we obtain a quadrilateral with four right angles and all sides equal to $2^n RT$. By choosing *n* large enough we can make the side of the last quadrilateral greater than any given line segment. A diagonal of this quadrilateral divides it into congruent isosceles right triangles whose legs are greater than the given line segment. Since the sum of the angles of both triangles is four right angles, the angle sum of each is two right angles.

L4. If there exists a triangle whose angle sum is two right angles, then the angle sum for every right triangle is two right angles.

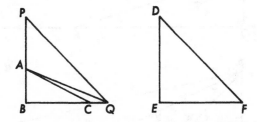

Proof. Let *ABC* be any right triangle, with right angle at *B*. By L3 there exists an isosceles right triangle *DEF* with right angle at *E* whose legs are greater than *AB* and *BC* and whose angle sum is two right angles. Extend *BA* to *P* and *BC* to *Q* so that *BP* = *BQ* = *ED* = *EF*, and draw *QA* and *QP*. Then triangle *PBQ* is congruent to triangle *DEF*, and hence its angle sum is two right angles. Then, by L1, the angle sum is two right angles in triangle *QAB* and hence also in triangle *ABC*.

We are now ready to prove that 5 implies 6. Let *ABC* be any triangle, and let its angle sum be *S*. Let *BD* be the altitude to the longest side of triangle *ABC*. Let S_1 and S_2 be the angle sums of triangles *ABD* and *BDC* respectively. Referring to the diagram

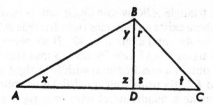

above we see that $S + z + s = S_1 + S_2$, so that $S = S_1 + S_2 - 2$
right angles. By L4, statement 5 implies that $S_1 = S_2 = 2$ right
angles. Therefore 5 implies that $S = 2$ right angles $+ 2$ right
angles $- 2$ right angles $= 2$ right angles. That is, 5 implies 6.

Proof that 6 implies 1. Given line a with P not on line a. Draw
PQ perpendicular to a, and draw straight line $S'PS$ perpendicular
to PQ. By Euclid's proposition 28 of Book I, PS does not intersect a. Assuming statement 6, we shall show that every other line
through P does intersect a. Let b be a line through P other than

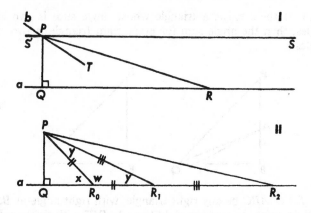

PS. Then a ray PT of b lies inside either angle SPQ or angle
$S'PQ$. Suppose PT lies inside angle SPQ. We show first that
there exists on line a a point R such that angle PRQ is less than
angle SPT. To do so we show that we can find a point R on a
such that angle PRQ is less than any given angle. Using diagram
II above, choose any point R_0 on a and draw PR_0. Then
choose R_1 on a so that $R_0R_1 = PR_0$. Let $x = $ angle PR_0Q,
$w = $ angle PR_0R_1, and $y = $ angle $R_0PR_1 = $ angle PR_1R_0. $x + w = $
2 right angles. By Legendre's theorem on page 197, $2y + w$
cannot be more than two right angles. Therefore $2y$ cannot be
more than x. That is, angle $PR_1Q \leq \frac{1}{2}x$. Now choose R_2 on a

so that $R_1R_2 = PR_1$. By similar reasoning we see that angle $PR_2Q \leq \frac{1}{2}(\frac{1}{2}x)$. By repeating this procedure, we find on a a sequence of points $R_1, R_2, R_3, \ldots, R_n, \ldots$ such that $R_{n-1}R_n = PR_{n-1}$ and angle $PR_nQ \leq \dfrac{1}{2^n}(x)$. If n is chosen large enough we can obtain an angle PR_nQ that is smaller than any given angle. Therefore, in diagram I, there exists a point R on a such that angle PRQ is less than angle SPT. If we assume statement 6, that is, if the sum of the angles of a triangle is two right angles, then angle PRQ + angle RPQ = a right angle = angle SPR + angle RPQ. Therefore angle SPR = angle PRQ. Consequently angle SPR is less than angle SPT. That is, the ray PT lies between PQ and PR. Therefore it must intersect a between Q and R. A similar argument shows that if PT lies in angle $S'PQ$, it must intersect a. Therefore 6 implies 1.

The following additional statements can also be shown to be equivalent to Euclid's fifth postulate:

7. If two lines do not intersect, there exists a line segment that is greater than the distance to one of the lines from any point on the other. (See page 235.)

8. There exists a rectangle. (See page 224.)

9. There exist similar triangles that are not congruent. (See page 228.)

10. Through any point in the interior of an angle there is a line which meets both sides of the angle (at points other than the vertex). (See page 218.)

11. There exists a pair of straight lines that are everywhere equally distant from one another. (See page 235.)

12. Given any three points not lying on a straight line, there exists a circle that passes through them. (See page 239.)

13. Given any area, there exists a triangle whose area is greater than the given area. (See page 232.)

14. The line joining the midpoints of the legs of an isosceles triangle passes through the midpoint of the median to the base. (See pages 193, 225 and 226.)

The Saccheri Quadrilateral

Saccheri's attempt to prove the fifth postulate was organized around a study of the properties of a special figure, a quadrilateral in which two equal sides are both perpendicular to a third side. Such a quadrilateral is now known as a *Saccheri quadrilateral*. The two equal sides that are both perpendicular to a third side are called the *legs* of the quadrilateral. The third side is called its *base*, and the fourth side, which lies opposite the base, is called the *summit*. The angles between the summit and the legs are called the *summit angles*.

Without using the fifth postulate Saccheri showed that this special kind of quadrilateral has the following properties:

The line joining the midpoints of the base and summit of a Saccheri quadrilateral is perpendicular to the base and summit. The summit angles of a Saccheri quadrilateral are equal.

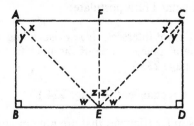

Proof. In the Saccheri quadrilateral *ABDC*, in which *AB* and *CD* are equal and are perpendicular to *BD*, and *E* and *F* are the midpoints of *BD* and *AC* respectively, draw *EF*, *AE* and *EC*. Triangle *ABE* is congruent to triangle *CDE*. Hence *AE = EC*, and it follows that triangle *AFE* is congruent to triangle *CFE*. Then the following angle equalities are obtained immediately: $x = x'$, $y = y'$, $z = z'$, $w = w'$. Adding the first two equalities, we find that the summit angles are equal. Adding the last two equalities, we find that angle *BEF* = angle *DEF*. Since their sum is two right angles, it follows that each of them is a right angle. From the congruent triangles we find, too, that angle *AFE* = angle *CFE*. Since these, too, add up to two right angles, each of them must be a right angle. Consequently *EF* is perpendicular to *BD* and *AC*.

After establishing this result, Saccheri then noted that there are three possible hypotheses that may be made about the summit angles: The summit angles of a Saccheri quadrilateral are either 1) right angles; 2) obtuse angles; or 3) acute angles. He was able to show that the first hypothesis, which he called the hypothesis of the right angle, implies the fifth postulate of Euclid. Therefore, to complete a proof of the fifth postulate, he had to eliminate the other two hypotheses. He succeeded in eliminating the second hypothesis, which he called the *hypothesis of the obtuse angle*, by a *reductio ad absurdum* argument. In the course of the argument, he tacitly assumed, as Euclid did, that a straight line is infinite in length. Then he tried to eliminate the third hypothesis, which he called the *hypothesis of the acute angle*. Hoping to show that it leads to a contradiction, he traced many of its implications. By doing so, he was in effect proving theorems of hyperbolic geometry, although he did not realize it. He finally arrived at this result: On the hypothesis of the acute angle, if P is a point not on line b, there are two lines r and s through P that are asymptotic to b towards the right and left respectively, which divide all the

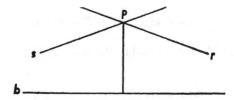

lines through P into two classes. One class consists of lines which intersect b. The other class consists of lines which have a common perpendicular with b. He then concluded that r and s, as limiting lines of the two classes, would partake of the nature of both: they would have a common perpendicular with b at a common point at infinity. Here he thought he had found the contradiction he was looking for, because it is impossible for two distinct lines to be perpendicular to a third line at the same point. His argument was erroneous, however, because he was assuming without justification that properties that lines have at finite points are also valid at infinity. Saccheri was a mathematical Sinbad who accidentally stumbled upon the hyperbolic valley of precious stones. But unlike Sinbad, who gathered the treasures he had discovered, Saccheri said, in effect, "You can't fool me. These jewels are only colored glass."

The Lambert Quadrilateral

Lambert, in his attempt to prove the fifth postulate, centered attention on another special figure, a quadrilateral three of whose angles are right angles. This figure, now known as a *Lambert quadrilateral,* is always obtainable as half of a Saccheri quadrilateral. For example, in the diagram on page 206, quadrilaterals *ABEF* and *FEDC* are both Lambert quadrilaterals.

Lambert listed three possible hypotheses concerning the fourth angle of a Lambert quadrilateral: the fourth angle may be a right angle, an obtuse angle, or an acute angle. The first hypothesis, like Saccheri's hypothesis of the right angle, leads to the fifth postulate and the geometry of Euclid. Lambert showed that the second hypothesis leads to a contradiction. Then he concentrated on trying to deduce contradictory results from the third hypothesis as well.

In the course of his exploration of the implications of the second and third hypotheses, Lambert arrived at this interesting result: If the second hypothesis holds, then the area of a triangle *ABC* can be computed from its angles by the formula

(1) $Area = k(A + B + C - 180°),$

and if the third hypothesis holds, the area can be computed from the formula

(2) $Area = k(180° - A - B - C),$

where *k* is some positive number. He recognized immediately that equation (1) has the form of the equation for the area of a triangle on a sphere,

(3) $Area = r^2(A + B + C - 180°),$

where *r* is the radius of the sphere and the unit of area is chosen so that the area of the sphere is $720r^2$. If in equation (3), *r* is replaced by *ir*, where $i = \sqrt{-1}$ is the imaginary unit of complex numbers, the equation becomes

(4) $Area = r^2(180° - A - B - C),$

which has the same form as equation (2). Probably because he may have observed this fact, he made the following significant observation: "From this I should almost conclude that the third hypothesis would occur in the case of an imaginary sphere." This hunch, as we shall see, contains an important element of truth.

However, pursuing the implications of the third hypothesis

further, he showed that it implies the existence of an absolute unit of length. He considered this conclusion absurd, so he rejected the third hypothesis. We shall see that there is indeed an absolute unit of length on a hyperbolic plane and it is not any more absurd than the absolute unit of length that occurs in the geometry on a sphere.

Lambert's recognition of the connection between the second hypothesis and spherical geometry was also prophetic. It foreshadowed the discovery of Riemann's elliptic geometry, which is, as we shall see, related to spherical geometry.

A Subtle Fallacy

Legendre's attempts to prove the fifth postulate revolved around an investigation of the sum of the angles of a triangle. There are three alternatives, analogous to the three hypotheses of Saccheri and Lambert. Given any triangle, the sum of the angles is either 1) equal to two right angles, or 2) greater than two right angles, or 3) less than two right angles. In the proof given on page 197, which tacitly assumes that a line is infinite, Legendre proved that the second alternative is impossible. In the proof given on page 203 he also showed that if there is a single triangle whose angle sum is two right angles, then every triangle has an angle sum of two right angles. It follows from this immediately that if there is a single triangle whose angle sum is less than two right angles, then every triangle has an angle sum that is less than two right angles. Thus the alternatives are sharply defined: either all triangles have an angle sum of two right angles, or all triangles have an angle sum that is less than two right angles.

Legendre made many attempts to prove that the latter alternative is impossible. He failed, of course, because he was trying to prove something that is not true. We reproduce one of his attempts, which, on first reading looks like a good proof. But, alas, second reading reveals that the proof rests on a fallacy.

Legendre's Fallacious Proof: Suppose the angle sum of triangle *ABC* is $180° - d$, where $d > 0°$. Construct triangle *ACD*, with *D* and *B* on opposite sides of *AC*, so that angle *DAC* = angle *ACB* and angle *DCA* = angle *CAB*. Then triangle *ADC* is congruent to triangle *ABC*. Therefore the angle sum of triangle *ADC* is also $180° - d$. Through *D* draw a straight line cutting the sides of angle *B* at *E* and *F* respectively. The angle sum of triangle *DCF* and also of triangle *AED* is less than or equal to

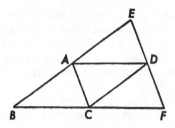

180°. Consequently the sum of the angles of all four triangles *ABC*, *BCD*, *DCF* and *BED* is less than or equal to *720° − 2d*. If we subtract from this sum the nine angles whose vertices are at *A*, *C* and *D* and whose sum is 3 × 180°, we find that the sum of the angles of triangle *EBF* is *180° − 2d*. If we now repeat this procedure, but start with triangle *EBF*, we can construct a triangle whose angle sum is *180° − 2(2d)*. If we repeat the procedure again and again, the *n*th triangle obtained has an angle sum of *180° − 2ⁿd*. If *n* is chosen large enough, *2ⁿd* is greater than 180°, and the *n*th triangle has an angle sum that is negative. This is obviously impossible. Hence the sum of the angles of a triangle cannot be less than two right angles.

The fallacy in this proof is in the step, "Through *D* draw a straight line cutting the sides of angle *B* at *E* and *F*." This step assumes that it is possible to draw a line through *D* that cuts both sides of angle *B*. This assumption is not justified, however. In fact, we shall see on page 218 that it is false in hyperbolic geometry. Hence, as we stated on page 205, this assumption is equivalent to Euclid's fifth postulate. By making this assumption, Legendre was in effect assuming what he was trying to prove.

The Timid Giant

At the end of the eighteenth century, trying to prove Euclid's fifth postulate became a popular pastime among mathematicians. It was inevitable that Gauss, who was probably the greatest mathematician of all time, should also turn his attention to this baffling problem. At first, like the others, he tried to prove the fifth postulate. However, with his deep mathematical insight and his clear grasp of the requirements of a rigorous proof, he did not fall into the trap of making any unjustified assumptions. In a letter written December 17, 1799 to his friend Wolfgang Bolyai (1775–1856) who was also trying to prove the fifth postulate and thought he had succeeded, Gauss said, "As for me, I have already made some progress in my work. However the path I have chosen does not lead at all to the goal which we seek, and which you assure me you have reached. It seems rather to compel me to doubt the truth of geometry itself.

"It is true that I have come upon much which by most people would be held to constitute a proof: but in my eyes it proves as good as *nothing*. For example, if one could show that a rectilinear triangle is possible whose area would be greater than any given area, then I would be ready to prove the whole of geometry absolutely rigorously.

"Most people would certainly let this stand as an Axiom; but I, no! It would, indeed, be possible that the area might always remain below a certain limit, however far apart the three angular points of the triangle were taken."

Gauss's doubt, hesitantly expressed in this letter, that (Euclid's) geometry is necessarily true, grew into a firmly held conviction that there is another geometry, which he called non-Euclidean geometry, which, though different from Euclid's, is equally deserving of consideration. Apart from mentioning the new geometry in letters to some trusted friends, Gauss kept his thoughts on the subject to himself. In a letter sent on November 8, 1824 to F. A. Taurinus, who had sent Gauss a report on his own efforts to prove the fifth postulate, Gauss wrote:

"I have not read without pleasure your kind letter of October 30th with the enclosed abstract, all the more because until now I have been accustomed to find little trace of real geometrical insight among the majority of people who essay anew to investigate the so-called Theory of Parallels.

"In regard to your attempt, I have nothing (or not much) to say except that it is incomplete. It is true that your demonstration of the proof that the sum of the three angles of a plane triangle cannot be greater than 180° is somewhat lacking in geometrical rigor. But this in itself can easily be remedied, and there is no doubt that the impossibility can be proved most rigorously. But the situation is quite different in the second part, that the sum of the angles cannot be less than 180°; this is the critical point, the reef on which all the wrecks occur. I imagine that this problem has not engaged you very long. I have pondered it for over thirty years, and I do not believe that anyone can have given more thought to this second part than I, though I have never published anything on it. The assumption that the sum of the three angles is less than 180° leads to a curious geometry, quite different from ours (the Euclidean), but thoroughly consistent, which I have developed to my entire satisfaction, so that I can solve every problem in it with the exception of the determination of a constant, which cannot be designated *a priori*. The greater one takes this constant, the nearer one comes to Euclidean Geometry, and when it is chosen infinitely large the two coincide. The theorems of this geometry appear to be paradoxical and, to the uninitiated, absurd; but calm, steady reflection reveals that they contain nothing at all impossible. For example, the three angles of a triangle become as small as one wishes, if only the sides are taken large enough; yet the area of the triangle can never exceed a definite limit, regardless of how great the sides are taken, nor indeed can it ever reach it. All my efforts to discover a contradiction, an inconsistency, in this non-Euclidean Geometry have been without success, and the one thing in it which is opposed to our conceptions is that, if it were true, there must exist in space a linear magnitude, *determined for itself* (but unknown to us). But it seems to me that we know, despite the say-nothing word-wisdom of the metaphysicians, too little, or too nearly nothing at all, about the true nature of space, to consider as *absolutely impossible* that which appears to us unnatural. If this non-Euclidean Geometry were true, and it were possible to compare that constant with such magnitudes as we encounter in our measurements on the earth and in the heavens, it could then be determined *a posteriori*. Consequently in jest I have sometimes expressed the wish that the Euclidean Geometry were not true, since then we would have *a priori* an absolute standard of measure.

"I do not fear that any man who has shown that he possesses a thoughtful mathematical mind will misunderstand what has been said above, but in any case consider it a private communication of which no public use or use leading in any way to publicity is to be made. Perhaps I shall myself, if I have at some future time more leisure than in my present circumstances, make public my investigations."

Gauss reveals in this letter why he kept his discovery of non-Euclidean geometry a closely guarded secret. Thinking in university circles at this time was dominated by the Kantian view that the properties of space supposedly expressed in Euclidean geometry were necessary properties imposed on reality by the process of perception. Gauss was afraid that his contemporaries, biased by the Kantian philosophy, and not sufficiently sophisticated mathematically to understand his ideas, would ridicule them. As he put it in a letter to Bessel on January 27, 1829, he withheld his discovery because he was afraid of the "clamor of the Boeotians."

However, Gauss did finally put in writing a summary of his

non-Euclidean geometry. In a letter he wrote to Schumacher on May 17, 1831, he said, "In the last few weeks I háve begun to put down a few of my own *Meditations*, which are already to some extent nearly 40 years old. These I had never put in writing, so that I have been compelled three or four times to go over the whole matter afresh in my head. Also I wished that it should not perish with me."

In January 1832 Gauss received a letter from his friend Bolyai that made it unnecessary for him to publish his non-Euclidean geometry. Bolyai's son Johann had independently made the same discovery and had already published it. A copy of Bolyai's paper was enclosed with the letter.

Creators of a Strange New Universe

Johann Bolyai (1802–1860) had acquired an interest in the theory of parallels from his father. Like his father, he first tried to prove Euclid's fifth postulate. Then he turned his attention to developing an absolute theory of space, in which the fifth postulate is neither affirmed nor denied. By 1823 he had discovered enough of hyperbolic geometry to be convinced that it was free of any inconsistencies, and hence was deserving of consideration as a possible mathematical model of physical space. In a letter he wrote to his father on November 3, 1823 he said: "I have now resolved to publish a work on the theory of parallels, as soon as I shall have put the material in order, and my circumstances allow it. I have not yet completed this work, but the road which I have followed has made it almost certain that the goal will be attained, if that is at all possible: the goal is not yet reached, but I have made such wonderful discoveries that I have been almost overwhelmed by them, and it would be the cause of continual regret if they were lost. When you will see them, you too will recognize it. In the meantime I can say only this: *I have created a new universe from nothing.* All that I have sent you till now is but a house of cards compared to the tower. I am as fully persuaded that it will bring me honor, as if I had already completed the discovery."

Replying to this letter, Johann's father urged him to complete his paper quickly and publish it as an appendix to the father's own forthcoming book. To explain the need for haste he made these interesting remarks: ". . . if you have really succeeded in the question, it is right that no time be lost in making it public, for two reasons: first, because ideas pass easily from one to another, who can anticipate its publication; and secondly, there is some truth in this, that many things have an epoch, in which they are found at the same time in several places, just as the violets appear on every side in spring. Also every scientific struggle is just a serious war, in which I cannot say when peace will arrive. Thus we ought to conquer when we are able, since the advantage is always to the first comer."

In 1825 Johann sent to his father an abstract of his discoveries. In 1829 he sent him the completed manuscript, and it was published, as planned, as an appendix to his father's book. When the appendix was sent to Gauss in 1832, Gauss made this comment on it in a letter to the elder Bolyai: "If I begin by saying that I am unable to praise this work, you will certainly be surprised for a moment. But I cannot say otherwise. To praise it would be to praise myself. Indeed, the whole contents of the work, the path taken by your son, the results to which he is led, coincide almost entirely with my meditations, which have occupied my mind partly for the last thirty or thirty-five years. So I find myself extremely surprised. So far as my own work is concerned, of which up till now I have put little on paper, my intention was not to let it be published during my lifetime. Indeed the majority of people have not clear ideas upon the questions of which we are speaking, and I have found very few people who could regard with any special interest what I communicated to them on this subject. To be able to take such an interest it is first of all necessary to have devoted careful thought to the real nature of what is wanted and upon this matter almost all are most uncertain. On the other hand it was my idea to write down all this later so that at least it should not perish with me. It is therefore a pleasant surprise for me that I am spared this trouble, and I am very glad that it is just the son of my old friend who anticipates me in such a remarkable manner."

Young Bolyai's joy at having created a new universe was dimmed when he heard the news that he was not the first to create it. It was dimmed even further when he found that he was not the first to publish it. Three years before Bolyai's paper was published, Lobatschewsky had already published in Russian a paper on the same subject. Bolyai did not learn of Lobatschewsky's work until 1848, when Lobatschewsky's book, *Geometrical Researches on the Theory of Parallels*, written in German, came to his attention. The honor that Bolyai hoped for had to be shared with others, because non-Euclidean violets were indeed appearing on every side. Bolyai's reluctant acceptance of this fact is recorded in this comment found among his notes on Lobatschewsky's book: "The nature of real truth of course cannot but be one and the same in Maros-Vasarhely as in Kamchatka and on the Moon, or, to be brief, anywhere in the world; and what one finite, sensible being discovers, can also not impossibly be discovered by another."

To give the reader a taste of the new geometry discovered by Gauss, Bolyai and Lobatschewsky, we present below an outline of some of its principal features.

The Axioms of Hyperbolic Geometry

The Hilbert axioms, arranged in groups I to VI on pages 74 to 77, serve as a foundation for Euclidean geometry. We obtain from the Hilbert axioms a set of axioms for hyperbolic geometry if we replace axiom III, the Playfair axiom, by the following contrary assumption:

III'. In a plane containing a given straight line and a given point that is not on the line, there is more than one straight line that does not intersect the given line. (Gauss-Bolyai-Lobatschewsky Axiom)

As we have noted on page 197, all theorems derived by using the Hilbert axioms except axiom III or III' are theorems of both hyperbolic geometry and Euclidean geometry. In addition, there are theorems whose proofs require the use of axiom III'. These are theorems of hyperbolic geometry alone. In the rest of this chapter up to page 232 all diagrams are assumed to be in a hyperbolic plane.

Parallels and Ultra-parallels

In a hyperbolic plane, let RS be any straight line, and let P be any point that is not on RS. Draw PQ perpendicular to RS, and draw AB perpendicular to PQ at P. Then, by Euclid's proposition 27 of Book I of the *Elements*, AB does not intersect RS. By axiom III' there is at least one more line CD through P

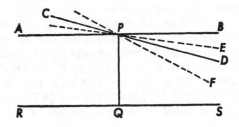

that does not intersect *RS*. Part of this line must lie in either angle *BPQ* or angle *APQ*. In the diagram we show it with the half-line *PD* inside the angle *BPQ*. We note first that any line such as *PE* which lies inside the vertical angles *BPD* and *APC* cannot intersect *RS*. For if it did, the points *P*, *Q*, and the intersection of *PE* and *RS* would be the vertices of a triangle, and the half-line *PD* would be inside angle *EPQ* of that triangle. Then, as a consequence of the Pasch axiom (see page 76), *PD* would intersect *RS*, which is contrary to our assumption. So, if we rotate *PB* clockwise to the position *PD*, in all intermediate positions it will still fail to intersect *RS*. If we continue to rotate it clockwise beyond *PD* we must ultimately come to a last position *PF* in which it fails to intersect *RS*. By reflecting *PF* across the line *PQ* we find another line *PF'* that has an analogous property: if the line *AB* is rotated counterclockwise, the position *PF'* is the last position in which it fails to intersect *RS*. Thus the lines *PF* and *PF'* divide all the lines through *P* into two parts. All the lines through *P* that lie inside angle *FPF'* and its vertical angle (the unshaded regions in the diagram below) intersect *RS*. All the rest of the lines through *P* (the lines *PF* and *PF'* and lines in the shaded region between them) do not intersect *RS*. We

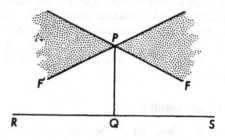

shall call the lines *PF* and *PF'* the lines through *P* that are *parallel* to *RS*. We shall refer to all the other lines through *P* that do not intersect *RS* as lines through *P* that are *ultra-parallel* to *RS*. Thus, while there are an infinite number of lines through *P* that do not intersect *RS*, there are exactly two lines through *P* that are parallel to *RS*. One of them lies on one side of *PQ* (the perpendicular from *P* to *RS*), and the other one lies on the other side of *PQ*. To distinguish them, we may say in reference to the preceding diagram that *PF* is parallel to *RS* on the right, while *PF'* is parallel to *RS* on the left. The angle *QPF* is called the angle of parallelism of *PF*. Similarly, angle *QPF'* which is equal to angle *QPF*, is called the angle of parallelism of *PF'*. Since angle $F'PF < 180°$, and angle QPF is half of angle $F'PF$, it follows that angle QPF, the angle of parallelism, is an acute angle.

It can be shown that parallel lines in the hyperbolic plane have the following properties: 1) If a line a is parallel to a line b in a given sense (to the right or to the left), and if P is a point

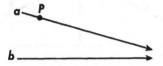

on a, then a is the line through P that is parallel to b in the given sense. 2) If one line is parallel to a second line, then the second line is parallel to the first line. 3) If two lines are both parallel to a third line in the same direction, then they are parallel to each other.

Ideal Points

In Euclidean geometry we found it convenient to add to each line an ideal point, or a point at infinity, so that we could say metaphorically that parallel lines "meet" at a point at infinity. We follow a similar procedure in hyperbolic geometry, but here we must introduce two ideal points on each line, one where it is "met" by lines that are parallel to it in one direction, and the other where it is "met" by lines that are parallel to it in the opposite direction. The ideal points are denoted by upper case Greek letters such as Ω (omega).

The introduction of ideal points permits us to extend the concept of a triangle to include triangles in which one or more vertices are ideal points. For example, if B is an ordinary point on a line l and Ω is an ideal point on l, and $A\Omega$ is the line through

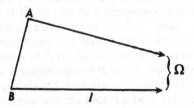

A parallel to l in the direction of Ω, then triangle $AB\Omega$ has two ordinary vertices and one ideal vertex. If RS is any line in the plane, and P is any point not on RS, we have already seen that

there are two lines PF and PF' that are parallel to RS. If we denote the ideal points where they meet RS by Ω_1, and Ω_2,

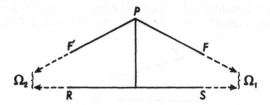

respectively, then triangle $P\Omega_1\Omega_2$ has one ordinary vertex and two ideal vertices. If we take a diagram like the latter with angle $F'PF$ equal to 120°, and place next to it on the other side of PF' and PF two more triangles $P\Omega_2\Omega_3$ and $P\Omega_1\Omega_3$ that are congruent to it, then we obtain a triangle $\Omega_1\Omega_2\Omega_3$ that has three ideal vertices.

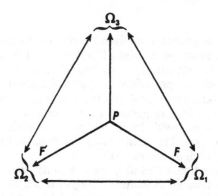

It can be shown that Pasch's axiom applies to triangles with one or more ideal vertices. (See exercises 9, 10, and 11 on page 258.) A significant consequence of this result is the fact that in the hyperbolic plane, if B is a point inside an angle, we cannot always find a line through B that cuts both sides of the angle. For example, in the diagram below, let B be a point inside angle $\Omega_1 A\Omega_2$, chosen so that A and B are on opposite sides of $\Omega_1\Omega_2$. If a line through B cuts the ray $A\Omega_1$ at P, then it must also cut $\Omega_1\Omega_2$, since it joins points B and P that are on opposite sides of $\Omega_1\Omega_2$. Then, since the Pasch axiom applies to this triangle, the line BP cannot cut the ray $A\Omega_2$. That is, there is no line through B that cuts both sides of angle A. This fact allows us to understand the fallacy of Legendre's proof on page 209.

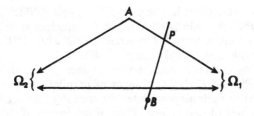

The Angle of Parallelism

If $P\Omega$ is parallel to $Q\Omega$ and PQ is the perpendicular distance from P to $Q\Omega$, then angle $QP\Omega$ is the angle of parallelism. One of the significant features of hyperbolic geometry is the fact that

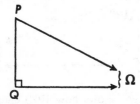

the angle of parallelism is related to the distance PQ. Two theorems which we now prove will show the nature of this relationship.

I. In triangles $PQ\Omega$ and $P'Q'\Omega'$ (in which $P\Omega$ is parallel to $Q\Omega$, and $P'\Omega'$ is parallel to $Q'\Omega'$), if $PQ = P'Q'$ and angle Q = angle Q', then angle $QP\Omega$ = angle $Q'P'\Omega'$.

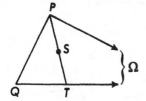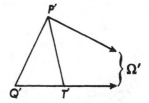

Proof: If angles $QP\Omega$ and $Q'P'\Omega'$ are not equal, then one of them, say angle $QP\Omega$, is larger than the other. Construct angle QPS equal to angle $Q'P'\Omega'$. The line PS will intersect $Q\Omega$ in some point T. On $Q'\Omega'$ draw $Q'T'$ equal to QT, and draw $P'T'$. Since $PQ = P'Q'$, angle Q = angle Q', and $QT = Q'T'$, triangle PQT is congruent to triangle $P'Q'T'$. Consequently angle

$Q'P'T'$ = angle QPT. But angle QPT = angle $Q'P'\Omega'$. Therefore angle $Q'P'T'$ = angle $Q'P'\Omega'$, and $P'T'$ coincides with $P'\Omega'$. But this is impossible, since $P'T'$ intersects $Q'\Omega'$ while $P'\Omega'$ does not. Consequently angle $QP\Omega$ = angle $Q'P'\Omega'$.

In the special case where angles Q and Q' are right angles, angles $QP\Omega$ and $Q'P'\Omega'$ are angles of parallelism. The theorem shows that to each perpendicular distance PQ there corresponds just one angle of parallelism. Then, if the length of PQ is d, the magnitude of the angle of parallelism that corresponds to it depends on d and on nothing else. To show this dependence, we shall denote by $\Pi(d)$ the magnitude of the angle of parallelism that corresponds to the distance d. (Read $\Pi(d)$ as "pi of d.")

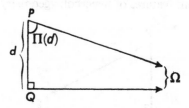

The second theorem shows how $\Pi(d)$ changes as d changes:

II. If $P\Omega$ is parallel to $Q\Omega$, and QP is extended to R, then the exterior angle $RP\Omega$ is greater than angle Q.

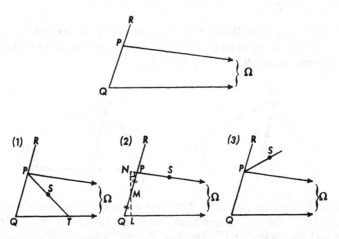

Proof: Draw angle RPS equal to angle Q. Then either (1) PS lies inside angle $QP\Omega$, or (2) PS coincides with $P\Omega$, or (3) PS lies inside angle $RP\Omega$. In case (1), PS intersects $Q\Omega$ at some point T

to form triangle PQT. Then exterior angle RPT is equal to the opposite interior angle Q. This is impossible, by proposition 16 of Euclid's Book I. In case (2), let M be the midpoint of PQ. Draw MN perpendicular to $P\Omega$, extended through P if necessary. On $Q\Omega$ draw QL equal to PN so that N and L lie on opposite sides of PQ, and then draw ML. Since angle Q = angle RPS = angle NPM, and $QM = PM$, triangles NPM and LQM are congruent. Then angle NMP = angle LMQ, and consequently NML is a straight line. Moreover, since angle MLQ = angle MNP, NL is perpendicular to $Q\Omega$. Then angle $LN\Omega$ is an angle of parallelism. But this is impossible, since angle $LN\Omega$ is a right angle, and an angle of parallelism is acute. Therefore case (3) is the only one that may arise, and in this case angle $RP\Omega$, being greater than angle RPS, is also greater than angle Q.

Let us apply this result to the diagram below, in which two lines are drawn parallel to $A\Omega$ from points P and Q on the line QPA which is perpendicular to $A\Omega$. If the distances of P and Q from $A\Omega$ are d and d' respectively, and $d' > d$, then the theorem just proved shows that $\Pi(d') < \Pi(d)$. If we picture the point P

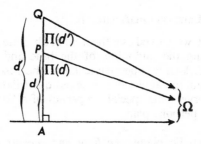

as moving toward Q, we see then that as the distance from $A\Omega$ increases, the corresponding angle of parallelism decreases. It can be shown that any acute angle, no matter how small, can be obtained as an angle of parallelism corresponding to some appropriate distance. Consequently, as the distance increases toward infinity, the corresponding angle of parallelism decreases toward zero. If P moves toward A, we see that as the distance from $A\Omega$ decreases toward zero the corresponding angle of parallelism increases toward 90°. The exact way in which $\Pi(d)$ varies with d is given explicitly by the following formula that was derived by Bolyai and Lobatschewsky:

(5) $$\tan \frac{\Pi(d)}{2} = e^{-\frac{d}{k}},$$

where k is a positive constant that is characteristic of the hyperbolic plane, and e is the sum of the infinite series $1 + \frac{1}{1} + \frac{1}{1 \cdot 2} + \frac{1}{1 \cdot 2 \cdot 3} + \frac{1}{1 \cdot 2 \cdot 3 \cdot 4} + \ldots$. Although the value of k is fixed for any given hyperbolic plane, it may be any positive real number.

Equation (5) relates lengths of line segments to the measures of the angles of parallelism that correspond to them. We have already observed that there is an absolute unit of angle measure, namely, the right angle. If we choose any specific multiple of a right angle that is within the range of values that an angle of parallelism may have (any value between 0° and 90°), equation (5) specifies a definite length that corresponds to it, and this length may serve as a unit of length. For example, we may use as a unit of length the distance d that corresponds to $\text{II}(d) = $ one-half of a right angle. Thus *in hyperbolic geometry there is an absolute unit of length*. In this respect it resembles the spherical geometry described on page 91, and differs from Euclidean geometry in which no absolute unit of length exists.

Saccheri and Lambert Quadrilaterals

On page 206 we proved as a theorem of absolute geometry that the line joining the midpoints of the base and summit of a Saccheri quadrilateral is perpendicular to the base and summit, and that the summit angles of a Saccheri quadrilateral are equal. We now observe some special properties of a Saccheri quadrilateral in a hyperbolic plane.

I. In a hyperbolic plane, *the base and summit of a Saccheri quadrilateral are ultra-parallel.*

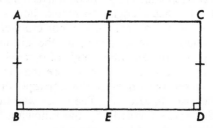

Proof: Let $ABDC$ be a Saccheri quadrilateral with base BD and summit AC. Let E and F be the midpoints of the base and summit respectively. Then, as stated above, FE is perpendicular to *BD* and *AC*. By Euclid's proposition 27 of Book I, *AC* does not intersect *BD*. Therefore, in a hyperbolic plane, *AC* and *BD* are either parallel or ultra-parallel. They cannot be parallel, for, if they were, the angle of parallelism, angle *CFE*, would have to be acute, and it isn't. Therefore they are ultra-parallel.

II. In a hyperbolic plane *the summit angles of a Saccheri quadrilateral are acute.*

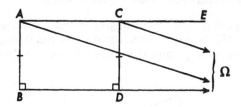

Proof: Let *ABDC* be a Saccheri quadrilateral with base *BD* and summit *AC*. Draw *A*Ω and *C*Ω parallel to *BD* in the same direction (to the right in the diagram above), and let *E* be any point on *AC* extended through *C*. Angle *ACD* + angle *DCE* = 180°. We shall show that angle *DCE* is greater than angle *ACD*. This inequality then implies at once that angle *DCE* > 90° and angle *ACD* < 90°, and hence the summit angle *ACD* is acute.

Since the summit angles *ACD* and *BAC* are equal, it will suffice to show that angle *DCE* is greater than angle *BAC*. Since *AC* is ultra-parallel to *BD* and *A*Ω is parallel to *BD*, it is clear from the discussion on page 216 that *A*Ω makes a smaller angle with *AB* than *AC* does. Consequently angle *BAC* = angle *BA*Ω + angle *CA*Ω. Similarly, *C*Ω makes a smaller angle with *CD* than *CE* does, and hence angle *DCE* = angle *DC*Ω + angle *EC*Ω. Since *AB* = *CD*, the angles of parallelism *BA*Ω and *DC*Ω are equal. Since *C*Ω is parallel to *A*Ω, the exterior angle *EC*Ω is greater than the interior angle *CA*Ω, as we proved on page 220. It follows by addition that angle *DCE* is greater than angle *BAC*.

This theorem shows that hyperbolic geometry is precisely what Saccheri was developing when he pursued the implications of the hypothesis of the acute angle.

Since a Lambert quadrilateral, in which three angles are right angles, is always obtainable as half of a Saccheri quadrilateral on one side of the line joining the midpoints of the base and summit, it follows that in a hyperbolic plane the fourth angle of a Lambert quadrilateral is acute. Consequently in a hyperbolic plane there are no quadrilaterals that have four right angles, that is, *there are no rectangles.* Since there are rectangles in a Euclidean plane and there are no rectangles in a hyperbolic plane, it follows that the statement that there exists a rectangle is equivalent to Euclid's fifth postulate. (See statement 8, page 205.)

In the Saccheri quadrilateral *ABDC* shown on page 222, which is longer, the leg *AB* or the line *FE* joining the midpoints of the summit and base? The answer to this question is supplied by the following theorem about a more general diagram:

III. If in quadrilateral *ABEF* angles *B* and *E* are right angles, then angle *AFE* is greater than, equal to, or less than angle *FAB* according as *AB* is greater than, equal to, or less than *FE*, and conversely.

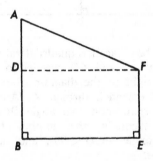

Proof: 1) If *BA* > *FE*, cut off on *BA* a segment *BD* = *FE*, and draw *DF*. Then quadrilateral *BDFE* is a Saccheri quadrilateral, and the summit angles *BDF* and *DFE* are equal. Since angle *AFE* > angle *DFE*, it follows that angle *AFE* > angle *BDF*. However, angle *BDF* is an exterior angle of triangle *ADF*. Therefore angle *BDF* > angle *BAF*. Consequently angle *AFE* > angle *BAF*. 2) If *BA* = *FE*, angle *BAF* = angle *AFE* since they are the summit angles of a Saccheri quadrilateral. 3) If *BA* < *FE*, an argument like that in 1) shows that angle *AFE* < angle *BAF*. 4) The reader can easily supply the proof of the converse, using a *reductio ad absurdum* argument. (See exercise 12 on page 259.)

In the Saccheri quadrilateral *ABDC* on page 222, *AB* and *FE* are both perpendicular to *BD*, so theorem III above applies to quadrilateral *ABEF*. Since angle *FAB* is acute, and angle *AFE* is a right angle, we see that *AB* > *FE*.

IV. If perpendiculars are drawn from the ends of one side of a triangle to the line passing through the midpoints of the other two sides, a Saccheri quadrilateral is formed.

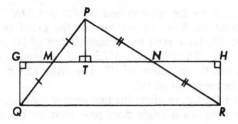

Proof: In triangle *PQR*, let *M* and *N* be the midpoints of *PQ* and *PR* respectively. Draw *QG*, *PT* and *RH* perpendicular to *MN*. Since *PM = MQ*, angle *PMT* = angle *GMQ*, and angle *QGM* = angle *PTM*, it follows that triangles *QGM* and *PTM* are congruent, and *QG = PT*. Similarly, triangles *PNT* and *RNH* are congruent, and *RH = PT*. Consequently *QG = RH*, and quadrilateral *QGHR* is a Saccheri quadrilateral with base *GH*.

V. In hyperbolic geometry, the line joining the midpoints of the legs of an isosceles triangle does not pass through the midpoint of the median to the base.

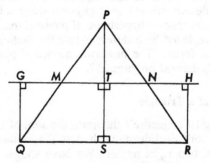

Proof: Let *PQR* be an isosceles triangle with *PQ = PR*. Let *M* and *N* be the midpoints of *PQ* and *PR* respectively. Draw *QG* and *RH* perpendicular to *MN*. Let *PS* be the median to *QR*. Since the median to the base of an isosceles triangle is perpendicular to the base, *PS* is perpendicular to *QR*. By IV above, quadrilateral *QGHR* is a Saccheri quadrilateral with base *GH* and summit *QR*. We proved on page 206 that the line joining the midpoints of the base and summit of a Saccheri quadrilateral is perpendicular to the base and the summit. This implies that the perpendicular bisector of the summit of a Saccheri quadrilateral is perpendicular to the base. Consequently *PS* is perpendicular to *GH*. Let *T* be the intersection of *PS* and *MN*. Then *PT* is perpendicular to *MN*. As we saw in the proof of IV above, *QG* = *PT*. We have seen as a consequence of III on page 224 that *QG* > *TS*. Therefore *PT* ≠ *TS*, and the intersection of *MN* and *PS* is not the midpoint of *PS*. Therefore *MN* does not pass through the midpoint of *PS*.

Since in a Euclidean plane the line joining the midpoints of the legs of an isosceles triangle does pass through the midpoint of the median to the base, while in a hyperbolic plane it does not, we see that statement 14 on page 205 is equivalent to Euclid's fifth postulate. Statement 14 is identical with statement A' on page 192. There we showed (using some physical assumptions that are listed on page 254) that statement A' is equivalent to the assumption A made by Archimedes. Consequently assumption A is equivalent to Euclid's fifth postulate. Therefore *if physical space is hyperbolic* rather than Euclidean, *assumption A is false*. Assumption A, restated in terms of forces rather than weights, is the familiar rule that the resultant of two equal forces of magnitude *F* acting in the same direction at the end of a line segment and at right angles to it is a force of magnitude *2F* acting at right angles to the segment at its midpoint. Consequently, while this rule is true if physical space is Euclidean, it is false if physical space is hyperbolic. If physical space is hyperbolic this rule is replaced by a different rule for finding the resultant of two equal forces. The new rule needed if physical space is hyperbolic is derived on page 257.

The Angles of a Triangle

According to Legendre's theorem, the sum of the angles of a triangle (in both Euclidean and hyperbolic geometry) cannot be greater than two right angles. We have shown on pages 203 to 205 that the existence of a triangle whose angle sum is two right angles implies that the space under consideration satisfies the Playfair axiom. Therefore in a hyperbolic plane, which does not satisfy the Playfair axiom, the sum of the angles of a triangle cannot be equal to two right angles. Thus we have proved that in a hyperbolic plane *the sum of the angles of a triangle is less than two right angles*.

The amount by which the sum of the angles of a triangle differs from two right angles is called the *defect* of the triangle. If *d* is the defect of triangle *ABC*, then

(6) $$d = 180° - A - B - C.$$

The Angles of a Polygon

A diagonal of a quadrilateral divides it into two triangles, and the sum of the angles of the quadrilateral is the sum of the angles of the two triangles. Consequently, in a hyperbolic plane, *the sum of the angles of a quadrilateral is less than four right angles.* Thus we see in another way that there are no rectangles in a hyperbolic plane.

In a polygon with n sides, the diagonals drawn from one vertex divide it into $n - 2$ triangles, and the sum of the angles of the polygon is the sum of the angles of these triangles. Consequently, in a hyperbolic plane the sum of the angles of a polygon with n sides is less than $(n - 2) \times (2$ right angles$)$. The amount by which the sum of the angles of the polygon differs from $(n - 2) \times (2$ right angles$)$ is called the defect of the polygon. If d is the defect of the polygon, and s is the sum of its angles, then

$$(7) \qquad\qquad d = (n - 2)180° - s.$$

A New Congruence Theorem

The methods for proving triangles congruent in a Euclidean plane are derived without the use of Euclid's fifth postulate. Therefore they apply to a hyperbolic plane. In hyperbolic geometry there is also another method for proving triangles congruent based on the following theorem:

In a hyperbolic plane, if the three angles of one triangle are equal respectively to the three angles of another triangle, then the triangles are congruent.

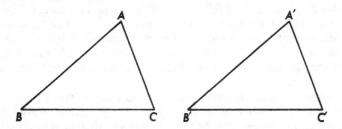

Proof: In triangles ABC and $A'B'C'$, let angle $A =$ angle A', angle $B =$ angle B', and angle $C =$ angle C'. We show that no pair of corresponding sides, such as AB and $A'B'$, can be unequal. Then it will follow at once that the triangles are congruent.

Suppose *AB* and *A'B'* are unequal. Then one of them, say *AB*, must be larger than the other. On *AB* measure off *AD* equal to *A'B'*, and on *AC* (extended if necessary) measure off *AE* equal to *A'C'*. Draw *DE*. Then triangle *ADE* is congruent to triangle *A'B'C'* (since side-angle-side = side-angle-side). Therefore angle

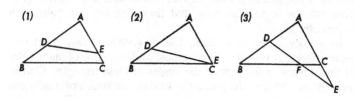

ADE = angle *A'B'C'* = angle *ABC*, and angle *AED* = angle *A'C'B'* = angle *ACB*. The rest of the proof hinges on a comparison of the lengths of *AE* and *AC*. Either 1) *AE* < *AC*, or 2) *AE* = *AC*, or 3) *AE* > *AC*. The three possibilities are shown in separate diagrams above. We show that neither one of them can occur, because each leads to a contradiction. In case 1), we have angle *ADE* + angle *BDE* + angle *AED* + angle *CED* = 4 right angles. Substituting angle *ABC* for angle *ADE*, and angle *ACB* for angle *AED*, we find that angle *ABC* + angle *BDE* + angle *ACB* + angle *CED* = 4 right angles. But this is impossible, because in a hyperbolic plane the sum of the angles of a quadrilateral is less than four right angles. In case 2), *E* and *C* coincide. Then angle *ADE* is an exterior angle of triangle *BDC*, and is therefore greater than angle *ABC*. But we have already established that angle *ADE* = angle *ABC*, so we have a contradiction. In case 3), *DE* does not cut segment *AC*, while it does cut segment *AB*. Then, by the Pasch axiom, *DE* cuts *BC* in some point *F*. Then angle *ADE* is an exterior angle of triangle *BDF* and is greater than angle *ABC*. Once again we have a contradiction of the statement already established that angle *ADE* = angle *ABC*. Since the assumption that *AB* and *A'B'* are unequal leads to a contradiction in every case, they cannot be unequal.

Because of this theorem, there are no similar triangles (that are not congruent) in a hyperbolic plane. Using everyday language to express this fact, we could say that in hyperbolic geometry, if two triangles have the same shape, they must also have the same size. This is another trait that hyperbolic geometry has in common with spherical geometry. (See page 90.)

The Concept of Area

The concept of *area* is a mathematical refinement of our intuitive notion of the "size" of a polygonal region. Underlying this concept is the concept of *equivalent* polygonal regions, which is a mathematical refinement of our intuitive notion of polygonal regions that have "the same size." Underlying this concept is the concept of *polygonal region* itself.

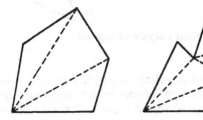

Examples of polygonal regions

A *polygonal region* in a plane is the figure obtained by putting side by side a finite number of triangular regions so that those which intersect at all intersect only on their boundaries. Suppose we divide a polygonal region into a finite number of triangular regions, and then rearrange the triangular regions to form a new polygonal region. Intuitively we think of the two polygonal regions as having the "same size" because they are made up of the same triangular pieces. This is the basis of the definition of equivalence: Two polygonal regions are *equivalent* if they can be divided into the same finite number of triangular regions and a one-to-one correspondence can be set up between the two sets of triangular regions so that corresponding triangular regions are congruent. For example, the two regions in the diagram below are equivalent because triangle I is congruent to triangle I′, triangle II is congruent to triangle II′, and triangle III is congruent to triangle III′. It can be shown that two polygonal regions that are equivalent to a third polygonal region are equivalent to each other.

A system of area measure is defined for polygonal regions if we assign to each polygonal region a positive number called its *area* in such a way that the system has these properties: 1) Equivalent polygonal regions have the same area; 2) polygonal regions that

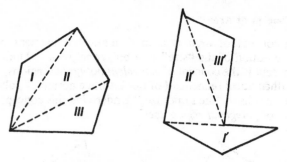

Equivalent polygonal regions

have the same area are equivalent; and 3) if a polygonal region is divided into a finite number of component polygonal regions, then the area of the original region is the sum of the areas of its components.

Area in a Euclidean Plane

In Euclidean geometry a system of area measure can be introduced in this way: first the area of a rectangular region is defined to be the product of the lengths of its base and height; then the definition is extended to triangular regions through the formula $A = \frac{1}{2}bh$; finally, it is extended to all polygonal regions through the rule that the area of a polygonal region is the sum of the areas of its component triangular regions.

Many different systems of area measure can be introduced in a Euclidean plane. However, they are all related to each other in a simple way: If A is the area of a polygonal region in one particular system, then the area of that region in some other system of area measure is r^2A, where r^2 is a positive constant. The change from one system to another is essentially a change of unit. For example, if we have a system of area measure in which the unit is a square foot, and we multiply each area measure by 144, we obtain a new system in which the unit is a square inch.

Area in a Hyperbolic Plane

To introduce a system of area measure in a hyperbolic plane, we cannot merely repeat the procedure used in a Euclidean plane, because there are no such things as rectangular regions in a hyperbolic plane. So we have to build a system of area measure in a hyperbolic plane on a different foundation. The foundation of the system consists of three theorems that are easily proved.

I. If two triangles have the same defect, then the triangular regions are equivalent.

II. If a triangle is divided into two triangles by a line drawn from a vertex to the opposite side, then the defect of the triangle is the sum of the defects of its two component triangles.

III. If two triangular regions are equivalent, then the triangles have the same defect.

We give here only the proof of II. The proofs of I and III may be found in any textbook on non-Euclidean geometry.

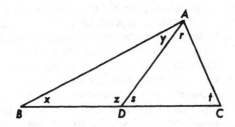

Proof of II. In triangle ABC, let D be any point on BC, and draw AD. Designate the angles of triangle ABD by x, y and z, and designate the angles of triangle ADC by r, s, and t, as shown in the diagram. Let d, d_1 and d_2 be the defects of triangles ABC, ABD and ADC respectively. Then, by equation (6), $d_1 = 180° - x - y - z$, and $d_2 = 180° - r - s - t$. Adding these two equations, we get $d_1 + d_2 = 360° - x - y - z - r - s - t$. If we take into account the fact that $z + s = 180°$, we see that $d_1 + d_2 = 180° - x - (y + r) - t = d$.

Because of theorems I, II, and III, we can introduce a system of area measure in a hyperbolic plane by defining the area of a polygonal region to be the sum of the defects of its component triangles. Any other system of area measure is obtainable from this one by multiplying the areas in this one by some positive constant r^2. Different choices of the value of r^2 correspond to different choices of the unit of area. The area of a triangle with defect d is therefore given by the formula $A = r^2 d$, or

$$(8) \qquad A = r^2(180° - A - B - C).$$

Triangle with Maximum Area

An immediate consequence of this formula is the fact that in hyperbolic geometry there is an upper limit to the area that a triangle may have, namely, $r^2(180°)$. It is the limit that is approached as the angles of a triangle all decrease to zero. This limit is the area of a triangle whose vertices are all ideal points.

Because there is an upper limit to the area of a triangle in a hyperbolic plane, while there is no such upper limit to the area of a triangle in a Euclidean plane, the assumption that there is no such upper limit is equivalent to Euclid's fifth postulate. (See statement 13 on page 205.)

Two Lines with a Common Perpendicular

We have already seen on page 223 that two lines in a hyperbolic plane that have a common perpendicular are ultra-parallel. The converse is also true: Two lines in a hyperbolic plane that are ultra-parallel have a common perpendicular. To prove this theorem it suffices to show that it is possible to construct a Saccheri quadrilateral whose base is on one of the ultra-parallel lines and whose summit is on the other one, because then the line joining the midpoints of the base and summit of the Saccheri quadrilateral would be the desired common perpendicular. To

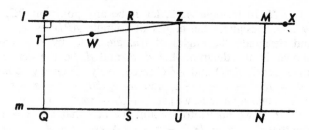

construct such a Saccheri quadrilateral bridging the space between two ultra-parallel lines l and m, we proceed as follows: Let P and R be any two points on l. Draw PQ and RS perpendicular to m. If $PQ = RS$, then $PRSQ$ is the desired Saccheri quadrilateral. If $PQ \neq RS$, then one of them, say PQ, is the greater segment. On PQ draw $QT = RS$. On the side of PQ on which R and S lie draw angle QTW equal to angle SRX, where X is a point on l chosen so that P and X are on opposite sides of R. It can be shown that the line TW will intersect l in some point Z. Draw ZU perpendicular to m. On l and m respectively, on the side of *RS* opposite *PQ*, draw *RM* = *TZ* and *SN* = *QU*, and draw *MN*. It is easy to prove with the help of appropriately drawn triangles that quadrilateral *TZUQ* is congruent to quadrilateral *RMNS*. It follows then that *MN* is perpendicular to *m*, *MN* = *ZU*, and *ZMNU* is the desired Saccheri quadrilateral.

Two ultra-parallel lines cannot have two common perpendiculars, because then the four lines would form a rectangle, which is impossible. Consequently two ultra-parallel lines have one and only one common perpendicular.

Ultra-ideal Points

Just as we say metaphorically that two parallel lines "meet" at an ideal point, we shall also say metaphorically that ultra-parallel lines "meet" at an *ultra-ideal point*. We shall say that all the lines which are perpendicular to a given line meet at the same ultra-ideal point. Under this convention, saying that several lines converge at an ultra-ideal point is equivalent to saying that they are all ultra-parallel to each other and have a common perpendicular.

The Extended Hyperbolic Plane

It is convenient to think of the ideal points and the ultra-ideal points as actual points that are adjoined to the ordinary points of a hyperbolic plane. The structure consisting of all three kinds of points, ordinary, ideal, and ultra-ideal, is called the *extended hyperbolic plane*. The relationships between the lines of a hyperbolic plane and the points of the extended hyperbolic plane can be shown in a simple diagram. Picture the extended hyperbolic plane as the set of points on the surface of a sheet of paper.

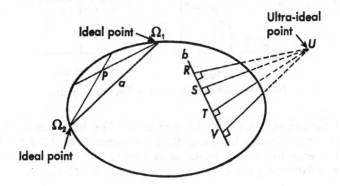

Draw an ellipse on the surface. Think of the interior of the ellipse as the hyperbolic plane; think of the points on the ellipse as ideal points; and think of the points outside the ellipse as ultra-ideal points. In this diagram, a line of the hyperbolic plane is represented by a chord of the ellipse without the endpoints of the chord. Two chords that intersect inside the ellipse represent intersecting lines in the hyperbolic plane. Two chords that intersect on the ellipse represent parallel lines in the hyperbolic plane. Two chords which, when extended, intersect outside the ellipse represent ultra-parallel lines in the hyperbolic plane. In the drawing above, Ω_1 and Ω_2 are the ideal points on line a. $P\Omega_1$ and $P\Omega_2$ are the lines through the ordinary point P that are parallel to a. R, S, T and V are ordinary points on the line b, and U is the ultra-ideal point where all the lines perpendicular to b converge. The ellipse whose points are the ideal points of the hyperbolic plane is called the *absolute* of the hyperbolic plane.

We shall see later that this diagram is more than a convenient visual aid. It has a deeper significance that will become evident on page 250 of this chapter and on page 351 of Chapter 10.

The Distance Between Two Lines

Given any two lines in a Euclidean plane or in a hyperbolic plane, if the perpendicular distance is drawn to one line from a point on the other, it is instructive to observe how the distance changes as the point moves along the line from which the perpendicular is dropped.

In a Euclidean plane there are two cases that arise:

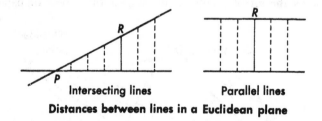

Intersecting lines   Parallel lines
Distances between lines in a Euclidean plane

1) *Intersecting lines.* The distance to one line from a point on the other decreases to zero as the point moves toward the point of intersection, and it increases to infinity as the point moves away from the point of intersection in either direction.

2) *Parallel lines.* The distance to one line from a point on the other remains the same as the point moves in either direction.

In a hyperbolic plane there are three cases that arise. The following propositions concerning them can be proved:

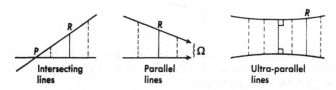

Intersecting   Parallel   Ultra-parallel
lines   lines   lines

Distances between lines in a hyperbolic plane

1) *Intersecting lines.* The distance to one line from a point on the other decreases to zero as the point moves toward the point of intersection, and it increases to infinity as the point moves away from the point of intersection in either direction.

2) *Parallel lines.* The distance to one line from a point on the other decreases to zero as the point moves in the direction of parallelism, and it increases to infinity as the point moves in the opposite direction.

3) *Ultra-parallel lines.* The distance to one line from a point on the other decreases to the length of the common perpendicular as the point moves toward the common perpendicular, and it increases to infinity as the point moves away from the common perpendicular in either direction.

If a plane satisfies the axioms of absolute geometry, and either statement 7 or statement 11 on page 205 is true about this plane, then it is clear from the properties listed above that the plane must be Euclidean, not hyperbolic. That is why statements 7 and 11 are both equivalent to Euclid's fifth postulate.

Corresponding Points

Consider two rays or half-lines *l* and *l'* that have a common vertex *O*. Let *P* be any point on *l*. There is obviously one and only one point *P'* on *l'* for which the distance *PO* equals the distance *P'O*. Let us call the point *P'* which has this property the point on *l'* that corresponds to the point *P* on *l*. Every ray whose vertex is *O* has a point that corresponds to *P*. Obviously any two points on rays from *O* that correspond to the same point

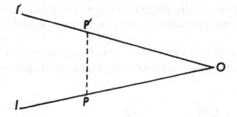

P also correspond to each other. All the points on the rays from O that correspond to a given point P lie on the circle whose radius is OP, and conversely.

In the paragraph above, corresponding points are defined on rays that converge to an ordinary point. We now consider the problem of extending the concept of corresponding points so that it makes sense for lines that "converge" to an ideal point or an ultra-ideal point as well. Suppose two lines l and l' meet at an ideal point Ω, and let P be any point on l. We cannot simply say that a point P' on l' corresponds to P if the distance $P\Omega$ equals the distance $P'\Omega$, since $P\Omega$ is infinite in extent. So, to be able to extend the concept of corresponding points we first have to modify our definition of it. The way in which we should modify the definition is easily seen if we re-examine the case of corresponding points on rays that converge to an ordinary point. If P and P' are corresponding points on the rays l and l' that have the same vertex O, so that $PO = P'O$, then angle $P'PO =$ angle $PP'O$. Conversely, if these two angles are equal then $PO = P'O$, and P' and P are corresponding points. This suggests that we redefine corresponding points as follows: If l and l' are two of the lines that meet at a given point which is ordinary, ideal or ultra-ideal, then a point P on l and a point P' on l' are corresponding points if the lines l and l' make equal angles with PP' on the same side.

Cousins of the Circle

In a Euclidean plane, suppose two lines l and l' meet at an ideal point Ω. That is, l and l' are parallel. If a point P on l corresponds to a point P' on l', then the interior angles on the same side of the transversal PP' are equal. But they are also supplementary. Therefore they are right angles, and PP' is perpendicular to l and l'. If for each line l' that is parallel to l we identify the point P' that corresponds to P, then the locus of P' is *the straight line that is perpendicular to l at P.*

This locus can also be obtained by a limiting process by first taking the lines l' that meet l at an ordinary point O, and allowing O to move toward the ideal point Ω on l. The locus of points on these lines that correspond to the point P on l is the circle whose center is O and whose radius is OP. If we allow O to move to infinity along l, the radius of the circle increases to infinity while the circle approaches as a limit the straight line perpendicular to l at P. For this reason it is customary to say that *in a Euclidean plane a circle with an infinite radius is a straight line.*

Now let us examine the corresponding locus in a hyperbolic plane. Consider all the lines that meet a given line l at the ideal point Ω on l. (That is, these lines are parallel to l in the direction of Ω.) If P is any point on l, it can be shown that there is one and only one point on each of these lines that corresponds to P. It can be shown, too, that if P' corresponds to P, then the perpendicular bisector of PP' passes through Ω, that is, it is parallel to l in the direction of Ω. To get some information about the nature of the locus of the points that correspond to P, let us examine any three corresponding points P, P' and P'' on the lines l, l' and l'' which meet at Ω. We show first that P, P' *and* P'' *do not lie in a straight line.*

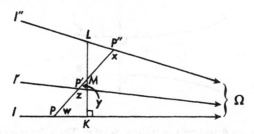

Proof: If P, P' and P'' lie in a straight line, then, using the notation of the diagram above, $y + z = 180°$. Since P' and P'' are corresponding points, $y = x$. Since P and P' are corresponding points, $z = w$. Making these substitutions, we see that $x + w = 180°$. Under these conditions, it is easy to show by congruent triangles that if M is the midpoint of PP'' and MK is perpendicular to l then it is also perpendicular to l'', so that l and l'' would have a common perpendicular, and hence would be ultra-parallel. But this is impossible, since l and l'' are parallel.

We show next that P, P' *and* P'' *do not lie on a circle.*

Proof: If P, P' and P'' lie on a circle whose center is the ordinary point O, then the perpendicular bisectors of PP', and $P'P''$

intersect at O. However, this is impossible since they are parallel to l and hence parallel to each other.

From these two results we conclude that the locus of points corresponding to P on lines parallel to l in the direction of Ω is neither a straight line nor a circle. It is a new kind of curve called a *horocycle*.

As in the case of the Euclidean plane, this locus can also be obtained by a limiting process by first taking the lines that meet l at an ordinary point O, and then allowing O to move toward the point at infinity Ω. Then we see that the horocycle described above is the limit approached by the circle with center O and radius PO when PO becomes infinite. Consequently, *in hyperbolic geometry, a circle with an infinite radius is not a straight line, but is a horocycle.*

In a hyperbolic plane we have one more case to consider that does not arise in a Euclidean plane, the case of lines that meet at an ultra-ideal point. Suppose l and l' meet at an ultra-ideal point. This means that they have a common perpendicular h. It can be shown that if P on l corresponds to P' on l', then they are equi-

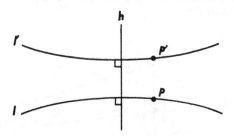

distant from h and on the same side of it. Then the locus of points corresponding to P that are on lines perpendicular to h is the locus of points on one side of h that have the same distance from h that P does. For this reason this locus is called an *equidistant curve*. To get information about the nature of this curve, let us examine any three corresponding points P, P' and P'' on the lines l, l' and l'' that have a common perpendicular h. We show first that P, P' and P'' do not lie in a straight line.

Proof: Let the distances of P, P' and P'' from h be PQ, $P'Q'$ and $P''Q''$ respectively. Since P, P' and P'' are corresponding points, $PQ = P'Q' = P''Q''$. Therefore quadrilaterals $PQQ'P'$ and $P'Q'Q''P''$ are Saccheri quadrilaterals. The summit angles x and y indicated in the diagram are acute. Consequently their

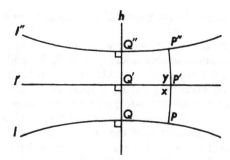

sum is less than two right angles, and the line $PP'P''$ is not straight.

We show next that P, P' and P'' do not lie on a circle.

Proof: If P, P' and P'' lie on a circle whose center is O, then the perpendicular bisectors of PP' and $P'P''$ meet at O. However, this is impossible, for the following reasons: Since PP' is the summit of a Saccheri quadrilateral, the perpendicular bisector of PP' is perpendicular to the base QQ'. Similarly the perpendicular bisector of $P'P''$ is perpendicular to $Q'Q''$. Therefore the perpendicular bisectors of PP' and $P'P''$ have a common perpendicular h, and therefore they do not intersect.

From these two results we conclude that an equidistant curve is neither a straight line nor a circle.

Thus we have in a hyperbolic plane two new kinds of curves, the horocycle and the equidistant curve, which share with the circle the property of being loci of corresponding points on lines with a given "common point."

In a Euclidean plane there are such things as pairs of straight lines that are everywhere equally distant, namely, parallel lines. However, there are no such things in hyperbolic geometry, since a locus of points equally distant from a straight line and on one side of it is an equidistant curve, not a straight line. That is why statement 11 on page 205 is equivalent to Euclid's fifth postulate.

In a Euclidean plane, there is a circle through any three non-collinear points P, P' and P''. We have just seen in effect that this is not true in a hyperbolic plane. There is a circle through P, P' and P'' if and only if the perpendicular bisectors of PP' and $P'P''$ meet at an ordinary point O, in which case O is the center of the circle through P, P' and P''. However, in a hyperbolic plane, the perpendicular bisectors of PP' and $P'P''$ may also meet at an ideal point or at an ultra-ideal point, and, as we have seen, in these cases there is no circle through P, P' and P''. That is why statement 12 on page 205 is equivalent to Euclid's fifth postulate.

In a hyperbolic plane, while we cannot always draw a *circle* through any three non-collinear points P, P', and P'', we can draw through them either a circle, or a horocycle or an equidistant curve, according as the perpendicular bisectors of PP' and $P'P''$ meet at an ordinary point, an ideal point or an ultraideal point. To prove the second and third parts of this assertion, it suffices to show that 1) if the perpendicular bisectors of PP' and $P'P''$ meet at an ideal point Ω then $P\Omega$ and $P'\Omega$ make equal angles with PP', and $P'\Omega$ and $P''\Omega$ make equal angles with $P'P''$; and 2) if the perpendicular bisectors of PP' and $P'P''$ have a common perpendicular h then P, P', and P'' are equidistant from h and on the same side of it. (See exercises 21 and 22 on page 260.)

A New Pythagorean Theorem

In a Euclidean plane the angles of a triangle ABC are related to each other by the angle-sum rule, $A + B + C = 2$ right angles, and if angle C is a right angle, the three sides of the triangle are related to each other by the Pythagorean theorem, $a^2 + b^2 = c^2$. It can be shown that these two rules are linked to each other. Since there is a deviation from the first of these rules in a hyperbolic plane, there is, in consequence, a deviation from the second of these rules as well. In a hyperbolic plane the relation connecting the three sides of a right triangle takes an entirely different form, expressed in the following equation:

$$\text{(9)} \qquad \cosh \frac{c}{k} = \cosh \frac{a}{k} \cosh \frac{b}{k},$$

where *cosh* is the abbreviation for *hyperbolic cosine*, defined on page 95, and k is a positive constant that is characteristic of the hyperbolic plane.

A New Plane Trigonometry

In a Euclidean plane, any two angles of a triangle are related to the opposite sides by the law of sines,

$$\frac{\sin A}{\sin B} = \frac{a}{b}.$$

This rule, too, is linked to the angle-sum rule. Since a hyperbolic plane has a different angle-sum rule it also has a different law of sines, which turns out to be

(10)
$$\frac{\sin A}{\sin B} = \frac{\sinh \dfrac{a}{k}}{\sinh \dfrac{b}{k}},$$

where *sinh* is the abbreviation for *hyperbolic sine*.

An Imaginary Sphere

On a sphere in Euclidean space the law of sines takes the form,

$$\frac{\sin A}{\sin B} = \frac{\sin \dfrac{a}{r}}{\sin \dfrac{b}{r}},$$

where r is the radius of the sphere. (See page 97.) For a real sphere, the radius r is a positive *real* number. It is interesting to see what happens to the formula above if we let r be an imaginary number ik, where $i = \sqrt{-1}$, and k is a positive real number. We then have

$$\frac{\sin A}{\sin B} = \frac{\sin \dfrac{a}{ik}}{\sin \dfrac{b}{ik}}.$$

Since $1 = -i^2$, it follows that $a = -i^2a$, and $b = -i^2b$. Making these substitutions into the formula, we obtain:

$$\frac{\sin A}{\sin B} = \frac{\sin \left(\dfrac{-i^2a}{ik} \right)}{\sin \left(\dfrac{-i^2b}{ik} \right)} = \frac{\sin \left[-i \left(\dfrac{a}{k} \right) \right]}{\sin \left[-i \left(\dfrac{b}{k} \right) \right]} = \frac{\sin \left[i \left(\dfrac{a}{k} \right) \right]}{\sin \left[i \left(\dfrac{b}{k} \right) \right]},$$

where the last substitution is based on the fact observed on page 94 that $\sin(-x) = -\sin x$, for all complex values of x. Finally, in view of equation (6) on page 95, we may replace sin [*i*(*a*/*k*)] by *i* sinh (*a*/*k*), and we may replace sin [*i*(*b*/*k*)] by *i* sinh (*b*/*k*). Then, canceling the *i* that occurs as a factor in both the numerator and the denominator of the last obtained fraction, we find that

$$\frac{\sin A}{\sin B} = \frac{\sinh \dfrac{a}{k}}{\sinh \dfrac{b}{k}}.$$

Comparing this result with equation (10) on page 241, we see that the law of sines on a sphere with an imaginary radius is the same as the law of sines on a hyperbolic plane. It is for this reason that hyperbolic geometry can be correctly described as the geometry of an imaginary sphere.

Spherical Geometry in Hyperbolic Space

A three-dimensional space is called hyperbolic if every plane in it is a hyperbolic plane. If a circle in such a space is rotated around a diameter it generates a sphere. Using great circles on the sphere as "lines," as we did on a sphere in Euclidean space, we can develop a spherical geometry in hyperbolic space. It turns out that it doesn't differ in any way from spherical geometry in Euclidean space. In both spaces, the sum of the angles of a spherical triangle is more than two right angles, and the law of sines for a spherical triangle takes the form of equation (8) on page 97.

The Horosphere and Its Geometry

We have seen that a horocycle may be thought of as a circle whose center is an ideal point. A ray from any point of a horocycle to its infinitely distant center is called a radius of the horocycle. If a horocycle in a three-dimensional hyperbolic space is rotated about one of its radii, it generates a surface called a *horosphere*. A horosphere may also be obtained as the limiting surface approached by a sphere as the radius of the sphere is increased to infinity.

On a horosphere, the shortest path between two points is always along a horocycle. So a horocycle is to a horosphere what a great circle is to a sphere or a straight line is to a plane. Consequently, there is a geometry on a horosphere in which horocycles play the role of lines. An examination of this geometry shows that it satisfies every one of the axioms of Euclidean geometry, if we read "horocycle" for the term "straight line" wherever this term occurs in an axiom. Therefore the geometry of a horosphere is Euclidean geometry.

Elliptic Geometry

In the construction of a set of axioms for plane geometry we have seen that there are three choices open to us for the selection of an axiom of parallels: Through a point not on a given line there is 1) one and only one line that does not intersect the given line; 2) more than one line that does not intersect the given line; 3) no line that does not intersect the given line. The first choice, the Playfair axiom, leads to Euclidean geometry, in which the sum of the angles of a triangle equals two right angles. The second choice leads to the hyperbolic geometry of Gauss, Bolyai and Lobatschewsky, in which the sum of the angles of a triangle is less than two right angles. The third choice leads to the elliptic geometry of Riemann, in which the sum of the angles of a triangle is more than two right angles.

We have already seen that the third choice is incompatible with the existence of straight lines of infinite extent. Therefore, if we wish to make this third choice we must modify those other axioms, the axioms of order, which imply that a line is infinite. We shall not discuss the details of this modification of the axioms other than to note that it is done in such a way that a straight line acquires the property of being *finite and unbounded*, just as a circle in Euclidean geometry is finite and unbounded. (See page 44.)

We shall now examine briefly some of the other consequences of making the third choice.

Let l be any line in an elliptic plane, and let P and Q be any two points on it. At each of these points draw a line perpendicular to l. Since there are no non-intersecting lines in an elliptic plane, these two lines intersect in some point O. Extend OP to O' so that $OP = O'P$, and draw $O'Q$. Triangle OPQ is congruent to triangle $O'PQ$, and as a consequence, angle PQO', being equal to angle PQO, is a right angle. Therefore OQO' is a straight line. Thus the same two lines PO and QO that intersect at O also intersect at O'. We now have two choices open to us. Either we may assume that O and O' are distinct points, or we may assume that O and O' coincide. If we make the first assumption, we must sacrifice the axiom that there is only one line between any two

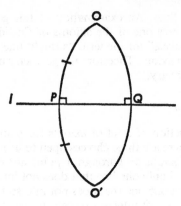

points, and the consequent property that two lines meet in at most one point. However, with this assumption we can obtain a plane that has the property of being orientable. (See page 98.) If we make the second assumption, we can preserve the axiom that there is only one line between any two points, but then we must sacrifice the orientability of the plane. We shall pursue each of these two alternatives far enough to show the meaning of these statements. Each of these alternatives leads to a different kind of elliptic plane. To distinguish them we shall call them the *Riemann plane* and the *Cayley-Klein plane* (Arthur Cayley, 1821–1895; Felix Klein, 1849–1925) respectively, after the mathematicians who first described them. The two kinds of elliptic plane are intimately related to each other, as we shall see.

The Riemann Plane

It is easy to describe what a Riemann plane is like because we have an exact model of it in a familiar surface in Euclidean space: The relations between points and lines in a Riemann plane are the same as the relations between points and great circles on a sphere in Euclidean space.

In making this comparison, we consider only the internal relationships of the two surfaces, and disregard the way the surfaces are related to any points that may exist outside the surface. Every true statement about a sphere can be converted into a true statement about a Riemann plane by merely substituting the term "straight line" for the term "great circle." Consequently the geometry of the Riemann plane is essentially the same as spherical geometry. (See pages 85 to 92.) In this geometry, any two lines meet in exactly two points. All the lines perpendicular to a given line meet at two particular points associated with the given line and known as its poles, just as all the meridian circles, which are perpendicular to the equator on the earth's surface, meet at the North and South Poles. A comparison of pages 90 and 227 shows that the congruence theorems for a Riemann elliptic plane are very much like the congruence theorems of a hyperbolic plane. The sum of the angles of a triangle in a Riemann plane is more than two right angles. The amount by which it exceeds two right angles is called the *excess* of the triangle. In the Riemann plane the excess of a triangle is a measure of its area just as in the hyperbolic plane the defect of the triangle is a measure of its area. In the Riemann plane as in the hyperbolic plane there is an absolute unit of length, there are no similar triangles, and there are no rectangles.

If a sphere is covered by a network of triangles, it is possible to assign a sense to each triangle in such a way that each pair of adjacent triangles induce opposite senses on their common side. For example, in the diagram below, the sphere is covered by a network of six triangles. If we assign to each triangle a clockwise sense as seen from a point outside the sphere and directly above the triangle then the sense of each triangle can be described by a sequence of its vertices as follows: *PSR*, *RST*, *PTS*, *PUT*, *PRU*, and *RTU*. The sense *PSR* means the sense from *P* to *S* to *R*. It induces on the line *SR* the sense from *S* to *R*. On the other hand,

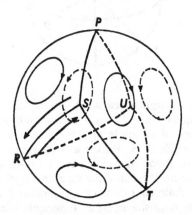

the sense *RST*, which means from *R* to *S* to *T*, induces on this same line the sense from *R* to *S*. Therefore the sphere satisfies the definition of an orientable surface. Since a Riemann plane, in its internal relationships, is the same as a sphere, a Riemann plane is orientable.

The Cayley-Klein Plane

A model of the Cayley-Klein elliptic plane can be derived very easily with the help of the spherical model of the Riemann plane. On a sphere, any two lines (that is, great circles) intersect in *two* diametrically opposite points. Consider the set S of pairs of diametrically opposite points on the sphere. Call each member of the set S a "point." Denote as a "line" in S the set of pairs in S obtained by taking all the pairs of points on the sphere that lie on any given great circle of the sphere. If $\{P, P'\}$ and $\{Q, Q'\}$ are distinct members of S, then P, P', Q and Q' all lie on the great circle determined by P and Q. Arc PQP' is half of a great circle. Either arc PQ or arc $P'Q$ is less than or equal to a quarter of a great circle of the sphere, while the other is greater than or equal to a quarter of a great circle. Call the smaller of these two

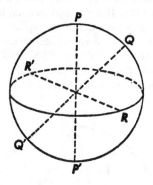

arc lengths the "distance" between $\{P, P'\}$ and $\{Q, Q'\}$. Each angle on the sphere determines a corresponding "angle" on S. Call the measure of an angle on the sphere the measure of the corresponding angle in S. With these definitions, the set S becomes an elliptic plane in which any two lines intersect in one and only one point. We can draw diagrams in S as diagrams on a sphere in the following manner: Any point P on a sphere or its diametrically opposite point P' represents the same point of S. Any great circle segment PQ on the sphere, or its diametrically opposite segment $P'Q'$, represents the same line segment in S. Any triangle PQR on the sphere, or its diametrically opposite triangle $P'Q'R'$, represents the same triangle in S.

To sum up the procedure described above for constructing a Cayley-Klein plane, a mathematician says that he is "identifying" each point P on a sphere with its diametrically opposite point P'. In this way, at one stroke, he converts the two points of intersection of two lines on a sphere into a single point of intersection in the Cayley-Klein plane. While the mathematician finds this procedure perfectly satisfactory, it may seem somewhat arbitrary to the layman. To remove from the mind of the reader any lingering doubts about the legitimacy of this procedure, we now carry it out in another way, as originally done by Klein. Consider the set of all straight lines that pass through the center O of a sphere. Call this set the "bundle" B. Any two diametrically opposite points on the sphere determine one and only one line in the bundle, and vice versa. In the bundle B let us use this terminology: Call every member of B (that is, every line through O), a "point." Call every set of coplanar lines through O a "line." Call the angle between any two lines through O the "distance" between the corresponding points of B. Call the dihedral angle between any two planes whose edge passes through O the "angle" between the corresponding lines in B. With point, line, distance, and angle defined in this way, the bundle B becomes a Cayley-Klein elliptic plane. That is, its "points" and "lines" have precisely the relationships that are characteristic of an elliptic plane in which any two lines meet in one and only one point. Obviously the elliptic plane B is completely equivalent to the elliptic plane S defined above. They differ only in the mental picture we associate with each point of the plane. In one case we picture the point as a pair of diametrically opposite points $\{P, P'\}$ on a sphere. In the other case we picture the point as the Euclidean line PP'.

To show that a Cayley-Klein plane is not orientable, let us use the set S as a model of the plane. In the diagram below, PQR is a triangle on a sphere, and PP', QQ', and RR' are diameters of the sphere. Consequently PQR and $P'Q'R'$ represent the same triangle in the Cayley-Klein plane. If we cover the plane S with a network of triangles, there is a corresponding network covering the sphere, with two diametrically opposite triangles on the sphere for every triangle in S. Part of one such network is shown in the diagram as a band of triangles connecting PQ to $P'Q'$. Let us see if we can assign a sense to each triangle in the network on S in such a way that adjacent triangles induce opposite senses on their common side. We begin by assigning the sense QPR to triangle PQR. Triangle PQR and triangle $P'Q'R'$ represent the same triangle in the elliptic plane S. P' represents the same point as P, Q' represents the same point as Q, and R' represents the

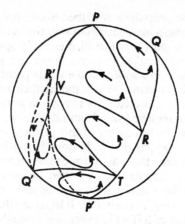

same point as R. Consequently the sense QPR shown in triangle PQR is the same as the sense $Q'P'R'$ shown in triangle $P'Q'R'$. The sense in each case is indicated by an arrow in the diagram. Now let us try to assign senses to the other triangles in the band in such a way that adjacent triangles induce opposite senses on their common side. In triangle PQR, the sense QPR induces the sense *from P to R* on the line PR. Since we want the sense of triangle VPR to induce the opposite sense on PR, namely from R to P, we must assign to triangle VPR the sense RPV. This sense induces on the line VR the sense *from V to R*. Since we want the sense of triangle VRT to induce the opposite sense on VR, *namely from R to V*, we must assign to triangle VRT the sense RVT. This sense induces on the line VT the sense *from V to T*. Since we want the sense of triangle VTQ' to induce the opposite sense on VT, namely *from T to V*, we must assign to triangle VTQ' the sense TVQ'. This sense induces on the line $Q'T$ the sense *from Q' to T*. Since we want the sense of triangle $TQ'P'$ to induce the opposite sense on $Q'T$, namely *from T to Q'*, we must assign to triangle $TQ'P'$ the sense $TQ'P'$. This sense induces on the line $Q'P'$ the sense *from Q' to P'*. But the sense $Q'P'R'$, already assigned to triangle $P'Q'R'$, also induces on the line $Q'P'$ the sense *from Q' to P'*. Thus we are compelled to have two adjacent triangles that do not induce opposite senses on their common side. Consequently the elliptic plane S is not orientable.

We have already encountered on page 99 another non-orientable surface, the Möbius strip. The Möbius strip has a single edge which is finite and unbounded and has the form of a closed loop. We can make use of the Möbius strip to help us visualize how the points on a Cayley-Klein plane are arranged with respect to each other. Take a sphere and draw a circle on it. Remove from the sphere the interior of the circle, so that you have a sphere with a circular hole. Now close up the hole by joining to its circular boundary the closed-loop edge of a Möbius strip. This joining cannot be done physically in a space of three dimensions, but can easily be carried out conceptually. All that is involved is setting up a one-to-one correspondence between the points on the boundary of the hole and the points on the edge of the Möbius strip, and then identifying each point on the boundary of the hole with the point to which it corresponds on the Möbius strip. The resulting surface is a closed, finite, unbounded surface. The arrangement of the points on this surface is like the arrangement of the points on a Cayley-Klein plane. Further discussion of this relationship is postponed until Chapter 12.

Trigonometry in an Elliptic Plane

A triangle on an elliptic plane of either kind is essentially a spherical triangle. Consequently trigonometry in an elliptic plane is the same as spherical trigonometry. For example, the law of sines in an elliptic plane takes the form of:

$$(11) \qquad \frac{\sin A}{\sin B} = \frac{\sin \dfrac{a}{r}}{\sin \dfrac{b}{r}},$$

which we have already met in equation (8) on page 97.

The Pythagorean Theorem in an Elliptic Plane

If ABC is a triangle in an elliptic plane, and angle C is a right angle, the relation connecting the three sides of the triangle has this form:

$$(12) \qquad \cos \frac{c}{r} = \cos \frac{a}{r} \cos \frac{b}{r},$$

where r is the radius of the sphere whose points, taken either singly or in diametrically opposite pairs, are the points of the elliptic plane.

The Three Geometries

The final outcome of two thousand years of investigation of
the theory of parallels was the discovery that there are three
different geometries, Euclidean, hyperbolic, and elliptic. While
these three geometries are distinct, they are also related to each
other. We shall note three ways in which they are related.

We have already observed that the geometry on a sphere in
Euclidean space is the same as the geometry in an elliptic plane.
In this sense, elliptic geometry exists within Euclidean geometry.
Hyperbolic geometry also exists within Euclidean geometry in
the same sense, because we can construct in a Euclidean plane a

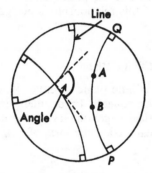

The Poincaré model of a hyperbolic plane

model of a hyperbolic plane, in the following way, due to Henri
Poincaré (1854–1912):

Denote as "points" of a "plane" all the points in the interior
of a fixed circle in a Euclidean plane. Denote as "lines" in the
"plane" the circular arcs in the interior of the fixed circle that
meet the fixed circle at right angles. Denote as the "angle" between
two "lines" in the "plane" the Euclidean angle formed at their
point of intersection by the tangents to these lines considered as
circular arcs in the Euclidean plane. To obtain a "length" for the
segment between any two points A and B in the "plane," we use
this definition: If the "line" AB cuts the fixed circle in the points
P and Q as shown in the diagram, denote as the "length of AB"
the number

$$d = \text{logarithm to the base } e \text{ of } \left(\frac{AQ}{AP} \middle/ \frac{BQ}{BP} \right),$$

where e is the number defined on page 222. With these definitions of "point," "line," "angle" and "length" in the "plane" it can be shown that the "plane" is a hyperbolic plane.

We have also seen that Euclidean geometry and elliptic geometry exist within hyperbolic geometry in the form of the geometries of a horosphere and a sphere respectively. Thus one way in which the three geometries are related is that Euclidean geometry and hyperbolic geometry each contains the other two geometries as the geometries of special configurations.

A second and third way in which the geometries are related can be observed by comparing the corresponding forms of the law of sines. The law of sines in elliptic geometry is given by equation (11) on page 249. On page 94 we observed that $\sin x = x - \dfrac{x^3}{6} + \dfrac{x^5}{120} - \dots$. For very small values of x, all terms after the first one in this infinite series are so small that the sum of the series is approximately equal to the first term alone, that is, $\sin x$ is approximately equal to x. Now, if triangle ABC is very small, the sides a and b are small, and so are the fractions a/r and b/r. Then $\sin a/r$ is approximately equal to a/r, and $\sin b/r$ is approximately equal to b/r. Making these substitutions in equation (11), we get the approximate equation

$$\frac{\sin A}{\sin B} = \frac{a/r}{b/r} = \frac{a}{b}.$$

Moreover, this equation becomes more and more exact as the triangle becomes smaller and smaller, with its longest side approaching length zero as a limit. But this equation is the law of sines for a Euclidean plane. Consequently we may say that in a small region of an elliptic plane the geometry is approximately Euclidean.

The law of sines in hyperbolic geometry is given by equation (10) on page 241. On page 95 we have the formula $\sinh x = x + \dfrac{x^3}{6} + \dfrac{x^5}{120} + \dots$. For very small values of x, $\sinh x$ is approximately equal to x. So, in hyperbolic geometry, if triangle ABC is small, $\sinh a/k$ is approximately equal to a/k, and $\sinh b/k$ is approximately equal to b/k. Making these substitutions in equation (10), we get the approximate equation

$$\frac{\sin A}{\sin B} = \frac{a/k}{b/k} = \frac{a}{b},$$

which again is the law of sines for a Euclidean plane. Consequently we may say that in a small region of a hyperbolic plane the geometry is approximately Euclidean.

The second way in which the three geometries are related, then, is this: *In small regions of an elliptic plane or a hyperbolic plane the geometry is approximately Euclidean.* This conclusion can be verified, too, by comparing the angle-sum rules for a triangle, or the Pythagorean theorems for the three geometries.

In the comparisons we made above, we made a/r and b/r small by making a and b small. Another way of making a/r and b/r small is to make r large. Similarly a/k and b/k may be made small by making k large. Consequently, when r and k become infinite, the law of sines in elliptic and in hyperbolic geometry is transformed into the law of sines of Euclidean geometry. Here then is the third way in which the three geometries are related: *Euclidean geometry is the limit approached by elliptic and hyperbolic geometry as r and k approach infinity.* The significance of this fact will become more apparent in Chapter 9.

The Consistency of Non-Euclidean Geometry

When Saccheri replaced Euclid's fifth postulate by his hypothesis of the acute angle he transformed the axioms of Euclidean geometry into the axioms of hyperbolic geometry. When he pursued the logical consequences of the new set of axioms, he did so in the hope of finding a contradiction. Had he found a contradiction, he would have proved that the axioms of hyperbolic geometry are inconsistent. However, neither he nor any of the later students of the theory of parallels ever found a contradiction in hyperbolic geometry. Consequently no one succeeded in proving that the axioms of hyperbolic geometry are inconsistent. This fact alone provides no answer to the question, "Are the axioms of hyperbolic geometry consistent or inconsistent?" The fact that no contradictions have yet been found in hyperbolic geometry is, by itself, no guarantee that contradictions will never be found. However, there is other evidence available to us that does answer the question.

In the preceding section we described the Poincaré model of a hyperbolic plane. This model was built out of points and arcs in the interior of a circle in a Euclidean plane. The relationships expressed in the axioms of hyperbolic geometry are properties of these points and arcs. As properties of these points and arcs in the Euclidean plane they are consequences of the axioms of Euclidean geometry. Consequently any contradiction that arises from using the axioms of hyperbolic geometry is a contradiction in Euclidean geometry. That is, if the axioms of hyperbolic geometry are inconsistent, so are the axioms of Euclidean geometry. The existence of the Poincaré model therefore proves that *if Euclidean geometry is consistent, so is hyperbolic geometry.*

On page 242 we noted that it has been proved that the geometry on a horosphere in a hyperbolic space satisfies the axioms of Euclidean geometry. Consequently any contradiction that might arise from using the axioms of Euclidean geometry would be a contradiction in hyperbolic geometry. The fact that the horosphere which is a model of the Euclidean plane exists in a hyperbolic space therefore proves *that if hyperbolic geometry is consistent, so is Euclidean geometry.* Combining this conclusion with the conclusion of the preceding paragraph we obtain this result: *Euclidean and hyperbolic geometry must stand or fall together as mathematical systems. If either is consistent, so is the other.*

Although we have not tried to do so in this book, it is possible to formulate a set of axioms for either the Riemann plane or the Cayley-Klein plane. Since models of these planes exist in Euclidean space in the guise of a sphere and the bundle of lines through the center of the sphere respectively, we can conclude that *if Euclidean geometry is consistent, so is elliptic geometry.* Conversely, since a Riemann plane, which is essentially a sphere, becomes a Euclidean plane when the radius of the sphere increases to infinity, we can conclude that if elliptic geometry is consistent, so is Euclidean geometry. Combining these two conclusions, we see that *Euclidean and elliptic geometry must stand or fall together as mathematical systems. If either is consistent, so is the other.*

Since we have already reached the same conclusion for Euclidean and hyperbolic geometry, we see that *the three geometries, Euclidean, hyperbolic, and elliptic, must stand or fall together as mathematical systems. If any one of them is consistent, then so are the other two.*

Thus, as mathematical systems, no one of the three geometries, Euclidean, hyperbolic, or elliptic, has any advantage over the other. All three are available to us as possible mathematical models of physical space. Which one of them is the best model of physical space is a problem for the physicist, not the mathematician.

Non-Euclidean Statics

We found on page 226 that the usual rule for adding two equal forces acting at the ends of a line segment is equivalent to

Euclid's fifth postulate, and is therefore valid only if physical space is Euclidean. Consequently if physical space is either hyperbolic or elliptic, the rule would have to be different. We outline here the procedure for deriving the form that the rule takes in each of these geometries. (The details of the derivation are found in the book *Non-Euclidean Geometry*, by Roberto Bonola.)

I. Six assumptions are made to serve as the foundation of the theory of statics:

1) Two or more forces, acting at the same point, have a definite resultant.

2) The resultant of two equal and opposite forces is zero.

3) The resultant of two or more forces, acting at a point, along the same straight line, is a force through the same point, equal to the sum of the given forces, and along the same line.

4) The resultant of two equal forces, acting at the same point, is directed along the line bisecting the angle between the two forces.

5) The magnitude of the resultant is a continuous function of the magnitude of the components.

6) The point of application of a force may be moved along its line of action.

II. Using these assumptions, and *without using Euclid's fifth postulate*, it can be shown that if two forces of magnitude F act at a point, and the angle between the two forces is 2α, then the magnitude of their resultant is $2F \cos \alpha$.

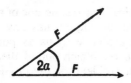

III. *Without using Euclid's fifth postulate*, it can be shown that a force R making angles α and β respectively with the perpendicular lines OX and OY may be resolved into components $R \cos \alpha$ and $R \cos \beta$ in the directions OX and OY respectively.

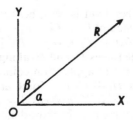

IV. Let F_1 and F_2 be equal forces of magnitude F acting at right angles to the segment AA' at the endpoints A and A' respectively. Let C be the midpoint of AA', and let B be any point on the perpendicular bisector of AA'. Draw BA and BA'. Denote by α the angle BAC, and denote by β the angle CBA. The sum $\alpha + \beta$ is equal to a right angle, or is less than a right angle, or is greater than a right angle, according as the space is Euclidean, hyperbolic, or elliptic. We shall prove that the resultant of F_1 and F_2 is a force acting at C along the line BC with magnitude $2F(\cos \beta / \sin \alpha)$.

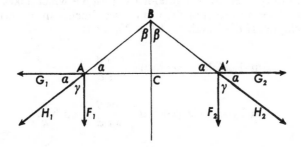

The force F_1 may be thought of as a component of a force H_1 acting along the line BA. Let G_1 be the component of H_1 in the direction $A'A$. Similarly, the force F_2 may be thought of as a component of a force H_2 acting along the line BA'. Let G_2 be the component of H_2 in the direction of AA'. The angle between H_1 and G_1 is α, and the angle between H_2 and G_2 is also α. If we denote by γ the angle between H_1 and F_1, we see that $\alpha + \gamma = 90°$. The angle between H_2 and F_2 is also equal to γ. Let H be the magnitude of H_1. By III above, $F = H \cos \gamma = H \sin \alpha$. Therefore $H = F/\sin \alpha$. Similarly, the magnitude of H_2 is also $H = F/\sin \alpha$. The magnitude of both G_1 and G_2 is $H \cos \alpha = (F/\sin \alpha) \cos \alpha$.

By assumption (6), the forces G_1 and G_2 may be moved until their points of application coincide. Since the forces are equal and opposite, their resultant is zero, because of assumption (2). Consequently the resultant $F_1 + F_2 = F_1 + F_2 + 0 = F_1 + F_2 + (G_1 + G_2) = (F_1 + G_1) + (F_2 + G_2) = H_1 + H_2$. By assumption (6), we may move the forces H_1 and H_2 along AB and $A'B$ respectively until their points of application coincide at B. Then by II above, the resultant $H_1 + H_2$ is a force acting along BC with magnitude $2H \cos \beta = 2(F/\sin \alpha) \cos \beta = 2F (\cos \beta/\sin \alpha)$. By assumption (6), this force may be moved along BC until its point of application is C. Thus we have proved that *the resultant of the forces F_1 and F_2, which have equal magnitude F, and act at right angles to the segment AA', each at one end of the segment, is a force with magnitude $2F(\cos \beta/\sin \alpha)$ acting at the midpoint of AA' and perpendicular to it*. Since this rule was derived without using Euclid's fifth postulate, it is valid whether physical space is Euclidean or not.

V. The ratio $\cos \beta/\sin \alpha$ occurs in the formula for the magnitude of the resultant of F_1 and F_2. The angles α and β are angles of triangle ABC in which angle BCA is a right angle. The value of this ratio is given by a trigonometric formula which takes on a different form in each of the geometries, Euclidean, hyperbolic, and elliptic. If the length of AA' is $2b$, then:

$$\text{in Euclidean geometry, } \frac{\cos \beta}{\sin \alpha} = 1;$$

$$\text{in hyperbolic geometry, } \frac{\cos \beta}{\sin \alpha} = \cosh \frac{b}{k};$$

$$\text{and in elliptic geometry, } \frac{\cos \beta}{\sin \alpha} = \cos \frac{b}{r}.$$

VI. If R is the magnitude of the resultant of F_1 and F_2, then by IV,

$$R = 2F \left(\frac{\cos \beta}{\sin \alpha} \right).$$

If we substitute for $\cos \beta/\sin \alpha$ the values given in V above, we get three different formulas for R:

$$R = 2F, \quad \text{if physical space is Euclidean;}$$

$$R = 2F \cosh \frac{b}{k} \quad \text{if physical space is hyperbolic;}$$

$$R = 2F \cos \frac{b}{r} \quad \text{if physical space is elliptic.}$$

Kant Refuted

According to the theory of space of Immanuel Kant, the properties of physical space are necessary properties imposed on reality by the process of perception. Since Euclidean geometry was the only geometry he knew of, he assumed that these necessary properties of space were the properties described in Euclidean geometry. This theory has been completely refuted by the discovery that there are two other geometries that are no less valid than Euclidean geometry. It is clear now that it is not at all necessary that physical space be Euclidean. It may be hyperbolic or elliptic. In Chapter 11 we shall see how the physicist may be able to find out from empirical data which of the three geometries most accurately describes the properties of physical space.

Axioms and Mathematical Systems

The refutation of the Kantian theory of space has also destroyed the notion that an axiom is a "self-evident" truth. It is now clear that an axiom is merely an assumption. A mathematical system is a set of assumptions and the logical implications that may be derived from them. When he constructs a mathematical system, a mathematician may begin with any arbitrary set of assumptions (axioms) about his undefined terms, subject only to the restriction that the assumptions should be consistent. For aesthetic reasons he may also impose the restriction that the assumptions be independent of each other, so that his set of axioms will not be redundant. For utilitarian reasons he usually constructs only mathematical systems that can serve as models for significant physical, biological or social systems, or for conceptual systems already in use.

EXERCISES FOR CHAPTER 8

1. *The fifth postulate.* State and prove the converse of Euclid's fifth postulate.

2. *Legendre's theorem.* In the proof of Legendre's theorem on page 197, why are triangles *ABD* and *CDE* congruent?

3. In the proof on page 199 that statement 1 implies statement 2, why is angle *AEF* + angle *CFE* more than two right angles?

4. a) In the proof of L1 on page 201, why does

$$x + y + z + r + s + t = S + z + s?$$

  b) Why does $S + z + s = 4$ right angles?

5. In the proof of L3 on page 203 we assert that by using four congruent quadrilaterals with four right angles and all sides equal to *RT* we can make a quadrilateral with four right angles and all sides equal to 2*RT*. (See the diagram on page 202.) Prove this assertion.

6. *Saccheri quadrilateral.* a) In the proof on page 206, why are triangles *ABE* and *CDE* congruent? b) Why are triangles *AFE* and *CFE* congruent?

7. Prove that the perpendicular bisector of the base of a Saccheri quadrilateral is the perpendicular bisector of the summit, and vice versa.

8. Prove that the line joining the midpoints of the legs of a Saccheri quadrilateral is perpendicular to the line joining the midpoints of the base and summit.

9. *Triangle with one ideal vertex.* In a hyperbolic plane, prove that if *AB*Ω is a triangle with one ideal vertex Ω, and half-line *AP* is inside angle *BA*Ω, then it intersects *B*Ω.

10. Using the result of exercise 9, prove that if a line intersects the side *A*Ω and does not pass through a vertex of triangle *AB*Ω, then it intersects one and only one of the other two sides.

11. Using the result of exercise 9, prove that if a line intersects the side *AB* and does not pass through a vertex of triangle *AB*Ω, then it intersects one and only one of the other two sides.

12. *Birectangular quadrilateral*
    Given: quadrilateral $ABEF$ is in a hyperbolic plane

    angle B = angle E = a right angle

    angle F > angle A

    Prove: $AB > FE$

13. In a hyperbolic plane prove that the summit of a Saccheri quadrilateral is greater than the base. (Hint: draw the line joining the midpoints of the summit and base, and prove that half the summit is greater than half the base.)

14. *Angles of a polygon.* a) Find the defect of a triangle whose angles are 60°, 70°, and 40°. b) Find the defect of a quadrilateral whose angles are 90°, 80°, 68°, and 95°.

15. In a hyperbolic plane, let $ABCD$ and $A'B'C'D'$ be Saccheri quadrilaterals with summits AD and $A'D'$ respectively. Prove that if $AD = A'D'$ and angle A = angle A', then $AB = A'B'$. (Hint: if $AB > A'B'$, draw AE on AB and draw DF on DC so that $AE = A'B'$ and $DF = D'C'$. Show that this leads to a contradiction.)

16. Use the conclusion of exercise 15 to show that two Saccheri quadrilaterals are congruent if they have equal summits and equal summit angles.

17. *Equivalent polygonal regions.* In the diagram on page 225, show that triangle PQR is equivalent to Saccheri quadrilateral $GQRH$.

18. Prove that two triangles with a side of one equal to a side of the other and having the same defect are equivalent. (Hint: use the results of exercises 16 and 17.)

19. *Common perpendicular.* In the proof on page 232, supply the details of the argument that shows that quadrilaterals $TZUQ$ and $RMNS$ are congruent.

20. *Corresponding points.* In the proof on page 237, supply the details of the argument that shows that triangles PMK and $P''ML$ are congruent.

21. *Horocycle.* If P, P' and P'' are distinct points in a hyperbolic plane, prove that if the perpendicular bisectors of PP' and $P'P''$ meet at an ideal point Ω, then $P\Omega$ and $P'\Omega$ make equal angles with PP', and $P'\Omega$ and $P''\Omega$ make equal angles with $P'P''$.

22. *Equidistant curve.* If P, P' and P'' are distinct points in a hyperbolic plane, prove that if the perpendicular bisectors of PP' and $P'P''$ have a common perpendicular h, then P, P' and P'' are equidistant from h.

23. *Three geometries.* On page 29 we proved that the sum of the angles of a triangle equals two right angles. a) What is the first step in the proof that is not valid in elliptic geometry? b) What is the first step that is not valid in hyperbolic geometry?

9

The Calculus and Geometry

Curved Figures

The axioms of Euclidean plane geometry express properties of straight lines and of figures, such as angles and triangles, that can be made from straight lines. However, there are important figures, such as the conic sections, which consist of curved lines. This fact confronts the mathematician with a challenging question: How can you use axioms about straight line figures to draw valid conclusions about curved line figures? The basic answer to this question was provided by the method of exhaustions of Eudoxus by which a curved line figure is obtained as the limit of a sequence of straight line figures that give better and better approximations to it. The elaboration of this method over the centuries ultimately led to the development of the specialized techniques of the calculus and the branch of geometry called *differential geometry* in which these techniques are employed.

In this chapter, without becoming involved in the specialized techniques of the calculus, we shall explain the geometric meaning of its basic concepts. Then we shall derive in a non-technical manner the concept of the curvature of space. Our reward will be a deeper understanding of the significance of the three geometries, Euclidean, hyperbolic, and elliptic, that emerged from the study of the theory of parallels.

Slope and Derivative

The fundamental problem of the *differential calculus* may be expressed geometrically as that of finding the slope of a straight line that is tangent to a given curved line at a given point. To solve this problem we first introduce a pair of rectangular coordinate axes OX and OY in the plane. Let P_1 be the given point on the curve, and let P_1T be the tangent to the curve at P_1. Let P_2 be another point on the curve, near P_1, and draw the straight line P_1P_2. A line like P_1P_2 which passes through two points of a curve

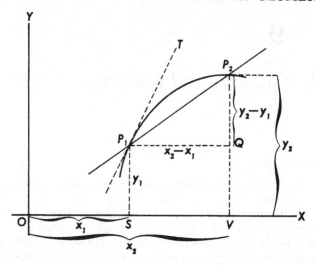

is called a *secant*. If P_2 is allowed to move toward P_1, the secant P_1P_2 rotates around the point P_1 and approaches the position of the tangent line P_1T. That is, P_1T is the limit approached by P_1P_2 as P_2 approaches P_1. Consequently *the slope of P_1T is the limit of the slope of P_1P_2 as P_2 approaches P_1.*

Let the coordinates of P_1 be (x_1, y_1), and let the coordinates of P_2 be (x_2, y_2). Then by equation (1) of page 109, the slope of P_1P_2 is

$$\frac{\Delta y}{\Delta x} = \frac{y_2 - y_1}{x_2 - x_1}.$$

As P_2 approaches P_1, the value of Δx approaches zero. Consequently we may say that *the slope of P_1T is the limit of $\dfrac{\Delta y}{\Delta x}$ as Δx approaches zero.* This limit is called the *derivative* of y with respect to x, and is denoted by the symbol $\dfrac{dy}{dx}$. When the equation of the curve is known, the value that the derivative $\dfrac{dy}{dx}$ has at the point P_1 can be computed by the methods of the calculus.

To give an example of the methods used, let us compute $\dfrac{dy}{dx}$ in the case where the equation of the curve is $y = x^2$. Since P_1 lies on the curve whose equation is $y = x^2$, its coordinates (x_1, y_1)

satisfy this equation. That is, $y_1 = (x_1)^2$. Since P_2 lies on the curve whose equation is $y = x^2$, its coordinates (x_2, y_2) satisfy the equation too. That is, $y_2 = (x_2)^2$. Since $\Delta y = y_2 - y_1$, $y_2 = y_1 + \Delta y$. Since $\Delta x = x_2 - x_1$, $x_2 = x_1 + \Delta x$. Making these substitutions for y_2 and x_2, we get the equation $y_1 + \Delta y = (x_1 + \Delta x)^2$, and multiplying out on the right-hand side of this equation, we find that

$$y_1 + \Delta y = x_1^2 + 2x_1(\Delta x) + (\Delta x)^2.$$

Subtracting from this equation the equation $y_1 = (x_1)^2$, we get

$$\Delta y = 2x_1(\Delta x) + (\Delta x)^2.$$

Dividing by Δx, we obtain a formula for the slope of the secant $P_1 P_2$:

$$\frac{\Delta y}{\Delta x} = 2x_1 + \Delta x.$$

As Δx approaches zero, the first term on the right-hand side of this equation remains unchanged, while the second term approaches zero. Consequently the slope $\dfrac{\Delta y}{\Delta x}$ approaches as a limit the value $2x_1 + 0$, or $2x_1$. So when the equation of the curve is $y = x^2$, the slope of the line tangent to the curve at the point with coordinates (x_1, y_1) is $2x_1$.

We have introduced the derivative $\dfrac{dy}{dx}$ here as the slope of a straight line tangent to a curve, thus relating it to a simple geometric picture. However, its significance goes beyond that of the geometric picture. The derivative is the basic mathematical tool used for studying rates of change. For example, suppose a particle is moving along a straight line, and we note its position on the line at each instant of time. Let t denote the elapsed time, as measured by a particular clock, and let s be the distance of the particle from some fixed point at the time t. Let t_1 and t_2 be particular values of t, and let s_1 and s_2 be the corresponding values of s. When the elapsed time changes from t_1 to t_2, the distance of the particle changes from s_1 to s_2. The change in time is $\Delta t = t_2 - t_1$. The corresponding change in distance is $\Delta s = s_2 - s_1$. The ratio $\dfrac{\Delta s}{\Delta t}$ is the average rate of change of the distance in the time interval Δt. We usually call it the average *speed* of the particle in that time interval. If we choose t_2 closer and closer to t_1, then, as Δt approaches zero, the ratio $\dfrac{\Delta s}{\Delta t}$ approaches a limit. This limit, denoted by the symbol $\dfrac{ds}{dt}$, is the rate of change of s with respect to t at the instant when $t = t_1$. For this reason we call it the *instantaneous* speed of the particle at the time $t = t_1$.

More generally, suppose we have a function defined for some interval of the real number system, which assigns a number denoted by y to each number x in the interval. Let x_1 and x_2 be particular values of x, and let y_1 and y_2 be the corresponding values of y. By analogy with the case of a moving particle, which defines the distance s as a function of the time t, we call the interval $\Delta x = x_2 - x_1$ the *change in x*, and we call the interval $\Delta y = y_2 - y_1$ the corresponding *change in y*. Then the ratio $\dfrac{\Delta y}{\Delta x}$ is the *average rate of change* of y in the interval Δx. If $\dfrac{\Delta y}{\Delta x}$ approaches a limit as x_2 approaches x_1, this limit, represented by the symbol $\dfrac{dy}{dx}$, is the *instantaneous rate of change* of y with respect to x when $x = x_1$.

Motion at an Instant

How can we tell whether a body is moving or is at rest? This is a fundamental question that involves many subtleties and has important physical implications. We shall consider now only one aspect of the question, which can be answered adequately with the help of the concept of the derivative. Other aspects of the question will be examined in Chapter 11.

Before we can answer this question it is necessary to divide it into two parts: 1) How can we tell whether a body is moving or is at rest during an *interval* of time? 2) How can we tell whether a body is moving or is at rest at an *instant* of time? This separation into two parts is made necessary by the fact that an *interval* of elapsed time is quite different from an *instant*, just as a *segment* of a line is quite different from a *point* on the line.

To answer the first question, we begin by measuring the passage of time, as indicated by a clock, and we determine for each instant the position of the body with respect to some object chosen as a point of reference. Then, *if the body is in the same position during every instant in a given interval of time, we say that the body is at rest during the interval.* However, *if there are two* *instants t_1 and t_2 at which the body is in different positions, then we say that the body has moved during the interval between t_1 and t_2.* Thus, the basic criterion for motion during an interval is that for at least part of the interval *a change in time is accompanied by a change in position.*

To answer the second question we cannot rely on this criterion at all, because an instant, being of duration zero, involves no change of time at all. Therefore we must develop another criterion for motion. The new criterion is derived from a re-examination of the first problem of determining whether or not a body has moved during an interval. Let us restrict ourselves to the simple case where the motion of the body is along a straight line. Then the position of the body can be indicated by its distance s from some fixed point on the line. If the distances that correspond to the instants t_1 and t_2 are s_1 and s_2 respectively, then the average speed of the body during the interval from t_1 to t_2 is, as we have seen, $\dfrac{\Delta s}{\Delta t}$, where $\Delta s = s_2 - s_1$, and $\Delta t = t_2 - t_1$. If the body has different positions at the instants t_1 and t_2, then its average speed during the interval is not zero. But if the body does not change its position during the interval, then its average speed during that interval is zero. Thus we have another criterion for determining whether or not the body has moved during an interval: *A body has moved during an interval if and only if its average speed during some part of the interval is not zero.* We can now answer the second question by allowing t_2 to approach t_1, so that the length of the interval shrinks to zero. As t_2 approaches t_1, the interval between t_1 and t_2 collapses to the instant t_1, and the average speed $\dfrac{\Delta s}{\Delta t}$ approaches the instantaneous speed $\dfrac{ds}{dt}$ as a limit. Thus we obtain the criterion for motion at an instant. *A body is moving at an instant t_1 if and only if its instantaneous speed at the time t_1 is not zero.* For example, when a ball is thrown into the air, it rises to a highest position and then falls again. As the ball rises, its instantaneous speed is directed upward, is different from zero, and is diminishing. That is, the ball slows down as it rises. In its highest position, the ball has an instantaneous speed of zero. As the ball falls again, its instantaneous speed is directed downward, is different from zero, and is increasing. That is, the ball falls faster and faster. At the instant when the ball is in its highest position it is at rest, because its instantaneous speed is zero. At every other instant of its flight the ball is in motion because its instantaneous speed is not zero.

The Paradox of the Arrow

Our analysis of the concepts of motion and rest make it possible for us to resolve the *paradox of the arrow*, one of the four famous paradoxes of Zeno. Zeno argued that an arrow in flight occupies a definite position at each instant of its flight. Since there is no change in position during any instant, he concluded that the arrow is at rest at each instant, and hence does not move at all. Zeno's paradox arises from the fact that he confused an instant with an interval. He assumed that where there is no change of position there is no motion. However, this assumption is not universally true. Where there is no change of position *accompanying a change in time*, there is no motion. However, there is no change in time during an instant. Zeno was applying to an instant a criterion for motion that is relevant only for an interval. The real test of motion at an instant, as we have seen, is to determine whether or not the instantaneous speed of the arrow at that instant is different from zero.

Area and Integral

The fundamental problem of the *integral calculus* may be expressed geometrically as that of finding the area bounded by an arc of a curve, the portion of the X axis under the curve, and the ordinates at the ends of the arc. In the diagram below, P_1 and P_2 are the ends of the arc, and P_1A and P_2B are the corresponding ordinates. The problem is to compute the area of the closed figure P_1ABP_2. Suppose the equation of the curve P_1P_2 is $y = f(x)$,

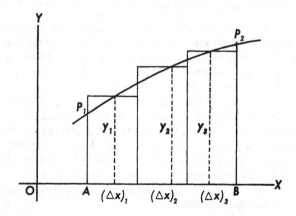

where the expression $f(x)$, read as "f of x," denotes some continuous function of x. We can obtain an approximation to the desired area in the following way: Divide the segment AB into n equal segments $(\Delta x)_1, (\Delta x)_2, \ldots, (\Delta x)_n$. The diagram shows such a division with $n = 3$. Erect over each of these segments a rectangle whose height is the ordinate of any point of the curve that lies directly over the segment. Denote by y_1 the height of the rectangle whose base is $(\Delta x)_1$, denote by y_2 the height of the rectangle whose base is $(\Delta x)_2$, and so on. Then the areas of these rectangles are $y_1(\Delta x)_1$, $y_2(\Delta x)_2$, etc. If we add the areas of these n rectangles we get an approximation to the area under the curve P_1P_2. Consequently the area under the curve is approximately $y_1(\Delta x)_1 + y_2(\Delta x)_2 + \ldots y_n(\Delta x)_n$. This sum may be represented by the symbol S_n. We can see intuitively that we get better approximations to the area under the curve by taking larger and larger values of n. The actual area under the curve is the limit approached by S_n as n increases to infinity. It is shown in the integral calculus that when $f(x)$ is a continuous function, this limit always exists and is unique. If a and b are the abscissas of P_1 and P_2 respectively, this limit is designated by the symbol $\int_a^b y\, dx$, and is called the *integral of y with respect to x*, over the interval from a to b. The value of the integral depends, of course, on the nature of the function $f(x)$ that occurs in the equation of the curve. It is shown in the integral calculus, for example, that if $y = f(x) = x^2$, then $\int_a^b y\, dx = \int_a^b x^2\, dx = (b^3 - a^3)/3$.

We have related the integral to a geometric picture by presenting it as the area under a curve. However, its significance extends beyond that of this geometric picture. What is essential to the concept of the integral is that it is the limit approached by a sum of small pieces when the size of each piece shrinks to zero and the number of pieces increases to infinity. The integral is the basic mathematical tool for computing quantities that may be decomposed into arbitrarily small pieces.

The Length of a Curve

An important geometric use of the integral is for finding the length of a curve. Suppose we want to determine the length of the curve extending from A to B in the diagram below. We first divide the curve into small pieces by inserting other points along the curve between A and B. Let P_1 and P_2 be the ends of any one

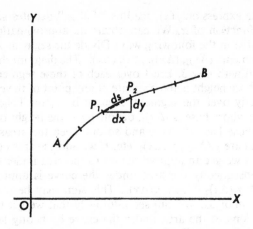

of these small pieces of the curve, and let (x_1, y_1) and (x_2, y_2) be their coordinates with respect to rectangular axes OX and OY. Denote by ds the distance from P_1 to P_2, and denote by dx and dy respectively the changes in x and y that arise when a point moves from P_1 to P_2. That is, $dx = x_2 - x_1$, and $dy = y_2 - y_1$. Then ds is the length of the hypotenuse of a right triangle whose legs have lengths dx and dy, respectively. Consequently, by the Pythagorean theorem,

(1) $$(ds)^2 = (dx)^2 + (dy)^2.$$

The length ds, computed by means of formula (1) is the length of the chord from P_1 to P_2. If P_2 is very close to P_1, the length of the arc P_1P_2 differs only slightly from the length of the chord P_1P_2. Consequently we get a good approximation to the length of the curve if we add the lengths ds of the chords joining successive points of division on the curve. We can improve the approximation by taking more points of division, dividing the curve into smaller pieces. The precise length of the curve, if it exists, is the limit of the sum of the lengths of the chords joining successive points of division, as the number of points of division increases to infinity and the lengths of the chords shrink to zero. This limit is denoted by the symbol $\int ds$. The integral calculus has developed techniques for computing this integral. One of the interesting discoveries made by the mathematicians who worked out the techniques for measuring the length of a curve is that not every curve has a length. Curves that have a length are known as *rectifiable* curves.

The Curvature of a Curve

The curvature of a curve is a measure of its deviation from a straight line. In the diagram below three paths are shown joining the two points A and B. One of them is part of a straight line. Each of the other two is part of a circle. We can see from the diagram that the path that is part of the larger circle deviates less from the straight line than does the path that is part of the smaller circle. For this reason we say that the larger a circle is,

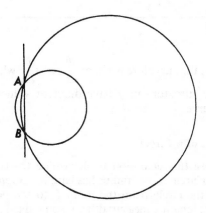

the less its curvature is. To give a precise meaning to our intuitive notion of curvature of a curve, we introduce a numerical measure of curvature. To do this, we first introduce the concept of the direction of a curve, and then the concept of the rate of change of the direction of a curve.

If a particle moves along a rectifiable curve, its direction of motion at any point of the curve is the direction of the line that is tangent to the curve at that point. We use as a measure of this direction the angle θ, measured in radians, that the tangent line makes with the coordinate axis OX.

Let F be any fixed point on a rectifiable curve. Denote by s the length of the arc of the curve extending from F to any point P. There is also associated with P the direction θ of the tangent line at P. When a particle moves along the curve, its direction of motion θ changes as the arc length s, measured from F to the particle's position P, changes. The *rate of change of direction* per unit length of arc at the point P is given by the derivative $\frac{d\theta}{ds}$ at P. The absolute value of $\frac{d\theta}{ds}$ is called the *curvature* of the

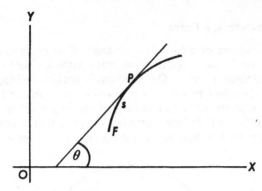

curve. That is, the curvature is either $\dfrac{d\theta}{ds}$ or $-\dfrac{d\theta}{ds}$, whichever is not negative. The curvature may have different values at different points of a curve.

The Curvature of a Circle

We shall use this definition to determine the curvature of a circle. Given a circle whose radius has length r. Suppose a particle moves along the circle from the point P to the point Q. As it moves, the arc length s measured from some fixed starting point F changes by an amount Δs which is equal to the length of the arc from P to Q. As the particle moves, its direction of motion, θ, given by the direction of the tangent line through its position, also changes. The total change in direction $\Delta \theta$ when the particle moves from P to Q is the angle through which the tangent line

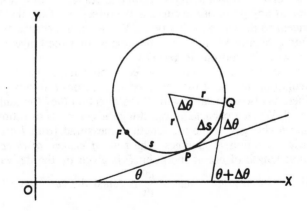

turns as the particle moves from P to Q. Since the radius to the point of contact of a tangent is perpendicular to the tangent, we may think of the radius as being rigidly attached to the tangent at a fixed right angle. Consequently, as the tangent turns through an angle $\Delta\theta$, the radius to the point of contact, being rigidly attached to the tangent, turns through an equal amount. Therefore the central angle subtended by the arc PQ is equal to $\Delta\theta$. We observed on page 97 that if an arc of a circle whose radius is r contains x' linear units and subtends a central angle of x radians, then $x = x'/r$, or $x' = rx$. Applying this formula, using Δs for x' and $\Delta\theta$ for x, we see that $\Delta s = r(\Delta\theta)$. Then the ratio $\dfrac{\Delta\theta}{\Delta s} = \dfrac{\Delta\theta}{r(\Delta\theta)} = \dfrac{1}{r}$. The magnitude of this ratio is the same no matter what the magnitude of Δs is. The curvature of the circle at the point P is the derivative $\dfrac{d\theta}{ds}$, which is the limit approached by $\dfrac{\Delta\theta}{\Delta s}$ as Δs shrinks to zero. Since, as Δs shrinks to zero the ratio $\dfrac{\Delta\theta}{\Delta s}$ is always equal to $\dfrac{1}{r}$, its limit $\dfrac{d\theta}{ds}$ is also equal to $\dfrac{1}{r}$. Therefore the curvature of the circle at the point P is $\dfrac{1}{r}$. The circle has the same curvature at every one of its points. For this reason a circle is often referred to as a curve with *constant curvature*.

If the radius r of a circle is allowed to increase to infinity, its curvature $\dfrac{1}{r}$ decreases to zero. In Euclidean geometry, a circle with infinite radius is a straight line. Consequently a straight line is a curve with constant curvature equal to zero. This fact can also be derived directly from the definition of curvature.

The Circle of Curvature

If a curve is not a circle or a straight line, its curvature varies from point to point. To visualize these changes of curvature from point to point it is helpful to imagine that there is attached to a curve at each of its points a circle that has the same curvature as the curve at that point. This is done in the following manner. To attach such a circle to a curve at a point P on the curve, first take two additional points of the curve Q and R chosen so that P, Q and R do not lie on the same straight line. Draw the circle determined by P, Q and R. Then let Q and R move along the curve toward P. As the points Q and R move toward P, the circle through P, Q and R moves with them and approaches a limiting position. It can be shown that this limiting circle is the circle that fits the curve most closely at P in the sense that it has the same curvature that the curve has at P. It is called the *circle of curvature* of the curve at P. Its radius is called the *radius of curvature* of the

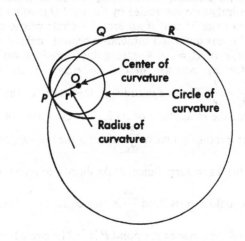

curve at P, and its center is called the *center of curvature* of the curve at P. If k is the curvature of the curve at P, and r is its radius of curvature, then, by the result of the preceding paragraph, $k = \dfrac{1}{r}$. It is not difficult to see that a curve and its circle of curvature at a point P have a common tangent at P. Since the radius of a circle drawn to the point of contact of a tangent is perpendicular to the tangent, it follows that the center of curvature of a curve at a point P is on the line that is perpendicular at P to the tangent to the curve at P. This line is called the *normal* to the curve at P.

The Curvature of a Surface

The curvature of a surface is a measure of its deviation from a plane. We introduce such a measure with the help of the concept of the curvature of a plane curve.

Let P be any point on a smooth surface, and draw the plane that is tangent to the surface at P. If any curve is drawn on the surface through P, it can be shown that the line tangent to the curve at P lies in this tangent plane. Draw the line through P that is perpendicular to the tangent plane. This line, known as the *normal* of the surface, is perpendicular to every line through P that is in the tangent plane. Consequently it is the normal of every curve that is on the surface and passes through P. The surface divides the normal into two halves. Let us arbitrarily designate one of these halves as the positive side of the normal, and the other half as the negative side.

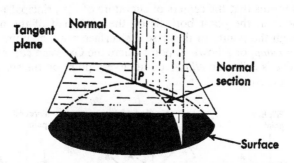

We now restrict our attention to a particular family of curves that lie on the surface and pass through P. Draw any plane that contains the normal to the surface. This plane intersects the surface in a curve called a *normal section* of the surface. We shall examine all possible normal sections of the surface. Each normal section has a definite curvature. If the curvature is not zero, the center of curvature of the normal section lies on the normal. We shall attach a sign, $+$ or $-$, to the curvature of the normal section according as its center of curvature lies on the positive or negative side of the normal. With this convention, the curvature of a normal section may be either positive, negative, or zero. It can be shown that among the curvatures of the normal sections through P there is a minimum curvature k_1 and a maximum curvature k_2, and that the normal sections whose curvatures have these extreme values meet at right angles. (See the diagram on page 274.) The two normal sections whose curvatures are k_1 and k_2 respectively are called the *principal* normal sections. It can be shown that the curvature of any other normal section depends only on k_1 and k_2 and on the angle it makes with one of the principal normal sections. Consequently the way in which the surface deviates from the tangent plane at P depends only on the values of k_1 and k_2. For this reason Gauss proposed the use of the product of these two numbers as a measure of the curvature of the surface. The product $k_1 k_2$ is known as the *total curvature* or the *Gaussian curvature* of the surface at P.

Bumps, Saddles and Ridges

The Gaussian curvature leads to a threefold classification of the points of a surface according as the Gaussian curvature at the point is positive, negative, or zero.

If the Gaussian curvature k_1k_2 at a point is positive, the principal curvatures k_1 and k_2 are either both positive or both negative. This means that the centers of curvature of the principal normal sections at the point both lie on the same half of the normal through the point. In this case the surface near the point looks like a *bump*, or a bowl. A point where the Gaussian curvature is positive is called an *elliptic* point. Every point on the surface of an egg is elliptic.

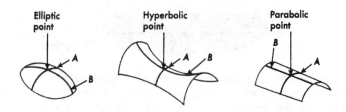

A — Normal section with maximum curvature
B — Normal section with minimum curvature

If the Gaussian curvature k_1k_2 at a point is negative, the principal curvatures k_1 and k_2 have opposite signs. This means that the centers of curvature of the principal normal sections at the point lie on opposite halves of the normal through the point. In this case the surface near the point looks *saddle shaped*. A point where the Gaussian curvature is negative is called a *hyperbolic* point. A

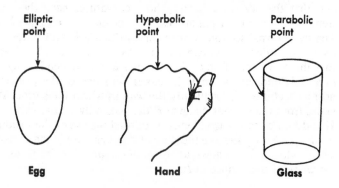

Egg   Hand   Glass

point on the surface of your hand between adjacent knuckles is a hyperbolic point.

If the Gaussian curvature k_1k_2 at a point is zero, then one of the principal curvatures, k_1 or k_2, is zero. In this case the surface near the point looks like a *ridge*. A point where the Gaussian curvature is zero is called a *parabolic* point. Every point on the curved surface of an ordinary drinking glass is a parabolic point.

At any point in a plane, every normal section of the plane is a straight line, with curvature zero. So, at each point of a plane, $k_1 = k_2 = 0$, and the Gaussian curvature is zero.

Intrinsic Properties of a Surface

Two points on a surface may be joined by many different paths drawn on the surface. The path that has the shortest possible length is called a *geodesic*. We have already noted on page 86 that on a spherical surface the geodesic paths are arcs of a great circle. If we imagine two-dimensional creatures living on a surface, they would naturally use as the distance between two points on the surface the length of a geodesic path between them, just as we who live on Earth use as the distance between two points on Earth the length of a great-circle arc between them. If two points are close enough on a surface, there is only one geodesic path between them, and so the length of the geodesic path is unique. We shall call this length the *geodesic distance* between the two points.

Any two points on a surface can also be joined by a straight line segment. In most cases this line segment will not lie on the surface. The length of the line segment is what we ordinarily call the "distance" between the two points. To distinguish it clearly from the geodesic distance, we shall call it the *linear distance* between the two points.

Let us now consider a surface from the point of view of two-dimensional creatures living on it. If the perception of these creatures is limited to two dimensions, they would be unaware that the surface is embedded in a three-dimensional space. If the surface on which they live were bent into a different three-dimensional shape, but the bending were done in such a way that all geodesic distances in the surface remained unchanged, the inhabitants of the surface would not be aware of any change. From their point of view, the surface before and after bending in the third dimension is the same, and has the same geometric properties. These observations lead us to the following definition: Properties of a surface that remain unchanged when the surface is bent without changing the geodesic distances between any two of its points are called *intrinsic* properties of the surface. Such properties reflect the internal relations within the surface and are independent of the way in which the surface is embedded in the surrounding three-dimensional space. One of the interesting and important facts about the Gaussian curvature of a surface is that it is an intrinsic property of the surface. That is, if the surface is bent in such a way that geodesic distances in it are unaltered, then the Gaussian curvature at each point also remains unaltered.

For example, a sheet of paper lying flat on your desk is part of a plane. The Gaussian curvature at each point of the sheet is zero. Now suppose you bend the sheet into a circular cylindrical surface with radius r. This bending does not alter geodesic distances on the sheet. Now draw the principal normal sections to the surface through any point that is not on the edge. One of these sections will be a circle with curvature $\frac{1}{r}$, and the other one

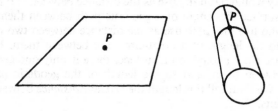

will be a straight line with curvature zero, as shown in the diagram. Consequently the Gaussian curvature is zero. That is, it is the same as it was before the sheet was bent.

Free Movement

One of the characteristics of physical space is that a rigid body can be freely turned in it or moved from place to place in it without undergoing any distortion. If a mathematical space is to be of any use as a model of physical space, it must have this property too. For this reason it is important to know which surfaces have the property of allowing free movement within the surface. Riemann pointed out that free movement is possible in a surface if and only if the Gaussian curvature of the surface has the same value at every point of the surface. A surface with the latter property is called a surface of *constant curvature*. Since the Gaussian curvature of a surface at a point may be either zero or positive or negative, there are three possible types of surfaces of constant curvature: The first type is a surface of *constant zero curvature*. A plane, and surfaces that can be obtained by bending a plane, are examples of this first type. The second type is a surface of *constant positive curvature*. A sphere is an example of this second type. The third type is a surface of *constant negative curvature*. The pseudosphere, shown in the drawing on page 286, is an example of this third type.

Three Intrinsic Geometries

A surface of constant curvature has its own intrinsic geometry in which geodesics play the role of straight lines. We already know that the intrinsic geometry of a surface of constant zero curvature is the same as the geometry of the plane, and this is the geometry we have called *Euclidean*. We know, too, that the intrinsic geometry of a surface of constant positive curvature, such as a sphere, is the geometry we have called *elliptic*. It will not be surprising then to learn that the intrinsic geometry of a surface of constant negative curvature, such as a pseudosphere, is the geometry we have called *hyperbolic*. This is indeed the case for the local properties of a pseudosphere, that is, properties observable in any small region, as was proved in 1868 by Eugenio Beltrami (1835–1900). Consequently by using the concept of the intrinsic geometry of a surface, we may describe in a new and meaningful way the three geometries that were discovered in the theory of parallels: Euclidean plane geometry is the intrinsic geometry of surfaces of constant zero curvature; elliptic plane geometry is the intrinsic geometry of surfaces of constant positive curvature; hyperbolic plane geometry is the intrinsic geometry of surfaces of constant negative curvature.

Riemann's Generalization of Geometry

The concept of curvature, defined by Gauss for two-dimensional spaces (surfaces), was extended to spaces of three or more dimensions by Riemann. This extension was made on the basis of Riemann's bold and far-reaching generalization of the concept of space itself which he presented in his famous 1854 lecture "On the Hypotheses Underlying Geometry."

We have already seen one generalization of the concept of space in the n-dimensional Euclidean space described on page 131. The essential features of this generalization of the notion of space were these: 1) A space of n dimensions is the set of all ordered *n*-tuples of real numbers $(x_1, x_2, \ldots, x_n)$; or we may think of the space as a set of objects called "points" to which these *n*-tuples, called the "coordinates" of the points, are attached. In either case the set of *n*-tuples is a system of coordinates for the entire space. 2) The structure of the space is determined by the definition given for the distance between two points $(x_1, x_2, \ldots, x_n)$ and $(x_1', x_2', \ldots, x_n')$:

$$d = \sqrt{(x_1' - x_1)^2 + (x_2' - x_2)^2 + \cdots + (x_n' - x_n)^2}.$$

Riemann extended this generalization of the concept of space to spaces that are not Euclidean by generalizing each of these two features. He generalized the first feature by assuming that the space is a smooth *manifold*. He generalized the second feature by using a more general formula to define the distance between two points. We explain the substance of these generalizations in the next two paragraphs.

The Concept of a Manifold

We shall introduce the concept of a smooth manifold by first producing and examining two examples.

Any smooth curve which consists of one *connected* set of points not necessarily in a plane, is an example of a one-dimensional smooth manifold. (The term *connected* means roughly consisting of one piece. For example, an ellipse is connected, but a hyperbola is not connected. However, one branch of a hyperbola is connected.) An ellipse, a parabola, and a single branch of a hyperbola are all examples of smooth manifolds. At each point on the curve there is a tangent line. Each tangent line, being a Euclidean space of one dimension, can be equipped with a coordinate system that assigns a real number to every point on the line. We can take advantage of this fact to introduce part of the coordinate system on each tangent line into a region of the curve that contains the point of tangency. We proceed in this way: Let *P* be any point

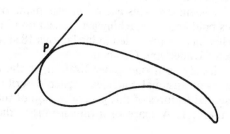

on the curve. Draw the straight line tangent to the curve at *P* and introduce a coordinate system on the line. Then take on the tangent line a small segment containing the point *P*. Every point on this segment has a coordinate attached to it. Now deform this segment of the tangent line by bending it and perhaps even stretching it in some parts and contracting it in others as if it were a rubber band, until the deformed segment has been made to coincide with a segment of the curve with *P* on it. Make the deformation one-to-one so that no two points of the tangent line

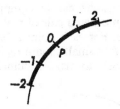

Coordinate system on a segment of the tangent at P....

....transferred to a region of the curve

land on the same point of the curve. Also, make the deformation continuous, so that points that are near each other on the tangent line are moved to points that are near each other on the curve. When each point of the segment is moved to the curve in this way, it carries its coordinate with it. In this way we have equipped with a coordinate system a segment of the curve that contains the point *P*. Such a segment of the curve that is equipped in this way with a coordinate system is called a *coordinate neighborhood* of the curve. Since there is a tangent line at every point of the curve, we can follow the same procedure at every point of the curve, and equip with a coordinate system a segment of the curve that contains that point. Then for every point on the curve there will be a coordinate neighborhood that contains it. We use this last property of the curve for defining a one-dimensional smooth manifold. A *one-dimensional smooth manifold* is a set of points such that each point is contained in a smooth coordinate neighborhood which is obtained by continuous deformation of a segment of a Euclidean straight line. We may describe a one-dimensional smooth manifold somewhat crudely but suggestively as the result of fusing smoothly end to end many continuously deformed segments of a straight line.

In general, two points of a one-dimensional smooth manifold need not be in the same coordinate neighborhood. However, if two points P and Q of the manifold are close enough to each other, there will be a coordinate neighborhood that contains them both. In that case both points have coordinates in the same coordinate system. If these coordinates are x and x' respectively, and if the difference between them is small, we shall use the symbol dx to stand for the difference $x' - x$.

Any smooth connected surface is an example of a two-dimensional smooth manifold. At each point on the surface there is a tangent plane. Each tangent plane, being a Euclidean space of two dimensions, can be equipped with a coordinate system that assigns an ordered pair of real numbers (x_1, x_2) to every point on the plane. For each point P on the surface take, on the tangent plane at P, a small square region that contains P, and deform it continuously to make it coincide with a small region around P

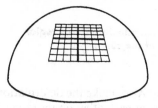   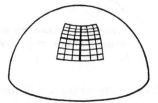

Coordinate system in a   ....transferred to the sur-
square region of the tan-   face
gent plane at P....

on the surface. In this way the coordinate system in the square region on the tangent plane is transferred to a region around P on the surface. On the surface, a region that is equipped with a coordinate system in this way is called a *coordinate neighborhood* of the surface. Then for every point on the surface there will be a coordinate neighborhood that contains it. We use this last property of the surface for defining a two-dimensional smooth manifold. A two-dimensional smooth manifold is a set of points such that each point is contained in a smooth coordinate neighborhood which is obtained by continuous deformation of a square region of a Euclidean plane. We may describe a two-dimensional smooth manifold somewhat crudely but suggestively as the result of fusing smoothly side by side many continuously deformed square pieces of a Euclidean plane.

If two points P and Q of a two-dimensional smooth manifold are close enough to each other, there will be a coordinate neighborhood that contains them both. In that case both points have coordinates in the same coordinate system. If these coordinates are (x_1, x_2) and (x_1', x_2') respectively, and if the differences of corresponding coordinates are small, we shall use the symbols dx_1 and dx_2 to stand for the differences $x_1' - x_1$ and $x_2' - x_2$ respectively.

Given any point P of a two-dimensional smooth manifold, and a point Q that is near it, there is a definite direction from P to Q. This direction can be represented by the ordered pair (dx_1, dx_2), which is a two-component vector and may be visualized as an arrow that is tangent to the manifold at P. All such arrows lie in the plane that is tangent to the manifold at P. For this reason the set of all the vectors (dx_1, dx_2) for the manifold at P is called the *tangent vector space* at P. In this case it is a vector space of two dimensions.

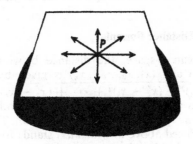

Tangent vector space at P

Taking our cue from these two examples, we can now define what is meant by an *n*-dimensional smooth manifold, where *n* is any positive integer. An *n-dimensional smooth manifold* is a set of points such that each point is contained in an *n*-dimensional smooth coordinate neighborhood, where an *n*-dimensional smooth coordinate neighborhood is the result of continuous deformation of a 2*n*-cell (the *n*-dimensional analogue of a square) of a Euclidean *n*-space. (See page 132.) We may think of an *n*-dimensional smooth manifold as the result of fusing smoothly side by side many continuously deformed 2*n*-cells cut out of a Euclidean *n*-space.

If two points P and Q of an *n*-dimensional smooth manifold are close enough to each other, there will be a coordinate neighborhood that contains them both. In that case both points have coordinates in the same coordinate system. If these coordinates are $(x_1, x_2, \ldots, x_n)$ and $(x_1', x_2', \ldots, x_n')$ respectively, and if the differences of corresponding coordinates are small, we shall use the symbols $dx_1, dx_2, \ldots, dx_n$ for the differences $x_1' - x_1$, $x_2' - x_2, \ldots, x_n' - x_n$ respectively.

Given any point P of an n-dimensional smooth manifold, and a point Q that is near it, there is a definite direction from P to Q. This direction can be represented by the ordered n-tuple $(dx_1, dx_2, \ldots, dx_n)$, which is an n-component vector. By analogy with the case of a two-dimensional manifold, we think of this vector as an arrow that is tangent to the manifold at P. The set of all such arrows is called the *tangent vector space* at P. We may picture it as the set of all arrows drawn from a fixed point in an n-dimensional Euclidean space. There are many planes through this fixed point in the n-dimensional Euclidean space, each with a different orientation. The set of arrows in any one of these planes is called a two-dimensional subspace of the tangent vector space.

A Generalized Distance Formula

In a Euclidean n-space, the distance between two points $(x_1, x_2, \ldots, x_n)$ and $(x_1', x_2', \ldots, x_n')$ is given by the formula $d = \sqrt{(x_1' - x_1)^2 + (x_2' - x_2)^2 + \cdots + (x_n' - x_n)^2}$. Let us assume that the two points are very close to each other, and let us use the symbol ds for the distance between them. Then, using the notation introduced above in which dx_1 stands for $x_1' - x_1$, etc., and squaring both sides, the formula takes this form:

$$(2) \qquad (ds)^2 = (dx_1)^2 + (dx_2)^2 + \cdots + (dx_n)^2,$$

which is the extension to n dimensions of equation (1) on page 268. In this equation, each term on the right-hand side is a product of two of the numbers $dx_1, dx_2, \ldots, dx_n$, but the two numbers multiplied in each term are always the same. Thus, $(dx_1)^2$ is the product of dx_1 and dx_1, $(dx_2)^2$ is the product of dx_2 and dx_2, etc. There are other possible products of these numbers, two at a time, in which the two numbers are different, such as $(dx_1)(dx_2)$, etc. We may think of terms like these as being present, too, on the right-hand side of the formula, except that their coefficients are all equal to 0, while the coefficients of $(dx_1)^2$, $(dx_2)^2$, etc. are all equal to 1. Then equation (2) may be written in this form:

$$(3) \quad (ds)^2 = 1(dx_1)^2 + \cdots + 1(dx_n)^2 + 0(dx_1)(dx_2) + \cdots ,$$

where the last set of three dots indicates other terms whose coefficient is zero.

Now suppose that the two points whose distance is given by equation (3) are in a $2n$-cell that has been continuously deformed to become a part of an n-dimensional smooth manifold. Since the deformation may involve stretching or contraction, the distance between the corresponding points in the manifold will not, in general, be given by equation (3). The distance itself will be either stretched or contracted. Moreover, since the stretching or contraction throughout the $2n$-cell need not be uniform, the amount of stretching near any point will depend on its coordinates. To take all these considerations into account, Riemann introduced as the distance formula for two neighboring points in the manifold the more general formula:

$$(4) \quad (ds)^2 = a_{11}(dx_1)^2 + \cdots + a_{nn}(dx_n)^2 + a_{12}(dx_1)(dx_2) + \cdots,$$

where the coefficient of each term is a function of the coordinates, and so may vary from point to point, instead of having only the fixed values 1 and 0 that occur in equation (3). (The subscripts of each coefficient are identification tags that show which term it belongs to.)

In the case of a three-dimensional manifold, equation (4) has this form:

$$(ds)^2 = a_{11}(dx_1)^2 + a_{22}(dx_2)^2 + a_{33}(dx_3)^2 + a_{12}(dx_1)(dx_2) \\ + a_{13}(dx_1)(dx_3) + a_{23}(dx_2)(dx_3).$$

A smooth manifold equipped with a distance formula in the form of equation (4) is called a *Riemannian space*.

Equation (4) makes it possible to determine the length of an arc of a rectifiable curve in the manifold in the same way that equation (1) made it possible to determine the length of an arc of a rectifiable curve in a plane. It suffices to divide the arc into small pieces ds and add them up, and then allow the pieces to shrink in size to zero while the number of pieces increases to infinity. The addition is done by means of the calculus by finding the integral, $\int ds$.

Geodesic Surface

Because a Riemannian space is equipped with a formula for measuring distance, it is possible to determine the length of any path between two points. The paths of shortest length, the geodesics, are of particular importance because they play the same role in a Riemannian space that straight lines play in a Euclidean space. We show next how the geodesics are used to generate certain special surfaces called geodesic surfaces. Then we shall show how the concept of the curvature of a Riemannian space is derived from the Gaussian curvature of its geodesic surfaces.

There is associated with each point P of a Riemannian space a tangent vector space. Each vector in this vector space specifies a definite direction in the Riemannian space. For any given vector in the tangent vector space at P there is one and only one geodesic in the Riemannian space that passes through P and is tangent to that vector. If the Riemannian space has three or more dimensions, then the tangent vector space at P contains many two-dimensional subspaces, each with a different orientation, just as there are many different planes through a point in three-dimensional Euclidean space. The set of all geodesics through P whose tangent vectors are in a two-dimensional subspace of the tangent vector space at P lie in a surface called a *geodesic surface*. There are many such geodesic surfaces, one for each two-dimensional subspace of the tangent vector space. Each of these geodesic surfaces has a different orientation in the Riemannian space. In the case of a three-dimensional Riemannian space there is a geodesic surface perpendicular to every direction in the space.

Curvature Tensor

The curvature at a point of a two-dimensional Riemannian space (a surface) is expressed by a single number, the Gaussian curvature of the surface at that point. For a Riemannian space of three or more dimensions, it is not, in general, possible to express the curvature of the space at a point by a single number. It is necessary to use as the measure of curvature at the space a set of numbers obtained in the following way: At each point of a Riemannian space of three or more dimensions there are many geodesic surfaces. Each of these geodesic surfaces has a Gaussian curvature. The Gaussian curvature of a geodesic surface through a point may be thought of as the curvature of the space for the particular orientation that the geodesic surface has. However, geodesic surfaces with different orientations may, in general, have different curvatures. Therefore, to characterize the curvature of the space for all possible orientations at a point, we must use the set of curvatures of all possible geodesic surfaces through the point. This set of curvatures of all possible geodesic surfaces through a point, one for each orientation, is called the *curvature tensor* through the point. The curvature tensor may vary from point to point in the space.

Spaces with Constant Curvature

The Riemannian spaces that we are particularly interested in are those of three dimensions that might serve as models for physical space. Fortunately, for these spaces it is possible to express the curvature of the space by a single number instead of by a curvature tensor that varies from point to point. We now show how this single number is obtained.

A Riemannian space of three dimensions is suitable as a model for physical space if it allows for free movement of rigid bodies. In particular, it should be possible to turn a body to give it any possible orientation. Free turning at a point is possible if every geodesic surface through the point has the same Gaussian curvature. When every geodesic surface through a point has the same Gaussian curvature, we say that the space is *isotropic* at that point. A Riemannian space will permit rotation at any point if it is isotropic at *every* point. It can be proved that if a Riemannian space is isotropic at every point, then *every* geodesic surface at *every* point has the same Gaussian curvature. In this case we say that the Riemannian space has *constant curvature*. In a space with constant curvature, the curvature may be expressed by a single number, namely the Gaussian curvature k of any geodesic surface through any point.

Three Space Geometries

If we assume that physical space is a Riemannian space of three dimensions with constant curvature, we have three possible models to choose from according as the Gaussian curvature of any of its geodesic surfaces is zero, or a positive number, or a negative number:

1) *A space with zero curvature*. In this case every geodesic surface has zero curvature. Then, on a geodesic surface, the sum of the angles of a triangle is two right angles. The geometry of a geodesic surface is Euclidean, and the space is infinite.

2) *A space with constant positive curvature*. In this case every geodesic surface has the same positive curvature. On such a geodesic surface, the sum of the angles of a triangle is more than

two right angles, as on a sphere. The geometry of a geodesic surface is elliptic, and the space is finite.

3) *A space with constant negative curvature.* In this case every geodesic surface has the same negative curvature. The sum of the angles of a triangle on a geodesic surface is less than two right angles, as on a pseudosphere. The geometry of a geodesic surface is hyperbolic, and the space is infinite.

Which of these three models best fits the known properties of physical space? We shall deal with this question in Chapter 11.

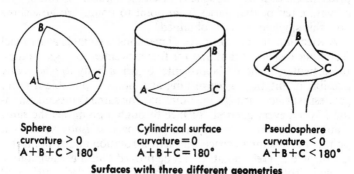

Sphere	Cylindrical surface	Pseudosphere
curvature > 0	curvature = 0	curvature < 0
A+B+C > 180°	A+B+C = 180°	A+B+C < 180°

Surfaces with three different geometries

10

Projective Geometry

The Reunification of Geometry

The study of the theory of parallels led to the pluralization of geometry by showing that there were three equally valid geometries instead of only one. Another independent line of geometric thought that developed side by side with the theory of parallels ultimately led to the reunification of geometry. This separate line of thought had its origin in some geometric problems in art and military engineering. Mathematicians, extending the investigations that were begun by artists and engineers, constructed out of their findings a new and simpler geometry known as *projective geometry*. A major achievement of the projective geometers was the discovery that Euclidean geometry, elliptic geometry and hyperbolic geometry are all branches of projective geometry.

The Problem in Art

Renaissance artists, who were trying to paint realistically, had to grapple with this problem of technique: How do you make a two-dimensional drawing of a three-dimensional object so that the drawing really looks like the object? To solve this problem, they first simplified it by assuming that the viewer of the object and the drawing uses only one eye. If the eye is represented by a point O, the rays of light that travel from the object to the eye may be represented by straight lines converging from the object to the point O. This set of straight lines joining the point O to every point of the object is called a *projection* of the object from the point O. If a glass window is interposed between the point O and the object, the plane of the window intersects each line of the projection at a point. The set of all points in which the plane intersects the projection is called a *section* of the projection. The eye at O sees the object as if it were a picture in the window, occupying the position of this section. Consequently the section

287

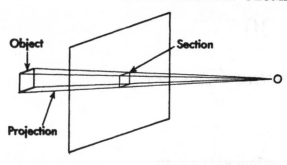

in the window of the projection of the object is a two-dimensional representation of the three-dimensional object. Thus the artist's problem is reduced to the purely geometric problem of determining the points of the section of the projection of the object from a point O for given positions of the object, the point O, and the plane of the window. Two different methods were used to solve the problem. One method was to locate the individual points of the section with the help of some mechanical device. The other method was to locate the points of the section by geometric reasoning from the given data about the form of the object and the relative positions of the object, the point O, and the plane of the section. The second method led to the development of the *theory of perspective*.

The famous Nuremberg painter Albrecht Dürer (1471–1528) used both methods. The frontispiece engraving shows a device he invented for locating one at a time the points of a perspective drawing of an object. A line in the projection of the object was produced physically as a thread stretched from the object to a nail representing the position of the eye. The thread passed through a frame that was standing between the object and the nail. The point where the thread crossed the plane of the frame was fixed by means of two movable threads that were mounted on the frame at right angles to each other. These threads were moved until they intersected at this point. Then the thread joining the object to the nail was removed, and a piece of drawing paper mounted on a board hinged to the frame was swung into the plane of the frame. Finally, the point fixed by the intersecting threads was transferred to the drawing paper.

Dürer's work on the theory of perspective was reported in his book called *A Course in the Art of Measurement*, devoted to geometry and its applications to architecture, engineering, decoration, typography, and art. In his preface to the book, Dürer said ". . . since geometry is the right foundation of all painting, I have decided to teach its rudiments and principles to all youngsters eager for art . . . ."

After Dürer, the next significant advances in the theory of perspective were made by Gerard Desargues (1593–1661).

The Problem in Engineering

A problem similar to that of the artist was also faced by the engineer: How do you make two-dimensional drawings of a three-dimensional object so that all the spatial properties of the object, such as size, form, and orientation can be reconstructed from the drawings? Gaspard Monge (1746–1818), mathematician and physicist, who encountered this problem as a designer of fortifications for the French Army, solved it by inventing *descriptive geometry*. Descriptive geometry is the theory and technique of representing a three-dimensional object by means of two views of it, a front view and a side view obtained by orthogonal projection of the points of the object onto two vertical planes that are perpendicular to each other. Each view is a plane section of a set of parallel projecting lines. This set of lines is the limit approached by a projection from a point when the point is allowed to move off to infinity. Consequently it may be thought of as a special case of projection from a point in which the lines "converge" to an infinitely distant point. For this reason the theory of descriptive geometry is essentially a subdivision of the theory of perspective.

The Underlying Geometry Problem

Two perspective drawings of the same object, drawn from different points of view, may have totally different forms. For example, a drawing of a horizontal square as seen from a point directly above its center is a square. But a drawing of the square as seen from a point that is off to one side of it is not even a parallelogram, since from that point of view one or both pairs of opposite sides seem to converge. Nevertheless, two drawings of an object, though they are different in form, look like that object to us. This means that the drawings and the object must have some geometric properties in common. This fact suggests the question, "What are these common properties?" Stated more specifically, the question is, "What properties of a geometric figure remain unchanged when it is transformed into another one by projection and section?" Investigations designed to answer this question led to the development of projective geometry as a separate branch of mathematics.

The first major contributions to projective geometry were made by Jean Victor Poncelet (1788–1867). The subject was developed further from different points of view by Jacob Steiner (1796–1863), Karl von Staudt (1798–1867), Julius Plücker (1801–1868), and Möbius, Cayley and Klein.

Projectivities

In Euclidean plane geometry a central role is played by those transformations of the plane into itself that we call isometries. (See pages 125 to 129 and Chapter 6.) In projective plane geometry the central role is played by transformations of another kind, called projectivities, which are defined by the repeated use of the operations of projection and section that occur in the theory of perspective. Before defining a projectivity of a plane into itself it is necessary first to define a projectivity of a line into another line.

Let l and l' be two lines in the plane and let O be a point in the plane that is not on either line. If A is any point on l, the line OA is the projection of A from O. If OA intersects l' at A', OA is also the projection of A' from O. That is, A and A' have a common projection from O. If we associate with each point on l the point on l' that shares with it a common projection from O, we establish a correspondence between the points on l and the points on l'. However, the correspondence is not perfect. If B is the point on l such that OB is parallel to l', there is no point on l' that shares with B a common projection from O. Similarly, if

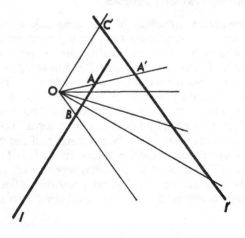

C' is the point on l' such that OC' is parallel to l, there is no point on l that shares with C' a common projection from O. To eliminate this troublesome imperfection in the correspondence we take our cue from Kepler (see page 84) and extend each line in the plane by adding to it an ideal point at infinity at which it is intersected by all the lines that are parallel to it. This time, however, we assume that it is done *actually*, not only metaphorically. We shall assume, too, that all the points at infinity lie on one straight line, *the line at infinity*. Then the ideal point on l' shares with B a common projection from O, and the ideal point on l shares with C' a common projection from O. Thus, by adding new points to the plane we have made the correspondence a one-to-one correspondence. The one-to-one correspondence that we obtain in this way between the points of the extended lines l and l' is called a *perspectivity*. The point O from which the lines are drawn that join corresponding points on l and l' is called the *center* of perspectivity.

A *projectivity* or a projective transformation between two lines is defined as the product of a finite sequence of perspectivities as follows: Let $l_1, l_2, \ldots, l_n$ be a finite sequence of lines such that every point on l_1 corresponds via a perspectivity to some point on l_2, every point on l_2 corresponds via a perspectivity to some point on l_3, etc. Then if P_1 on l_1 corresponds to P_2 on l_2, which in turn corresponds to P_3 on l_3, etc., and finally P_{n-1} reached in this way on l_{n-1} corresponds to P_n on l_n, then the one-to-one correspondence which associates in this way each point P_1 on l_1 with a point P_n on l_n is called a *projectivity* between l_1 and l_n. For example, in the diagram below, a perspectivity with center O connects A and B on l with A' and B' on l' respectively. Also a perspectivity with P as center connects A' and B' on l' with A'' and B'' on l'' respectively. The product of these two perspectivities is a projectivity which connects A and B on l with A'' and B'' on l'' respectively.

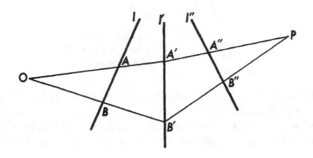

In the definition of a projectivity it is possible for the first line l_1 and the last line l_n in the sequence $l_1, \ldots, l_n$ to be the same line. In this case the projectivity that is defined is a projectivity of the line l_1 into itself.

A projectivity of a plane into itself is defined as a one-to-one onto transformation of the plane into itself such that the image of every straight line is a straight line and the correspondence between the points of any straight line and the points of its image is a projectivity as defined above.

A one-to-one onto transformation of a plane into itself such that the image of every straight line is a straight line is called a *collineation*. It is clear from the definition of a projectivity of a plane that it is a collineation. It can be shown that the converse is also true if we consider only continuous transformations (see page 56). That is, every continuous collineation of a plane is a projectivity.

The central problem of projective plane geometry is to determine those properties of figures that remain unchanged when the plane is subjected to a projective transformation.

The concepts of perspectivity and projectivity are defined in a three-dimensional space in a similar manner: If P and Q are two distinct planes, and O is a point that is not on either of them, a perspectivity of the two planes with center O is the correspondence that associates with each point of P the point of Q that

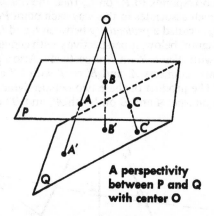

**A perspectivity
between P and Q
with center O**

shares with it a common projection from O. In order to make every perspectivity between two planes a one-to-one correspondence between their points, we add to each plane a line at infinity as we did above. In addition, we assume that parallel planes share the same line at infinity, and that all the lines at infinity lie on one plane, *the plane at infinity*. A transformation between two planes that can be obtained as the product of a finite sequence of perspectivities is called a projectivity between the two planes. (See exercise 2 on page 352.) A one-to-one onto transformation of three-dimensional space into itself is called a projectivity if the image of every straight line is a straight line, and the correspondence between the points of any straight line and the points of its image is a projectivity.

The Real Projective Plane

In order to be sure that a perspectivity between two lines in a plane would be a one-to-one correspondence between their points, we have extended the plane by adding to it the points at infinity. One of the axioms of the Euclidean plane is the axiom of completeness (axiom VI on page 77) which says that no additional points can be added to the plane without violating one of the axioms I and V. Since we *have* added points, one or more of these axioms are violated. Consequently the extended plane used in projective geometry does not satisfy the axioms of Euclidean geometry and therefore it is not a Euclidean plane. It is a different kind of plane which we call the *real projective plane*. By seeing which of the Hilbert axioms are violated and which are not when we add the points at infinity to the Euclidean plane, we can identify a set of axioms for the real projective plane.

Let us examine the Hilbert axioms (see pages 74–77) one at a time, to see which of them are valid in the real projective plane. Axioms I,1 and I,2 may be combined into one statement: *There is one and only one straight line that contains two distinct points.* To see if this statement is valid for the real projective plane, let A and B be any two points in the real projective plane. We must consider three cases that may arise: both points may be ordinary points of the Euclidean plane that we extended; one point may be an ordinary point while the other point is an ideal point; or, both points may be ideal points. In the first case, there is indeed one and only one line between the two points, namely the line that joins them in the Euclidean plane. In the second case, suppose A is the ordinary point, and B is the ideal point on line l in the Euclidean plane. If A is on l, then l is the unique line containing A and B. If A is not on l, then the line through A that is parallel to l is the unique line that contains A and B. In the third case, where both points are ideal points, the line at infinity is the unique line that contains them both. Consequently the statement that combines axioms I,1 and I,2 is valid for the real projective plane.

We may skip axioms I,3–6 because they refer to a Euclidean space of three dimensions.

Axiom I,7 asserts that on every straight line there are at least two points. Since we have just added to each line another point, namely the point at infinity, this axiom carries over to the real projective plane in the following modified form: *On every straight line there are at least three points.*

Axiom I,7 also asserts that in a Euclidean plane there are at least three points that are not on the same straight line. Consequently we are sure that in the real projective plane *not all points lie on the same line.* Moreover, the existence of at least two points assures us *that there exists at least one line,* namely the line that joins them.

We may skip the rest of axiom I,7 since it refers to a space of three dimensions.

Axiom II deals with the relations that are characteristic of linear order. When we add an ideal point to a Euclidean line this linear order is destroyed. What takes its place can be seen by examining the projection of the points on a line *l* from a point *O*. Let *A* be an ordinary point on *l*, and let Ω be the ideal point on *l*. Allow the line *OA* to rotate counterclockwise around *O*. As the line rotates, its intersection with *l* moves to the right on the half-line of *l* to the right of *A*, until the rotating line reaches the

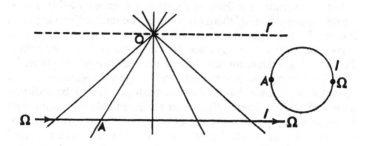

position of *l'* which is parallel to *l*. In this position it intersects *l* at Ω. As it continues rotating counterclockwise, its intersection returns to *A* by moving to the right along the half-line of *l* that is to the left of *A*. The sequence of points on the extended line is like the sequence of points on a closed loop, since the two half-lines whose vertex is *A* have another common point at Ω. Consequently the points on the extended line have circular order, not linear order. If we remove the point Ω, what is left is an ordinary Euclidean line, whose points can be put into one-to-one correspondence with the real number system. If we remove from the extended line any other point instead of Ω, what is left also has the property that its points have linear order and can be put into one-to-one correspondence with the real number system. (See exercise 3 on page 352.) Therefore the property of the extended line that replaces axiom II may be expressed as follows: *There is an order-preserving one-to-one correspondence between the real numbers and all but one point of a line.*

Axiom III is the axiom of parallels. By introducing ideal points at which lines that are parallel in the Euclidean plane meet in the real projective plane, we guarantee that in the latter any two lines meet. Consequently in the real projective plane we have this property: *Two distinct lines meet in one and only one point.*

Axiom IV deals with congruence. Congruent segments are segments with the same lengths. In projective geometry we are concerned only with properties of a figure that are preserved by a projectivity. The diagram on page 291 shows that a segment AB and its image under a projectivity need not have the same length. Consequently the length of a segment is not preserved by all projectivities, and the concept of congruence is not relevant to projective geometry. Therefore axiom IV does not apply to the real projective plane.

Axiom V is the axiom of continuity. Since it involves the concepts of linear order and congruence, it does not apply to the real projective plane. Its essential content does occur, however, in modified form, as part of the property we have already noted, that there is a one-to-one correspondence between the real numbers and all but one point of a line. (This is so because the axiom of Archimedes is a property of the real number system.)

By our survey of the Hilbert axioms, we have identified six properties of the real projective plane, indicated by italics in the text above. These six properties are the *axioms for the real projective plane*. All other properties of the real projective plane will be deduced from these axioms. We recapitulate the axioms here for convenient reference using the word *line* instead of *straight line*.

1. There exists at least one line.
2. On each line there are at least three points.
3. Not all points lie on the same line.
4. Two distinct points lie on one and only one line.
5. Two distinct lines meet in one and only one point.

6. There is an order-preserving one-to-one correspondence between the real numbers and all but one point of a line.

Notice that in the formulation of these properties no distinction is made between ordinary points and ideal points. There is no second-class citizenship in projective geometry. All points of the real projective plane have the same privileges whether they were "native born" ordinary points of the original Euclidean plane or were "immigrant" ideal points introduced from the outside. Consequently, in the development of theorems of projective geometry, no further reference to this distinction need be made.

Notice that, except for axiom 6, the axioms of projective geometry deal only with the existence of points and lines and their relations of connection. We shall rarely use axiom 6. (See pages 319 and 320.) For this reason it is correct to say that projective geometry is the geometry of connection, or as von Staudt called it, the geometry of "position."

Duality

There is an obvious similarity of form in axioms 4 and 5 for the real projective plane. To emphasize this similarity we introduce a slight change in language. To express the fact that a line contains a point or passes through the point we shall say that the line lies *on* the point. In this *on* terminology, axiom 5 takes the following form: Two distinct lines lie on one and only one point. Comparing axiom 5 with axiom 4, we see that each can be obtained from the other by interchanging the words "point" and "line." Two propositions that have this relationship to each other are said to be *dual* to each other, and each is called the dual of the other. Axiom 4 is a valid statement in real projective geometry. Its dual, since it is also an axiom, is also a valid statement. This suggests the following general question: If a proposition is valid in projective geometry, is its dual also valid? Let us try to answer this question first for the six axioms of the real projective plane. We list the six axioms again below in the *on* terminology, and write the dual of each one next to it in the column on the right.

1. There exists at least one line.

1'. There exists at least one point.

2. On each line there are at least three points.

2'. On each point there are at least three lines.

3. Not all points lie on the same line.

3'. Not all lines lie on the same point.

4. Two distinct points lie on one and only one line.

4'. Two distinct lines lie on one and only one point.

5. Two distinct lines lie on one and only one point.

5'. Two distinct points lie on one and only one line.

6. There is an order-preserving one-to-one correspondence between the real numbers and all but one point on a line.

6'. There is an order-preserving one-to-one correspondence between the real numbers and all but one line on a point.

Propositions 4' and 5' are valid because they are identical with axioms 5 and 4 respectively. The reader can easily verify that propositions 1', 2', 3' and 6' are also valid. (See exercise 4 on page 352.) Consequently the duals of all the axioms of the real projective plane are valid. This fact has far-reaching consequences. Any time we prove a theorem about the real projective plane, we present an argument that shows that the theorem is a logical consequence of axioms 1 to 6. If in this argument we interchange the words "point" and "line" we obtain a dual argument that shows that the dual of the theorem is a logical consequence of propositions 1' to 6'. But since propositions 1' to 6' are valid, so is any logical consequence of them. Therefore *the dual of any theorem about the real projective plane is also a theorem about the real projective plane.* This statement which we have just proved is known as the *principle of duality* for the projective plane. Because of this principle, any time we prove a theorem about the projective plane we can immediately get another valid one by interchanging the words "point" and "line."

Projective Three-Space

When a Euclidean space of three dimensions is extended by adding to each line a point at infinity, with the assumption that all the points at infinity lie in one plane, the plane at infinity, the result is a real projective space of three dimensions, or a projective three-space. This space can be characterized by a set of axioms like those we have listed for the real projective plane. In a three-dimensional projective space, as in a projective plane, the relation of parallelism is excluded. Any two lines in the same plane intersect at one and only one point, and any two planes intersect in one and only one line. There is a principle of duality in a projective space of three dimensions analogous to the principle of duality we have established for the projective plane. If two propositions about a projective three-space have the relation that each can be obtained from the other by interchanging the words "point" and "plane," then each of them is called the "space dual" of the other. The space dual of any theorem about the real projective three-space is also a theorem about that space. This principle of duality for a projective three-space was foreshadowed by the duality noted on page 28 in our discussion of the regular solids.

In the pages that follow, in which we explore some properties of the real projective plane, we shall assume that it is part of a real projective space of three dimensions, unless we state explicitly that this assumption is excluded.

Desargues' Theorem

To show the kind of theorem that arises in projective geometry we shall prove one known as the theorem of Desargues. In order to state the theorem briefly we shall need a few definitions first.

Triangle. A triangle is the configuration consisting of three points that are not on the same line and the three lines determined by pairs of these points. The points are called the *vertices*, and the lines are called the *sides* of the triangle. (Note that a side of a triangle in projective geometry is the line through two vertices, and not the segment between them. This is made necessary by the absence in projective geometry of the concept of *segment* which depends on the linear order relations that we have excluded.)

The dual of this definition defines a figure called a *trilateral*: A trilateral is the configuration consisting of three lines that are not on the same point and the three points determined by pairs of these lines. Obviously a trilateral is the same as a triangle.

Perspective triangles. Two triangles are *perspective from a point*, called the *center of perspectivity*, if their vertices can be put into one-to-one correspondence in such a way that the lines on corresponding vertices lie on the center of perspectivity. The dual of this definition gives us another definition: Two trilaterals (triangles) are *perspective from a line*, called the *axis of perspectivity*, if their sides can be put into one-to-one correspondence in such a way that the points on corresponding sides (the points of intersection of corresponding sides) lie on the axis of perspectivity.

In the terminology introduced by these definitions, Desargues' theorem takes the form shown on the next page.

If two triangles are perspective from a point, they are perspective from a line, and conversely.

Proof: Let triangles ABC and $A'B'C'$ be perspective from a point O. (That is, the lines AA', BB' and CC' are concurrent at O.) We shall show that the triangles are perspective from a line. That is, the points of intersection of AB and $A'B'$, of AC and $A'C'$, and

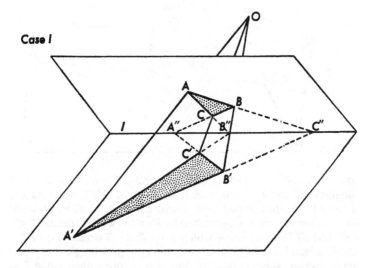

of BC and $B'C'$, are collinear. We consider two cases that may arise:

Case I. Assume that triangles ABC and $A'B'C'$ are not in the same plane. Then the intersection of planes ABC and $A'B'C'$ is some line l. The lines OAA' and OBB' determine a plane. The lines AB and $A'B'$ lie in this plane. Hence, since they are coplanar, they meet at some point C''. C'' is in both planes ABC and $A'B'C'$. Therefore it is on their intersection l. Similarly, the lines AC and $A'C'$ meet at some point B'', and B'' is on l; and the lines BC and $B'C'$ meet at some point A'', and the point A'' is on l. Therefore the triangles ABC and $A'B'C'$ are perspective from the line l.

Case II. Assume that triangles ABC and $A'B'C'$ are in the same plane α. Draw a line through O that is not in α, and let P and Q be two points on this line that are distinct from O. The lines $OA'A$ and OPQ determine a plane. In this plane the lines PA and QA' intersect at some point A''. Similarly the lines $OB'B$ and OPQ determine a plane, and in this plane, PB and QB' intersect at

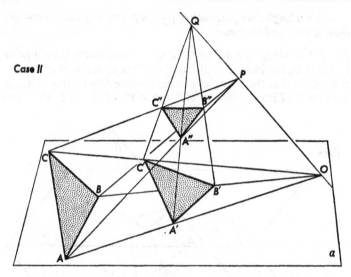

some point B''; and in the plane determined by $OC'C$ and OPQ, the lines PC and QC' intersect at some point C''. The points A'', B'' and C'' cannot be collinear, because if they did all lie on some line m, then the plane determined by P and m would contain PA'', PB'' and PC'', and also A which is on PA'', B which is on PB'', and C which is on PC''. Thus it would contain triangle ABC, and hence would coincide with α. But this is impossible, since P is not in α. Therefore A'', B'' and C'' determine a triangle. Since P is not in α, triangle $A''B''C''$ is not in α. Triangles ABC and $A''B''C''$ are perspective from P. Therefore, by case I, they are also perspective from a line, and this line is the intersection n (not shown in the diagram) of α and the plane of $A''B''C''$. Therefore AB and $A''B''$ intersect at some point C''' on n. That is, AB, $A''B''$ and n intersect at C'''. Triangles $A'B'C'$ and $A''B''C''$ are perspective from Q. Therefore, by case I, they are also perspective from the line n. Consequently $A'B'$ and $A''B''$ also meet at a point on n. This point must be C''', the intersection of $A''B''$ and n. Therefore AB and $A'B'$ meet at a point on n. Similarly, AC and $A'C'$ meet at a point on n, and BC and $B'C'$ meet at a point on n. That is, triangles ABC and $A'B'C'$ are perspective from the line n.

Thus we have proved the proposition that, if two triangles are perspective from a point, they are perspective from a line. The converse of this proposition is its dual. Therefore it is also valid, by the principle of duality.

Using Desargues' Theorem

Desargues' theorem applies to the real projective plane. Since the real projective plane includes all the points of the Euclidean plane from which it was derived, Desargues' theorem can sometimes be used to draw conclusions about figures in the Euclidean plane. As an illustration, we shall give a third proof of a Euclidean theorem that we have already proved twice before by other methods. On page 80 we proved by the methods of Euclidean geometry that the medians of a triangle are concurrent. On page 144 we proved the same theorem by vector methods. Now we

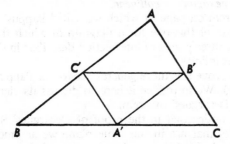

shall prove it once again by projective methods, using Desargues' theorem:

In any triangle ABC, let A', B' and C' be the midpoints of BC, CA, and AB respectively. We want to prove that AA', BB' and CC' are concurrent. BC and $B'C'$ are parallel. Therefore in the extended plane they meet at a point on the line at infinity. Similarly AC and $A'C'$ meet on the line at infinity, and AB and $A'B'$ meet on the line at infinity. Therefore triangles ABC and $A'B'C'$ are perspective from a line, namely the line at infinity. Consequently, by Desargues' theorem, they are also perspective from a point. That is, AA', BB' and CC' are concurrent.

Pappus' Theorem

Another important theorem of projective geometry is this one, named after Pappus:

If A, B and C, and A', B' and C' are triples of distinct points on two distinct lines that lie in the same plane, then the points of intersection of AB' and $A'B$, of AC' and $A'C$, and of BC' and $B'C$ are collinear.

We shall find it convenient to reformulate the statement of this theorem in another way. Think of the six points A, B, C and

A', B', C' as the vertices of a hexagon $AB'CA'BC'$ inscribed on the two lines ABC and $A'B'C'$. With the six vertices listed in the cyclical order $AB'CA'BC'$, where it is understood that C' is followed by A, any two consecutive vertices determine a side of the hexagon, and any two sides separated by only one vertex that they do not share are opposite sides. Thus AB' and $A'B$ are opposite sides. So are AC' and $A'C$, and so, too, are BC' and $B'C$. Using this convention, we can state Pappus' theorem in the following form:

If a hexagon is inscribed in two lines, so that three vertices are on each line, then the intersections of the three pairs of opposite sides of the hexagon are collinear.

The theorem on page 81 which we called Pappus' theorem is a special case of this one for a diagram in which the line that contains the three points of intersection described in the theorem is the line at infinity.

We shall prove this more general version of Pappus' theorem on page 320. We introduce it here to discuss its significance in relation to Desargues' theorem.

You will notice that in the proof of Desargues' theorem for two triangles that are in the same plane we assumed that the plane was embedded in a projective space of three dimensions, and we used points and lines that were not in the plane. An examination of the details of the proof shows that in addition to using some relations of points, lines and planes in projective three-space, we used in the proof only axioms 1 to 5 of the real projective plane. This fact suggests the question, "Can we prove Desargues' theorem on the basis of axioms 1 to 5 alone, without assuming that the plane is embedded in a space of three dimensions?" It has been shown that the answer to this question is "No." However, we can dispense with the assumption that the plane is embedded in a projective three-space if we assume Pappus' theorem in its place. This is the essential meaning of the following theorem, which we shall now prove:

In any plane that satisfies axioms 1 to 5 on page 295, Pappus' theorem implies Desargues' theorem.

Given: Triangles ABC and $A'B'C'$ are in the same plane and are perspective from a point O. That is, AA', BB' and CC' are distinct lines that meet at O. AB and $A'B'$ meet at C'', AC and $A'C'$ meet at B'', and BC and $B'C'$ meet at A''.

Prove (with the help of Pappus' theorem): A'', B'' and C'' are collinear

Procedure: We shall show that A'' is on the line $B''C''$.

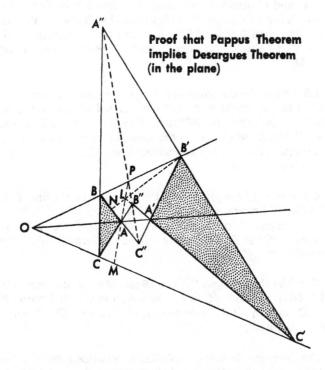

Proof that Pappus Theorem implies Desargues Theorem (in the plane)

Proof: First we establish the existence of certain points of intersection.

(1) $B''C''$ cannot be the same line as OB, because if it were, then O, B, B'' and C'' would be on a line. This line would also include B' (which is on OB) and A (which is on BC'') and A' (which is on $B'C''$). That is, A, A', B and B' would be on the same line, contrary to the hypothesis that AA' and BB' are distinct lines. Therefore $B''C''$ meets OB at some point P.

(2) PA cannot be the same line as $A'C'$, because if it were, then P, A, A', C' and B'' would be on a line. This line would also include C (which is on AB''). Then A, A', C and C' would be on

the same line, contrary to the hypothesis that AA' and CC' are distinct lines. Therefore PA meets $A'C'$ at some point L.

(3) PA cannot be the same line as OC, because if it were, then P, A, O and C would be on a line. This line would also include C' (which is on OC) and A' (which is on OA). Then A, A', C and C' would be on the same line, contrary to the hypothesis that AA' and CC' are distinct lines. Therefore PA meets OC at some point M.

(4) LB' cannot be the same line as AB, because if it were, then L, B', A and B would be on the same line. This line would also include C'' (which is on AB) and A' (which is on $B'C''$). Then A, A', B and B' would be on the same line, contrary to the hypothesis that AA' and BB' are distinct lines. Therefore LB' meets AB in some point N.

(5) Consider the hexagon $ONALA'B'$ inscribed on the lines OAA' and NLB'. NA meets $A'B'$ at C'', and AL meets $B'O$ at P. Therefore, by Pappus' theorem, ON meets LA' at a point on PC''. But LA' meets PC'' at B''. Therefore ON passes through B''. Therefore ON meets AC at B''.

(6) Consider the hexagon $ONMACB$ inscribed on the lines OMC and NAB. ON meets AC at B'', as shown in (5). MA meets BO at P. Denote by Q the intersection of NM and BC. Then, by Pappus' theorem, Q is on PB''.

(7) Consider the hexagon $ONMLC'B'$ inscribed on the lines OMC' and NLB'. ON meets LC' at B'', as shown in (5). ML meets $B'O$ at P. Therefore, by Pappus' theorem, the intersection of NM and $C'B'$ lies on PB''. But by (6), NM meets PB'' at Q. Therefore Q is on $C'B'$. But Q is on BC. Therefore Q is the intersection of BC and $B'C'$. Therefore $Q = A''$. But Q is on $PB''C''$. Therefore A'', B'' and C'' are collinear.

It can be shown that any theorem of real projective geometry that deals only with relations of connection between points and lines can be derived from the theorems of Desargues and Pappus. But Desargues' theorem itself can be derived from Pappus' theorem, as we have shown above. For this reason it may be said that the hexagonal configuration that occurs in Pappus' theorem is the most important configuration in projective geometry.

Analytic Projective Geometry

In Chapter 4 we outlined the elements of analytic geometry, which uses algebraic methods to study the Euclidean plane. There is an analogous analytic geometry of the projective plane which makes it possible to study projective geometry, too, by algebraic methods. Analytic projective geometry can be derived from a modified version of analytic Euclidean geometry. We shall trace this derivation now, in order to get acquainted with some significant concepts that we shall have to use later in this chapter. To do so, we once again picture the real projective plane as an extension of the Euclidean plane obtained by adding to it the ideal points. On page 310 we shall see that analytic projective geometry can also be derived in another way, without beginning with the Euclidean plane.

Homogeneous Coordinates

In the analytic geometry described in Chapter 4, each point of the Euclidean plane is represented by its coordinates, the ordered pair of real numbers (x, y), where x and y are the directed distances of the point from two fixed perpendicular lines OY and OX respectively. In this system of coordinates, every ordered pair of real numbers is assigned to a point of the Euclidean plane, that is, to an *ordinary* point of the real projective plane. Therefore there are no ordered pairs of real numbers left unassigned that may be used to represent the *ideal points* in the real projective plane. Consequently we cannot use the old system of coordinates to represent all the points in the real projective plane.

To provide coordinates for all points in the real projective plane, we first introduce a new system of coordinates in which only *some* of the coordinates are assigned to the points of the Euclidean plane. From the remaining unassigned coordinates we shall then select some that can be used to represent the ideal points. In the new system of coordinates, known as *homogeneous coordinates*, we use ordered triples of real numbers (X_1, X_2, X_3) rather than ordered pairs. We assign some of these ordered triples to each point of the Euclidean plane according to the following rule: An ordered triple (X_1, X_2, X_3) belongs to the point whose old coordinates are (x, y) if

(1) $$x = \frac{X_1}{X_3} \quad \text{and} \quad y = \frac{X_2}{X_3}.$$

We note first that this rule assigns more than one ordered triple to each point. In fact, if (X_1, X_2, X_3) is assigned to a particular point, and k is any real number except 0, then (kX_1, kX_2, kX_3) is assigned to the same point. Consequently, under this rule, each point is represented not by one ordered triple but by a whole family of ordered triples whose members are proportional to each other.

We note, secondly, that rule (1) leaves some ordered triples unassigned: Since division by 0 has no meaning, every ordered triple in which $X_3 = 0$ fails to satisfy condition (1) and therefore is not assigned to any point of the Euclidean plane. We therefore have a supply of triples of the form $(X_1, X_2, 0)$ which are not assigned to ordinary points in the real projective plane. We shall now look for a sensible rule for assigning some of them to ideal points in the real projective plane.

Consider any line through the origin whose slope is m. The equation of this line, in old-style coordinates, is

(2) $$y = mx.$$

Let (a, b), with $a \neq 0$, be the old coordinates of a fixed point on the line. Then $m = \dfrac{b}{a}.$ Now consider the ordered triple (a, b, X_3), where $X_3 > 0$. This ordered triple belongs to the point whose old coordinates are

(3) $$x = \frac{a}{X_3}, \qquad y = \frac{b}{X_3}.$$

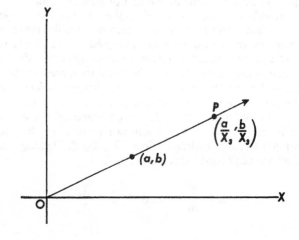

These coordinates satisfy equation (2), so they belong to a point P on the same line. If we allow X_3 to shrink to 0, the absolute values of x and y will become larger and larger without limit, so that as X_3 approaches 0 as a limit the point P will move along the line toward the point at infinity on that line. This suggests that we assign the ordered triple $(a, b, 0)$ in which $a \neq 0$ to the ideal point on the line through the origin whose slope is b/a. For similar reasons we assign every ordered triple of the form $(0, b, 0)$ in which $b \neq 0$ to the ideal point on the line through the origin that has no slope, namely the vertical line OY. The ordered triple $(0, 0, 0)$ remains unassigned and is not used at all.

Equation of a Straight Line

In the old-style system of coordinates, every straight line in the Euclidean plane is represented by an equation of the form

$$(4) \qquad ax + by + c = 0 \qquad \text{(with } a \text{ or } b \text{ not equal to 0),}$$

which is satisfied by those and only those ordered pairs (x, y) that represent points on the line. If we substitute into this equation the values of x and y given by (1) and then multiply by X_3, we obtain the equation of the line in homogeneous coordinates:

$$(5) \qquad aX_1 + bX_2 + cX_3 = 0.$$

Thus this equation is satisfied by the homogeneous coordinates of every ordinary point on the line that is represented by equation (4). In addition it is satisfied by the homogeneous coordinates of the ideal point on this line. (See exercise 9 on page 353.) So, in homogeneous coordinates, every *ordinary* line has an equation of form (5). Moreover, the line at infinity also has an equation of this form, with $a = 0$, $b = 0$, $c \neq 0$. Conversely, every equation of form (5) in which a, b and c are not all zero belongs to a straight line in the real projective plane.

Equation of a Conic Section

In the old-style coordinates, every conic section has an equation of the form:

$$(6) \qquad Ax^2 + Bxy + Cy^2 + Dx + Ey + F = 0.$$

By substituting for x and y the values given by (1), and multiplying by $X_3{}^2$, we obtain the equation of the conic section in homogeneous coordinates:

(7) $\quad AX_1{}^2 + BX_1X_2 + CX_2{}^2 + DX_1X_3 + EX_2X_3 + FX_3{}^2 = 0.$

Equation of a Circle

If a circle in the Euclidean plane has center (a, b) and radius r, its equation, as shown on page 123 in Chapter 4, is

(8) $\qquad\qquad (x - a)^2 + (y - b)^2 = r^2.$

If we expand this equation, substitute the values of x and y given by (1), and then multiply by $AX_3{}^2$, we get the equation of the circle in homogeneous coordinates:

(9) $\quad AX_1{}^2 + 0X_1X_2 + AX_2{}^2 - 2AaX_1X_3 - 2AbX_2X_3$
$$+ A(a^2 + b^2 - r^2)X_3{}^2 = 0.$$

Comparing equation (9) with equation (7), we see that if a conic section with equation (7) is a circle, then $A = C$, and $B = 0$. The converse is also true. That is, if $A = C$, and $B = 0$, then the conic section given by equation (7) is a circle. We shall use this fact on page 312.

Homogeneous Coordinates on a Line

A coordinate system can be provided for a Euclidean line by putting a real number scale on the line. Then each point on the line can be represented by a single coordinate, namely the real number that is attached to it. We can introduce homogeneous coordinates for the points on a line, just as we have introduced homogeneous coordinates for the points on a plane: Assign the ordered pair of real numbers (X_1, X_2) to the point whose coordinate is x if

(10) $\qquad\qquad x = \dfrac{X_1}{X_2}.$

In addition, assign to the ideal point on the line the ordered pairs of the form $(X_1, 0)$ where $X_1 \neq 0$. The ordered pair $(0, 0)$ remains unassigned and is not used. Notice that while homogeneous coordinates for the points of a plane consist of ordered triples, homogeneous coordinates for the points of a line taken all by itself are ordered pairs.

Equations of a Projectivity

The use of homogeneous coordinates to represent the points of the real projective plane makes it possible to represent every projectivity of the plane into itself by a simple set of equations that relate the coordinates of each point of the plane to the coordinates of its image under the projectivity. If (X_1, X_2, X_3) are the homogeneous coordinates of an arbitrary point of the plane, and (X_1', X_2', X_3') are homogeneous coordinates of its image under a given projectivity, then there are real numbers a, b, c, d, e, f, g, h, j such that these coordinates are related by equations of the form

(11)
$$kX_1' = aX_1 + bX_2 + cX_3,$$
$$kX_2' = dX_1 + eX_2 + fX_3,$$
$$kX_3' = gX_1 + hX_2 + jX_3,$$

where k is real and is not zero, and the determinant

$$D = \begin{vmatrix} a & b & c \\ d & e & f \\ g & h & j \end{vmatrix} \neq 0.$$

Conversely, every set of equations of form (11) with $k \neq 0$ and $D \neq 0$ represents a projectivity that moves (X_1, X_2, X_3) to (X_1', X_2', X_3').

Similarly, a projectivity of a line into itself can be expressed by a set of equations. If (X_1, X_2) are the homogeneous coordinates of an arbitrary point on the line, and (X_1', X_2') are the homogeneous coordinates of its image under the projectivity, then there are real numbers a, b, c and d such that these coordinates are related by equations of the form

(12)
$$kX_1' = aX_1 + bX_2,$$
$$kX_2' = cX_1 + dX_2,$$

where k is real and not zero, and the determinant

$$\begin{vmatrix} a & b \\ c & d \end{vmatrix} \neq 0.$$

Projective Geometry as a Branch of Algebra

In the preceding pages, to derive our outline of the analytic geometry of the real projective plane, we began with the plane and then associated a set of proportional ordered triples of real numbers with each point. Taking our cue from the discussion on page 129, we could also introduce analytic projective geometry in another way, in which a set of such ordered triples is not merely *associated* with a point but *is* the point.

Take the set of all ordered triples of real numbers except $(0, 0, 0)$ and divide it into subsets such that each subset consists of all the ordered triples that are proportional to any one of the members of the subset. Call every such subset a *point*, and call any ordered triple in the subset *homogeneous coordinates* of the point. Denote as a *line* the set of all points whose coordinates satisfy an equation of form (5) with real coefficients. Then it is possible to prove that the points and lines defined in this way satisfy axioms 1 to 6 for the real projective plane. Thus the set of all points defined in this manner is entitled to be called a real projective plane. So we see that projective geometry, like Euclidean geometry, can be developed by the use of numbers alone, and hence may be considered a branch of algebra.

The Complex Projective Plane

The development of projective geometry via ordered triples of numbers indicates a way in which projective geometry may be generalized. In the definition of the real projective plane given in the preceding paragraph, we defined a point to be a certain kind of set of ordered triples of real numbers, and we defined a line as the set of points satisfying an equation of form (5) with real coefficients. If we modify the definition by substituting the word *complex* for the word *real*, we obtain a new set of points with properties that are analogous to those of the real projective plane. In fact, it is possible to prove that this new set of points satisfies axioms 1 to 5 on page 295. It does not satisfy axiom 6, but it does have an analogous property, namely that there is a one-to-one correspondence between the *complex* numbers and all but one point of a line. For this reason we call this new set of points the *complex projective plane*.

Since every real number a is a complex number of the form $a + 0i$, the real projective plane is part of the complex projective plane. We may thus distinguish two kinds of points in the complex projective plane: Those points of the complex projective plane that are in the real projective plane are called *real points*. All the rest are called *imaginary points*. A point of the complex projective plane whose homogeneous coordinates are not real is not necessarily an imaginary point. If these coordinates are proportional to an ordered triple of real numbers, then the point is a real point. For example, the ordered triple (i, i, i) represents a real point, since it is proportional to the ordered triple $(1, 1, 1)$. However, the ordered triple $(i, 1, 0)$ represents an imaginary point, since there is no ordered triple of real numbers to which it is proportional. (See exercise 11 on page 353.)

We may also distinguish two kinds of lines in the complex projective plane: A line whose points satisfy an equation of form (5) with real coefficients is called a *real* line. All the rest are called *imaginary lines*. The points on a real line in the complex plane include both real points and imaginary points. For example, the real line whose equation is $X_1 + X_2 - X_3 = 0$ contains the real point whose homogeneous coordinates are $(1, 1, 2)$, and also contains the imaginary point whose homogeneous coordinates are $(1, i, 1 + i)$. (See exercise 12 on page 353.)

The real points on a real line, since they satisfy an equation of form (5) with real coefficients, constitute a line in the real projective plane. Thus every real line in the complex projective plane contains a line in the real projective plane, and every line in the real projective plane is part of a real line in the complex projective plane. A real line in the complex projective plane that contains a given line of the real projective plane is said to be its extension in the complex projective plane. For the sake of brevity we shall say that an imaginary point is on a given line of the real projective plane if its coordinates satisfy the equation of the line, so that it is on the extension of the line in the complex projective plane. Similarly, we shall say an imaginary point is on a curve in the real projective plane if its coordinates satisfy the equations of the curve.

Extension of a Projectivity

Any projectivity of the real projective plane into itself or of a line in that plane into itself can be represented by a set of equations of form (11) or (12) respectively. Ordinarily we apply the projectivity to real points only, by substituting real numbers for X_1, X_2 and X_3. However, if we wish, we may extend the projectivity to imaginary points as well, by substituting imaginary values of X_1, X_2 and X_3 that are the coordinates of imaginary points. We shall find it useful to follow this procedure later in this chapter.

Example: The following equations represent a projectivity of the real projective plane into itself:

$$X_1' = .6X_1 + .8X_2 + X_3,$$
$$X_2' = -.8X_1 + .6X_2 + X_3,$$
$$X_3' = X_3.$$

This is so because the determinant

$$\begin{vmatrix} .6 & .8 & 1 \\ -.8 & .6 & 1 \\ 0 & 0 & 1 \end{vmatrix} \neq 0.$$

Substitution of $(1, i, 0)$ for (X_1, X_2, X_3) shows that the extension of this projectivity to the complex projective plane moves the imaginary point with homogeneous coordinates $(1, i, 0)$ to the point with the homogeneous coordinates $(.6 + .8i, -.8 + .6i, 0)$. But the latter coordinates are proportional to $(1, i, 0)$, as may be seen by multiplying 1, i and 0 by $.6 + .8i$. Consequently they represent the same point as $(1, i, 0)$. That is, the point with coordinates $(1, i, 0)$ is a fixed point of the given projectivity.

The Circular Points at Infinity

In the example given above, we identified the image under a given projectivity of the imaginary point $(1, i, 0)$. We chose this particular point to use as an example because this point and a companion point $(1, -i, 0)$ will play an important role in our discussion later in this chapter. (See page 345.) These points are clearly points at infinity, because for each of them $X_3 = 0$. They have the interesting property that a conic section in the Euclidean plane is a circle if and only if it contains these two points. To prove this fact, we make use of the observation already made on page 308: A conic section described by equation (7) on page 308 is a circle if and only if $A = C$, and $B = 0$. If the conic section is a circle, its equation has the form

$$AX_1^2 + AX_2^2 + DX_1X_3 + EX_2X_3 + FX_3^2 = 0.$$

This equation is obviously satisfied by the values $X_1 = 1$, $X_2 = i$, $X_3 = 0$, and also by the values $X_1 = 1$, $X_2 = -i$, $X_3 = 0$.

Consequently the imaginary points with coordinates $(1, i, 0)$ and $(1, -i, 0)$ lie on every circle. Conversely, suppose these two points lie on a conic section whose equation is equation (7). Substituting the coordinates $(1, i, 0)$ into equation (7), we find that $A + Bi - C = 0$. Substituting the coordinates $(1, -i, 0)$ into equation (7), we find that $A - Bi - C = 0$. By adding these two equations, we find that $A = C$, and by subtracting them we find that $B = 0$. Consequently any conic section that contains the two points $(1, i, 0)$ and $(1, -i, 0)$ is a circle. Because of the theorem we have just proved, the points $(1, i, 0)$ and $(1, -i, 0)$ are known as the *circular points at infinity*.

Complete Quadrangle

After this brief excursion into the complex projective plane, let us return to the real projective plane for some unfinished business. We shall now outline a chain of ideas that culminates in a proof of Pappus' theorem. The first link in the chain is the concept of a complete quadrangle. A *complete quadrangle* is the configuration consisting of four points, no three of which are collinear, and the six lines determined by pairs of these points. The points are called the *vertices* of the quadrangle, and the lines are called the *sides* of the quadrangle. Sides of the quadrangle that do not lie on a common vertex are called *opposite sides*. The point that lies on a pair of opposite sides of a quadrangle is called a *diagonal point* of the quadrangle. There are three such diagonal points. It can be shown that they do not lie on the same line. In the complete quadrangle shown below, A_1, A_2, A_3 and A_4 are the vertices, and P, Q and R are the diagonal points.

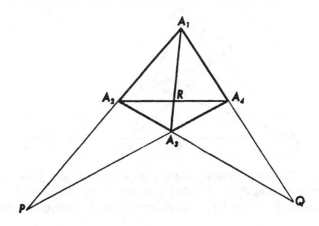

The dual of a complete quadrangle is called a *complete quadrilateral*. It is defined by the dual of the definition of a complete quadrangle. (See exercise 15 on page 353.) The dual of a diagonal point of a complete quadrangle is a diagonal line of a complete quadrilateral. (See exercise 16 on page 353.)

Harmonic Sequences

The concept of a complete quadrangle is used to define the concept of a harmonic sequence of points on a line. Four points A, B, C and D on a line are said to form a *harmonic sequence* if A and B are diagonal points of a complete quadrangle, and C and D are on the sides of the quadrangle that meet at the third diagonal point. By taking the dual of this definition we obtain the definition of the dual concept of a harmonic sequence of lines on a point. (See exercise 17 on page 354.) If A, B, C and D form a harmonic sequence, we say that C and D are *harmonic conjugates* with respect to A and B. It is clear from the definition of a harmonious sequence that the first and second members of the sequence may be interchanged and the third and fourth members may be interchanged. That is, if A, B, C, D is a harmonic sequence, so are B, A, C, D, and A, B, D, C. It can be shown, too, that the first pair and the second pair of a harmonic sequence may be interchanged. That is, if A, B, C, D is a harmonic sequence, then so is C, D, A, B.

If A, B and C are any three distinct points on a line, there is a point D on the line such that the four points A, B, C and D form a harmonic sequence.

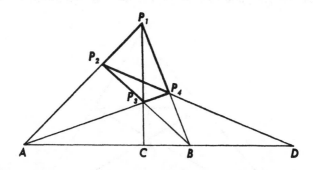

Proof: Choose a point P_1 that is not on line AB, and draw AP_1. Choose a point P_2 on AP_1 that is distinct from A and P_1. Draw BP_1 and BP_2. Draw CP_1, and denote by P_3 its intersection with

BP_2. Draw AP_3, and denote by P_4 its intersection with BP_1. Draw P_2P_4, and denote by D its intersection with AB. The completed diagram is on page 314. In this diagram, P_1, P_2, P_3 and P_4 are the vertices of a complete quadrangle. A and B are diagonal points of the quadrangle, and C and D are on the sides of the quadrangle that meet at the third diagonal point. Therefore the points A, B, C and D form a harmonic sequence. It can be shown with the help of Desargues' theorem that the point D obtained above is unique. Consequently, *for any three distinct points A, B and C on a line, there is one and only one point D on the line such that A, B, C and D form a harmonic sequence.* By taking the dual of this theorem, we obtain another: *For any three distinct lines a, b and c on a point, there is one and only one line d on the point such that a, b, c and d form a harmonic sequence.*

A harmonic sequence of lines can be obtained from a harmonic sequence of points in a very simple way. Suppose A, B, C and D form a harmonic sequence of points on a line. Let P_1 be a point that is not on the line, and denote AP_1, BP_1, CP_1 and DP_1 as a, b, c and d respectively. Then the lines a, b, c and d, which are all on the point P_1, form a harmonic sequence.

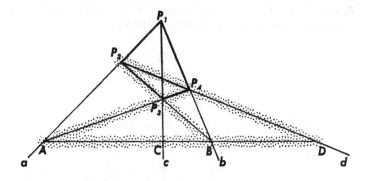

Proof: Starting with A, B, C and P_1, repeat the construction of the preceding theorem by choosing P_2 on AP_1, and determining P_3 and P_4 as described on page 314. Then, because of the uniqueness of the harmonic conjugate of C with respect to A and B, the line P_2P_4 meets AB at D. Now consider the complete quadrilateral whose four sides are AB, AP_4, BP_2, and P_2P_4, indicated by shading in the diagram above. Each of the six pairs of sides determines a vertex of the quadrilateral. AB and AP_4 meet at A. AB and BP_2 meet at B. AB and P_2P_4 meet at D. AP_4 and BP_2 meet at P_3. AP_4 and P_2P_4 meet at P_4. BP_2 and P_2P_4 meet at P_2.

Consequently the six vertices of the quadrilateral are A, B, D, P_3, P_4 and P_2. The vertices A and P_2 are opposite vertices of the complete quadrilateral, since they do not lie on a common side. This can be seen by the fact that the line AP_2 is not shaded in the diagram, and hence is not a side of the quadrilateral. Similarly, B and P_4 are opposite vertices, and D and P_3 are opposite vertices. Consequently, the line a, which joins the opposite vertices A and P_2, is a diagonal line of the quadrilateral. The line b, which joins the opposite vertices B and P_4, is also a diagonal line of the quadrilateral. Moreover, the lines c and d are on P_3 and D respectively, which are the vertices that determine the third diagonal line. Therefore the lines a, b, c and d form a harmonic sequence. (See exercise 17 on page 354, and the answer on page 399.)

What we have proved above is that *a projection of a harmonic sequence of points on a line is a harmonic sequence of lines on a point*. The converse of this theorem is its dual, so it is also a valid theorem: *A section of a harmonic sequence of lines on a point is a harmonic sequence of points on a line.*

By using these two theorems we can establish with ease the following important result: If two sequences of points A, B, C, D and A', B', C', D' are perspective from a point, and one of the sequences is harmonic, then so is the other.

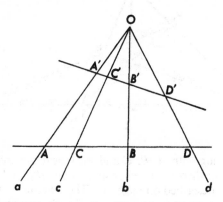

Proof: Let O be the center of perspectivity, and denote OA, OB, OC and OD by a, b, c and d respectively. If A, B, C and D form a harmonic sequence, then so do their projections, a, b, c and d. But then, since a, b, c and d form a harmonic sequence, so do their sections A', B', C' and D'.

Because a projectivity is the product of a sequence of perspectivities, we can extend this result immediately to obtain the following important theorem: *If A, B, C, D is a harmonic sequence of points on a line and A', B', C', D', on another line or on the same line, is the image of this sequence under a projectivity, then A', B', C', D' is also a harmonic sequence.*

The Fundamental Theorem of Projective Geometry

The next link in the chain of ideas leading to Pappus' theorem emerges from a consideration of this question: In the construction of a projectivity of a line into a line, how much freedom of choice do we have for selecting the images of individual points under the projectivity? This question is answered in part when we show that *if A, B and C are any three distinct points on a line l, and A', B' and C' are any three distinct points on a line l', then there is a projectivity of l into l' that moves A, B and C to A', B' and C' respectively.*

Proof: Assume first that *l* and *l'* are distinct lines. There are two cases that may arise.

Case I

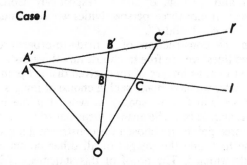

Case I: One of the pairs of corresponding points, say *A* and *A'*, coincide. Let *O* be the intersection of *BB'* and *CC'*. Then a perspectivity with center *O* moves *A*, *B* and *C* to *A'*, *B'* and *C'* respectively.

Case II: *A* is not *A'*, *B* is not *B'*, and *C* is not *C'*. Draw the line *AA'* and let *O* be any point on it distinct from *A* and *A'*. Draw a line *l''* that is distinct from *l* and *l'*, that passes through *A'*, and meets *l*. Construct the perspectivity of *l* into *l''* with center *O*. It moves *A* to *A'*, *B* to some point *B''*, and *C* to some point *C''*. Call this perspectivity R. Let *P* be the intersection of *B'B''* and *C'C''*. The perspectivity of *l''* into *l'* with center *P* moves *A'* to *A'*, *B''* to

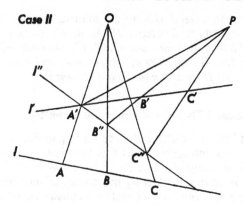

B', and C'' to C'. Call this perspectivity S. Then the projectivity RS obtained by using first R and then S moves A to A', B to B' and C to C'. If l' is the same line as l, let m be a line distinct from l. Use any perspectivity T to move the points of l into m. Let X, Y and Z be the images of A, B and C respectively under this perspectivity. As shown above, two perspectivities will suffice to move X, Y and Z to A', B' and C' respectively. Consequently there is a sequence of three perspectivities whose product moves A, B and C to A', B' and C'.

Thus, in the construction of a made-to-order projectivity between two lines, we are free to choose any three distinct points on one line to use as images of three given distinct points on the other line. After we have made such a choice of images for three points, are we still free to specify at will what the images of other points should be? The answer turns out to be *no*. Once the images of three points are chosen for constructing a projectivity from a line to a line the images of all other points are automatically determined. The proof of this statement is based on three key ideas: 1) the theorem proved above that a projective image of a harmonic sequence of points is a harmonic sequence of points, 2) axiom 6 for the real projective plane, and 3) the fact that a projectivity is a continuous transformation. We shall outline the main steps in the proof.

Suppose A, B and C are three distinct points on a line l, A', B' and C' are three distinct points on a line l', and T is a projectivity from l to l' that moves A to A', B to B' and C to C'. We want to show that if D is any other point on l, its image under T is uniquely determined. To begin with we shall consider only certain special points on l. Suppose D is the harmonic conjugate of C with respect to A and B. Then, by the theorem on page 317, its image under T must be the harmonic conjugate of C' with respect to A' and B'. Since there is one and only one point on l' that is the harmonic conjugate of C' with respect to A' and B', the image of D is uniquely determined. Let us call it D'. Similarly, if E is the harmonic conjugate of B with respect to A and C, its image E' is uniquely determined as the harmonic conjugate of B' with respect to A' and C'; and if F is the harmonic conjugate of A with respect to B and C, its image F' is uniquely determined as the harmonic conjugate of A' with respect to B' and C'. We thus have six points, A, B, C, D, E and F whose images under T are determined. We can obtain more points whose images are determined in the same way by finding the point which forms a harmonic sequence with any three of these six points and adding it to the list of points whose images are determined. We can continue this procedure indefinitely, extending the list of points whose images are determined by adding to the list every point that forms a harmonic sequence with any three points that are already on the list. The set of all such points obtained in this way is called the *net of rationality* on l determined by the three points A, B and C. The image under T of each point in the net of rationality on l is uniquely determined, and the set of all these images is the net of rationality on l' determined by the points A', B' and C'.

The completion of the proof depends on the fact that the projectivity T is a continuous transformation. On page 56 we defined a continuous transformation somewhat crudely as one that carries points that are near each other into points that are near each other. The meaning of this statement, expressed in technical language, is this: If a sequence of points in the domain of a continuous transformation T approaches P as a limit, then the images of these points under T approach some point P' as a limit, and P' is the image under T of P. We now use this property of T to show that if P is any point on l that is not in the net of rationality, then its image under T is also uniquely determined. It can be shown with the help of axiom 6 that there is a sequence of points in the net of rationality on l that approaches P as a limit. In the preceding paragraph we have shown that the image of each point in the sequence is uniquely determined as a member of the net of rationality on l'. Because T is a continuous transformation, these images approach a limit P' which is the image of P under T. Thus the image of P is also uniquely determined.

Suppose two projective transformations S and T of a line l into a line l' both carry three distinct points A, B and C into A', B' and C' respectively. We have just shown that if P is any other point on *l*, there is a unique point *P'* that must be the image of *P* under either S or T. Consequently the image of any point of *l* under S is the same as its image under T. That is, the projectivities S and T are the same. This completes the proof of what is known as the *fundamental theorem of projective geometry: There is one and only one projectivity of a line into a line that assigns specified distinct images to three given distinct points.*

Using the fundamental theorem, it is easy to prove the following corollary: *A projectivity between two distinct lines that moves a point into itself must be a perspectivity.* (See exercise 18 on page 354.) This corollary is the last link in the chain of ideas that leads to the proof of Pappus' theorem.

Proof of Pappus' Theorem

We are now ready to give the proof.

Given: *A, B, C* and *A', B', C'* are triples of distinct points on two distinct lines in a plane.

Prove: The intersections of *AB'* and *A'B*, of *AC'* and *A'C*, and of *BC'* and *B'C* are collinear.

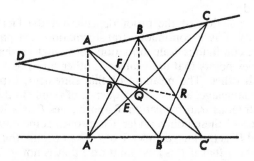

Proof: Let *P* be the intersection of *AB'* and *A'B*, and let *Q* be the intersection of *AC'* and *A'C*. We shall prove that the intersection of *BC'* and *B'C* lies on *PQ*. Let *R* be the intersection of *PQ* and *BC'*. Let *D* be the intersection of *AB* and *PQ*, let *E* be the intersection of *AB'* and *A'C*, and let *F* be the intersection of *A'B* and *AC'*. Consider the perspectivity with center *A'* that projects the points of *AB'* into *AC'*. This perspectivity moves *A, P, E* and *B'* to *A, F, Q* and *C'* respectively. Now consider the perspectivity with center *B* that projects the points of *AC'* into *PQ*. This perspectivity moves *A, F, Q* and *C'* to *D, P, Q,* and *R*

respectively. The product of these two perspectivities is a projectivity of the line AB' into the line PQ that moves A, P, E and B' to D, P, Q and R respectively. Since this projectivity moves P into itself, it must be a perspectivity, according to the corollary of the fundamental theorem. The center of this perspectivity is the point where the lines joining corresponding points intersect. Since AD and EQ intersect at C, C is the center of this perspectivity, and $B'R$ passes through C. That is, R is on $B'C$. But R is also on BC' and on PQ. Therefore the intersection of BC' and $B'C$ lies on PQ.

Involutions

Among the projectivities of a line into itself there are some that are involutions. (See the definition of an involution on page 161.) We recall that if an involution moves a point A to A', then it also moves A' to A. That is, it interchanges A and A'. We shall now examine the question, "How many fixed points may a projectivity that is an involution have?" Suppose that a projectivity T of a line into itself has three fixed points A, B, and C. That is, T moves A, B and C to A, B and C respectively. The identity transformation also moves A, B and C to A, B and C respectively. Then, by the fundamental theorem of projective geometry, T must be the identity transformation, and therefore cannot be an involution. Consequently an involution has at most two fixed points.

We now prove that if a projective involution of a line into itself has one fixed point, then it has two fixed points. Suppose A is a fixed point of a projective involution of a line into itself. Let B be another point on the line. If B is also a fixed point, then there are two fixed points. If B is not a fixed point, then its image under the involution is a third point B'. Let C be the harmonic conjugate of A with respect to B and B', and let C' be its image under the involution. The points B, B', A and C form a harmonic sequence. Since projectivities preserve harmonic sequences, the images of these points, namely B', B, A and C' form a harmonic sequence. We saw on page 314 that we may interchange the first two members of a harmonic sequence. Consequently, the sequence B, B', A, C' is harmonic. Thus both C and C' form a harmonic sequence with B, B' and A. Therefore, by the theorem stated on page 315, $C = C'$. That is, C is a fixed point, and the involution has two fixed points.

The argument of the preceding paragraph also shows incidentally that if an involution has two fixed points, such as A

and C, corresponding points under the involution, such as B and B', are harmonic conjugates with respect to the fixed points.

We know, therefore, that a projective involution of a line in the real projective plane into itself has either exactly two fixed points or none. A line in the real projective plane is, as we have seen, part of a line in the complex projective plane. A projective involution of a line in the real projective plane can be extended into a projective involution of the line in the complex projective plane of which it is a part. Now it can be shown that the extended involution always has two fixed points, and that they are either both real points or both imaginary points. The fixed points of the extended involution are real if and only if these same two points are the fixed points of the original involution on the line in the real projective plane. The fixed points of the extended involution are imaginary if and only if the original involution on the line in the real projective plane has no fixed points. For this reason we may classify projective involutions of a line in the real projective plane into two kinds: those that have two (real) fixed points, and those whose extensions to the complex projective plane have two imaginary fixed points. In the latter case it is customary to attach the property of the extended involution to the original one by saying that the original involution has two imaginary fixed points.

The Conic Sections

The curve called a conic section in the Euclidean plane is a set of ordinary points whose homogeneous coordinates satisfy an equation of the second degree, like equation (7) on page 308. If we add to the curve those ideal points whose homogeneous coordinates satisfy the same equation, we obtain the curve called a *point conic* in the real projective plane. There are three different kinds of conic sections in the Euclidean plane, namely, the ellipse, the parabola, and the hyperbola. When we extend each of them (by the addition of the appropriate ideal points) to form a point conic in the real projective plane, we find that the distinction among the three curves disappears. While there are three kinds of conic sections in the Euclidean plane, there is only one kind of point conic in the real projective plane. Let us see why.

The ellipse, the parabola, and the hyperbola may be distinguished from each other by these properties: the ellipse is a closed curve; the parabola is an open curve with only one branch; the hyperbola is an open curve with two branches. However, a parabola may be related to an ellipse in this way: In any ellipse, draw the longest diameter. The end points of this diameter are called the vertices of the ellipse. Keep one vertex of the ellipse fixed and make the ellipse longer and longer, so that the other vertex moves off to infinity. Then the ellipse can be shown to

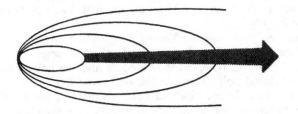

A parabola is an ellipse with one vertex at infinity.

approach a parabola as a limit. Consequently a parabola may be thought of as an ellipse that has one point on the line at infinity. If we add this ideal point to the ordinary points of a parabola, the resulting point conic becomes a closed curve, like an ellipse.

A hyperbola is also related to a closed ellipse-like curve. There is associated with every hyperbola a pair of straight lines called its *asymptotes*, as shown in the diagram below. If a point moves in either direction along a branch of the hyperbola, as it goes off

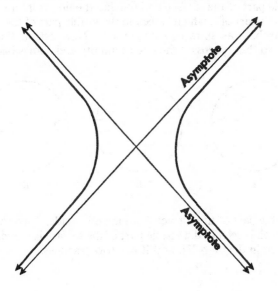

to infinity it comes closer and closer to one of the asymptotes. So each branch may be thought of as meeting each asymptote at its ideal point. If we add these two ideal points to the hyperbola, its two branches are joined at these points, both of which are on the line at infinity. The resulting point conic becomes a closed curve, like an ellipse, with two points on the line at infinity.

Thus *the point conic obtained from either an ellipse, or a parabola or a hyperbola is an ellipse-like closed curve.*

It is helpful to reverse the procedure and see how from one kind of closed point conic in the real projective plane we can obtain three different kinds of conic sections in the Euclidean plane. When we add the points at infinity to a Euclidean plane to form the real projective plane, it becomes a closed surface. It can be shown that this closed surface is like a sphere in which a hole has been punched and then sealed with a Möbius strip. (See pages 99 and 249.) Imagine that you are looking at this closed surface from the outside, and that the troublesome part of the surface, the part that contains the line at infinity, is turned away from you. Then you see only the ordinary points of the real projective plane, while the ideal points are out of sight on the "other side." Now draw on the surface a point conic whose points are all ordinary points. This conic is an ordinary ellipse in the visible part of the projective plane, as shown in diagram I. Now extend the conic to the right until it reaches around to the invisible part of the surface and touches the line at infinity there. Then that part of it which is seen in the visible part of the surface is a parabola as shown in diagram II. Now extend the conic further so that it crosses the line at infinity and comes back into

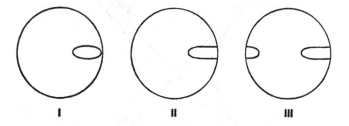

I　　　　　　　　II　　　　　　　　III

the visible part of the surface from the other side. Then that part of it which is seen in the visible part of the surface is a hyperbola, as shown in diagram III, and it has two branches.

Projective Definition of a Point Conic

There is an interesting way of defining a point conic in terms of purely projective relationships. To be able to give the projective definition of a point conic, we shall first dualize the definition of a perspectivity given on page 291, and repeated here in modified form: Let A be a fixed point, and let l and l' be two distinct lines that are not on A. The perspectivity with center A between the points on l and the points on l' is the one-to-one correspondence in which corresponding points have a common projection on A. The dual of this definition, obtained by interchanging the words point and line, and using an appropriate notation for points and lines, reads as follows: Let a be a fixed line, and let L and L' be two distinct points that are not on a. The *perspectivity with axis* a between the lines on L and the lines on L' is the one-to-one correspondence in which corresponding lines have a common section on a. Thus, in the perspectivity shown in the diagram below, p corresponds to p', q corresponds to q', r corresponds to r', and s corresponds to s'. Similarly, by dualizing the definition

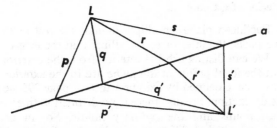

given on page 291 for a projectivity between the points on one line and the points on another line, we obtain the definition of a projectivity between the lines on one point and the lines on another point. Every such projectivity is obtainable as the result of a sequence of alternate projections and sections.

We can now give the projective definition of a point conic: In a plane, if the lines on one point are related by a projectivity that is not a perspectivity to the lines on another point, the set of intersections of corresponding lines is a point conic. (See exercise 20 on page 354.) If the projectivity is a perspectivity, or if the two points whose lines are related by the projectivity coincide, the set of intersections of corresponding lines is called a degenerate point conic. A degenerate point conic may consist of the entire plane, or two lines, or a single line, or a single point.

Order Relations on a Point Conic

We have seen that every point conic in the real projective plane is a closed curve, like an ellipse. The points of a point conic therefore have the same kind of order relations as the points of an ellipse. These are the relations of circular order, which are rather different from those of linear order. Two distinct points *A* and *B* on a point conic divide it into two arcs whose endpoints are *A* and *B*. Two other points *C* and *D* on the conic are said to separate *A* and *B* if *C* lies on one of these arcs and *D* lies on the other one. Obviously if *C* and *D* separate *A* and *B*, then *A* and *B* separate *C* and *D*. So we may say that in this case the pairs *A*, *B* and *C*, *D* separate each other. If *C* and *D* are on the same arc with endpoints *A* and *B*, then we say that the pairs *A*, *B* and *C*, *D* do not separate each other. The idea of separation of point pairs is the basic concept in the study of circular order just as the idea of betweenness of points is the basic concept in the study of linear order.

The Interior of a Conic

In a Euclidean plane, an ellipse divides the rest of the plane into two parts, the interior of the ellipse and the exterior of the ellipse. We can distinguish the interior from the exterior by the fact that the ideal points of the plane are in the exterior. In the projective plane defined by axioms 1 to 6 of page 295 we cannot use this distinction, because there is no distinction there between the ideal points and the ordinary points. So, in projective geometry we need a different criterion for distinguishing the interior from the exterior of a point conic. This criterion is easily obtained if we make use of the order relations on the point conic. Let *A*, *B* and *C*, *D* be two pairs of points on a point conic. We say that the point of intersection of the lines *AB* and *CD* is an *interior point* of the conic if the point pairs *A*, *B* and *C*, *D* separate each other. We say it is an *exterior point* of the conic

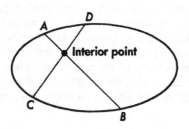

if the point pairs do not separate each other. The set of all interior points is called the *interior* of the conic. We shall use this concept on page 351.

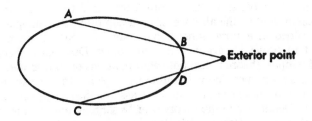

Pascal's Theorem

If six points are chosen on a point conic and numbered from 1 to 6, so that no three consecutive points in the cyclic order 123456 are collinear, then the polygon determined by the six points and the six lines joining each pair of consecutive points is called a simple hexagon inscribed in the conic. Such an inscribed hexagon has the interesting property that the points of intersection of its opposite sides are collinear. This theorem was discovered by Blaise Pascal (1623–1662) when he was 16 years old. In the case of a degenerate point conic consisting of two straight lines rather than a closed curve, the theorem is also true, and is in fact the theorem of Pappus which we have already encountered. For this reason, Pappus' theorem is sometimes called Pascal's theorem.

General Projective Geometry

We observed on page 310 that the real projective plane is not the only projective plane. There is also a complex projective plane that shares many of the properties of the real projective plane. This fact suggests that there may be other projective planes as well. We turn now to the problem of identifying these other possible projective planes.

First we must choose the properties that we want a plane to have to entitle it to be called a projective plane. Roughly, these properties are all those properties of the real projective plane that involve only relations of connection. At first sight, it may seem that we can obtain all the properties involving relations of connection by simply dropping axiom 6 on page 296, and deducing the consequences of axioms 1 to 5. This procedure would be analogous to what we did in Chapter 8, where, by using all the Hilbert axioms except the parallel postulate, we identified properties that are shared by the Euclidean and the hyperbolic plane. However, by dropping axiom 6, we drop too much. We have seen that axiom 6 was used to prove Pappus' theorem, and in the absence of an assumption that the plane is embedded in a three-dimensional space, Pappus' theorem was used to prove Desargues' theorem. Since Desargues' theorem and Pappus' theorem deal only with relations of connection, they express properties that we would like to retain. In order to retain them in the absence of axiom 6, we assume them as axioms in its place. Denote by D the theorem of Desargues, and denote by P the theorem of Pappus. We define a *projective plane* to be any set of objects that satisfies axioms 1 to 5 and D and P. The seven axioms 1, 2, 3, 4, 5, D, P are called the axioms of *general projective* (*plane*) *geometry*. The properties that all projective planes have in common are those that can be derived from the axioms of general projective geometry.

We could, if we wish to, omit D as an axiom, since it is a consequence of P, as we proved on page 302. However, we list it as a separate axiom in order to see how much we can prove using the weaker property D without invoking the stronger property P. In this way we shall develop some insight into the peculiar significance of axiom P.

The Algebra of Points on a Line

Axiom 6 for the real projective plane asserts that if we delete one point from a line in the plane, we can attach a real number scale to the remaining points, so that the rest of these points acquire the structure of the field of real numbers. We shall now show that in general projective geometry we can do *almost* as much: if we delete one point from a line in a plane that satisfies axioms 1 to 5 and D and P, then we can give a *field structure* to the set of remaining points. We do so by defining operations of addition and multiplication for these points, and then showing that these operations satisfy the axioms of a field given on pages 32–33.

Addition of Points

Let l be any line in the projective plane. Choose any point on the line and call it ∞ (infinity). Choose any other point on the line and call it 0 (zero). The reason for choosing these peculiar names for these two points will become clear as we go along. The point ∞ is the one we shall exclude from the line, while we define addition of any two other points on the line as follows: Draw two more lines m and n through the point ∞, and let A

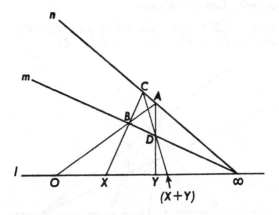

be any point on n except infinity. If X and Y are any two points on l other than ∞, we determine in succession a sequence of points which we call B, C, D, and $X + Y$ by the following sequence of steps:

1) OA meets m at B.
2) XB meets n at C.
3) YA meets m at D.
4) CD meets l at $X + Y$.

Notice that the two points being added are the points mentioned first in steps 2) and 3). Notice, too, that the line that meets l at their sum is the line determined by the points mentioned last in steps 2) and 3). These observations will be helpful when you use this definition of addition to find the sum of two particular points.

In this sequence of steps leading to the point $X + Y$ we used two arbitrarily chosen lines m and n through ∞, and an arbitrarily chosen point A on n. If we made a different choice of the lines m and n, and a different choice of the point A, would we, perhaps, arrive at a different point $X + Y$? The answer to this question turns out to be *no*. It can be proved with the help of axiom D that no matter which lines m and n we choose through ∞, and no matter which point A we choose on n, the sequence of steps 1) to 4) leads to the same point $X + Y$. Thus the point $X + Y$ is *uniquely determined*. We now show that the operation $+$ defined in this way satisfies axioms 1 to 4 of a field. (See page 32.)

The Associative Law of Addition

Let X, Y and Z be any three points on l other than ∞. We shall show that $(X + Y) + Z = X + (Y + Z)$.

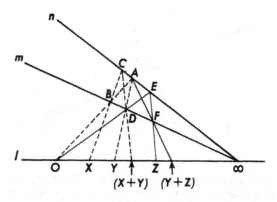

Choose lines m and n through ∞, choose a point A on n, and use steps 1) to 4) above to determine the point $X + Y$. Using the point D determined in step 3), draw OD, and let E be its intersection with n. Now using E instead of A, and Y and Z instead of X and Y, follow the same sequence of four steps to determine the point $Y + Z$ as follows:

> OE meets m at D.
> YD meets n at A.
> ZE meets m at F.
> AF meets l at $Y + Z$.

Using E again as the chosen point on n, apply steps 1) to 4) to the addition of $(X + Y)$ and Z:

> OE meets m at D.
> $(X + Y)D$ meets n at C.
> ZE meets m at F.
> CF meets l at $(X + Y) + Z$.

In order to keep the diagram uncluttered, we have not drawn the line CF and the point $(X + Y) + Z$ in the diagram.

Using A again as the chosen point on n, apply steps 1) to 4) to the addition of X and $(Y + Z)$:

> $0A$ meets m at B.
> XB meets n at C.
> $(Y + Z)A$ meets m at F.
> CF meets l at $X + (Y + Z)$.

Notice that both $(X + Y) + Z$ and $X + (Y + Z)$ are determined as the intersection of CF and l. But CF and l intersect at only one point. Therefore $(X + Y) + Z = X + (Y + Z)$.

The Commutative Law of Addition

Let X and Y be any two points on l other than ∞. We shall show that $X + Y = Y + X$.

Choose m, n, and A, and determine $X + Y$ by steps 1) to 4) using the diagram on page 329. $X + Y$ is the intersection of CD and l.

Now let m and n interchange roles in the construction, and use B as the chosen point on m, to determine the point $Y + X$:

> $0B$ meets n at A.
> YA meets m at D.
> XB meets n at C.
> DC meets l at $Y + X$.

Thus $X + Y$ and $Y + X$ are both determined as the intersection of CD and l. Therefore $X + Y = Y + X$.

The Law of Zero

Let X be any point on l other than ∞. We shall show that $X + 0 = X$.

Draw m and n through ∞ and choose A on n. Then using 0 instead of Y in steps 1) to 4) on page 329, determine $X + 0$:

> $0A$ meets m at B.
> XB meets n at C.
> $0A$ meets m at B.
> CB meets l at $X + 0$.

But CB meets l at X. Therefore $X + 0 = X$. A similar sequence of steps shows that $0 + X = X$. (See exercise 23 on page 354.)

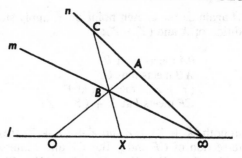

The Negative of a Point

Let X be any point on l other than ∞. We show that there exists on l a point Y that has the property that $X + Y = 0$.

Draw m and n through ∞ and choose A on n. We determine Y by the following sequence of steps:

> OA meets m at B.
> XB meets n at C.
> OC meets m at D.
> AD meets l at Y.

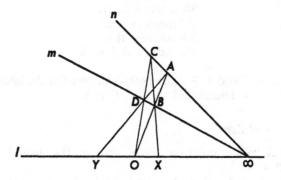

Now let us add X and Y by applying steps 1) to 4) as on page 329:

> OA meets m at B.
> XB meets n at C.
> YA meets m at D.
> CD meets l at $X + Y$.

But, in this diagram, CD meets l at 0. Therefore $X + Y = 0$. Following the usual custom, we designate the point Y determined above by the symbol $-X$. Then $X + (-X) = 0$. By the commutative law already established, we also have $(-X) + X = 0$.

Multiplication of Points

To define multiplication of points on the line l, we first choose any point on it other than 0 and ∞, and call it 1 (one). Draw a line m through 0 and a line n through ∞ so that neither coincides with l. Choose a point A that is on n but not on m. If X and Y are any two points on l other than ∞, we determine in succession a sequence of points which we call B, C, D and XY by the following sequence of steps:

    a) $1A$ meets m at B.
    b) XB meets n at C.
    c) YA meets m at D.
    d) CD meets l at XY.

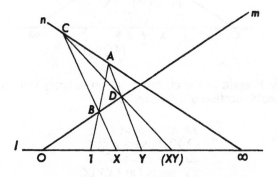

As in the case of addition, it can be shown with the help of axiom D that the choice of the lines m and n and the point A does not affect the result of the last step. The point XY is uniquely determined. We now show that the multiplication operation defined in this way satisfies axioms 5 to 9 of a field. (See page 32.)

The Associative Law of Multiplication

Let X, Y and Z be any three points on l other than ∞. We shall show that $(XY)Z = X(YZ)$.

Choose lines m and n through 0 and ∞ respectively, choose a point A on n, and use steps a) to d) above to determine the point

XY. Using the point *D* determined in step c), draw *1D*, and let *E* be its intersection with *n*. Now using *E* instead of *A*, and *Y* and *Z* instead of *X* and *Y*, follow the same sequence of four steps to determine the point *YZ* as follows:

> *1E* meets *m* at *D*.
> *YD* meets *n* at *A*.
> *ZE* meets *m* at *F*.
> *AF* meets *l* at *YZ*.

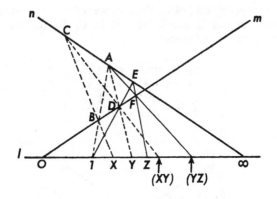

Using *E* again as the chosen point on *n*, apply steps a) to d) to the multiplication of (*XY*) and *Z*:

> *1E* meets *m* at *D*.
> (*XY*)*D* meets *n* at *C*.
> *ZE* meets *m* at *F*.
> *CF* meets *l* at (*XY*)*Z*.

Using *A* again as the chosen point on *n*, apply steps a) to d) to the multiplication of *X* and (*YZ*):

> *1A* meets *m* at *B*.
> *XB* meets *n* at *C*.
> (*YZ*)*A* meets *m* at *F*.
> *CF* meets *l* at *X*(*YZ*).

Thus both (*XY*)*Z* and *X*(*YZ*) are at the intersection of *CF* and *l*. Therefore (*XY*)*Z* = *X*(*YZ*).

The Commutative Law of Multiplication

Let X and Y be any two points on l other than ∞. We show that $XY = YX$.

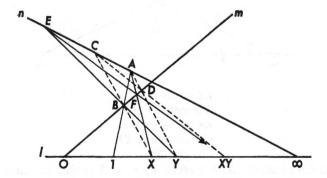

Choose m, n and A, and determine XY by steps a) to d) as we did above. Then apply steps a) to d) to determine YX as follows:

$$1A \text{ meets } m \text{ at } B.$$
$$YB \text{ meets } n \text{ at } E.$$
$$XA \text{ meets } m \text{ at } F.$$
$$EF \text{ meets } l \text{ at } YX.$$

To complete the argument, we now invoke axiom P (Pappus' theorem). Consider the hexagon $XACDBY$ inscribed on the two straight lines XBC and DAY, with three vertices on each. The lines XA and DB meet at F.

BY and AC meet at E.

CD and XY meet at the point (XY).

Then, by axiom P, the points F, E and (XY) are collinear. Thus, (XY) lies on the line FE as well as on the line l. But we have already found that the intersection of FE and l is the point (YX). Therefore $XY = YX$.

The Law of One

Let X be any point on l other than ∞. We show that $X1 = X$.

Draw m, n and A as we did above, and use 1 instead of Y in steps a) to d) to determine $X1$:

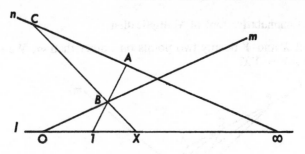

1A meets m at B.
XB meets n at C.
1A meets m at B.
CB meets l at X1.

But CB meets l at X. Therefore X1 = X. A similar sequence of steps shows that 1X = X. (See exercise 24 on page 354.)

The Reciprocal of a Point

Let X be any point on l other than 0 and ∞. We show that there exists on l a point Y that has the property that XY = 1.

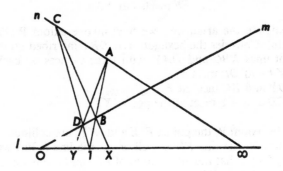

Draw m, n, and A as we did above, and determine Y by the following sequence of steps:

1A meets m at B.
XB meets n at C.
1C meets m at D.
AD meets l at Y.

Now let us multiply X and Y by applying steps a) to d) as on page 333:

$1A$ meets m at B.
XB meets n at C.
YA meets m at D.
CD meets l at XY.

But, in this diagram, CD meets l at 1. Therefore $XY = 1$. By the commutative law of multiplication, Y also has the property that $YX = 1$. The point Y determined above is called the reciprocal of X and is represented by the symbol $\frac{1}{X}$. Consequently

$$X\left(\frac{1}{X}\right) = \left(\frac{1}{X}\right)X = 1.$$

The Distributive Law

Using constructions analogous to those given above, and using only axioms 1 to 5 and D, it is possible to prove that if X, Y and Z are any three points on l other than ∞, $X(Y + Z) = XY + XZ$, and $(Y + Z)X = YX + ZX$.

The Meaning of Pappus' Theorem

Since the addition and multiplication operations defined in the preceding pages obey all the axioms of a field, the set of points on the line l after the point ∞ has been deleted is a field with respect to these operations. The reader will notice that we had to invoke axiom P only once, in order to prove the commutative law of multiplication. It is possible to prove that conversely, if the commutative law of multiplication is assumed together with axioms 1 to 5 and D, then Pappus' theorem can be proved as a consequence. This fact reveals the essential meaning of Pappus' theorem: In a plane that satisfies axioms 1 to 5 and D, *Pappus' theorem is equivalent to the commutative law of multiplication.* That is, Pappus' theorem, which is a geometric statement, and the commutative law of multiplication, which is an algebraic statement, have the same content.

Algebra in Geometry

On page 310 we observed that real projective geometry can be introduced by purely algebraic means, by defining a point as a set of proportional ordered triples of real numbers (homogeneous coordinates), and defining a line as the set of points whose homogeneous coordinates satisfy an equation of the first degree. That is, by using the algebraic structure called the real number field, we are able to construct the geometric structure called the real projective plane. We concluded then, as we did in Chapter 4, that geometry is a branch of algebra. We have now turned the tables. Using purely geometric means, we have shown that, with appropriate definitions of addition and multiplication of points, the set of all points but one on a line in a projective plane is a field. That is, by using the geometric structure called a projective plane, we are able to construct the algebraic structure called a field. For this reason we may conclude with equal validity that algebra is a branch of geometry. Thus we see that arguments about which is primary, algebra or geometry, make about as much sense as arguments about which came first, the chicken or the egg. Neither is primary. Each implies the other. They have a common content, but they approach this content from different points of view. The way in which the algebraic viewpoint and the geometric viewpoint differ from each other will be discussed in more detail in Chapter 13.

Many Projective Planes

The axioms of general projective geometry imply that, with addition and multiplication of points defined as they were in the preceding paragraphs, the set of points on a line that remain after one point is deleted is a field. But there is nothing in the axioms of general projective geometry that specifies which field it is. There are many different kinds of field, such as the field of real numbers, the field of complex numbers, the field of rational numbers, finite fields, etc. The field on a line in a projective plane may be any one of these fields. To each different kind of field there corresponds a different kind of projective plane. If the field is the field of real numbers, the plane is the real projective plane. If the field is the field of complex numbers, the plane is the complex projective plane. If the field is the field of rational numbers, the plane is a third kind of projective plane called the rational projective plane. If the field is a finite field, the plane is one of many finite projective planes which contain only a finite number of points. Thus there are many different kinds of projective planes that satisfy the axioms of general projective geometry.

Miniature Geometries

The finite projective planes are miniature geometric structures that exhibit all the properties that are deducible from the axioms of general projective geometry. The number of points on a finite projective plane can be calculated easily from the number of elements in the finite field on each of its lines.

Suppose l is a line in a finite projective plane, and P is a point in the plane that is not on l. Let Q be any other point in the plane. Then the line PQ meets the line l, since any two lines in a projective plane intersect. Consequently every point in the plane lies on a line through P that intersects l. This fact gives us a way of counting all the points in the plane. All we have to do is count the points that lie on all the lines that join P to points of l:

Let N be the number of points in the finite field on l after the point ∞ is deleted. Then the number of points on l is $N + 1$. It is not difficult to prove that every line in the plane contains the same number of points. (See exercise 30 on page 355.)

There is a line from P to each point of l. Consequently there are $N + 1$ lines on P. Each of these lines contains $N + 1$ points, including P. Each of these lines contains N points, not counting P. Therefore, not counting P, the number of points in the plane is $N(N + 1) = N^2 + N$. If we now include P, we see that the total number of points in the plane is $N^2 + N + 1$.

To complete the count for any particular finite plane, all we need do is substitute into this formula the appropriate value of N. As we noted on page 37, the number N always has the form p^n, where p is a prime number and n is any positive integer.

Seven-Point Geometry

If we use the smallest possible values of p and n, namely 2 and 1, $N = p^n = 2$, and $N^2 + N + 1 = 7$. Consequently the smallest possible projective plane has exactly seven points. By the principle of duality, it also has exactly seven distinct lines. In this seven-point plane every line has three points on it, since $N + 1 = 3$, and every point has three lines through it. If the points of the plane are denoted by A, B, C, D, E, F and G, then the lines may be represented by the columns in the following rectangular array:

A	B	C	D	E	F	G
B	C	D	E	F	G	A
D	E	F	G	A	B	$C.$

Coordinates in the Plane

The fact that every line in a projective plane, after deletion of one point, has the structure of a field makes it possible to use purely geometric means to assign coordinates to each point in the plane.

The first step is to introduce a system of old-style *non-homogeneous coordinates* in which each point of the plane is represented by an ordered pair of members of the field:

Let *l* be a line in the plane, with points, *0, 1* and ∞ chosen on it as on page 333, and with the corresponding field structure on it defined. Draw any other line *n* through ∞ and choose another point on *n* and denote it by the symbol ∞'. Draw the line through *0* and ∞' and call it *m*. On *m* choose any point that is not *0* or ∞', and call it *1'*. Let *U* be the intersection of the line through *1* and

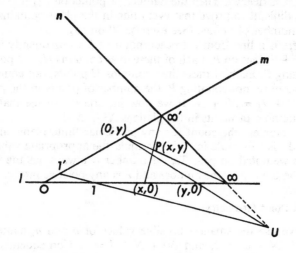

1' and the line through ∞ and ∞'. We assign non-homogeneous coordinates to all points of the plane except those on the line through ∞ and ∞' as follows:

1) If the point is on *l*, it is a member *x* of the field on *l*. Assign to it the coordinates (*x, 0*).

2) If the point is on *m*, the line through this point and *U* intersects *l* at some point *y*, whose coordinates by rule 1) are (*y, 0*). Assign to the point on *m* the coordinates (*0, y*).

3) If the point is some point P that is not on l or m, draw $P\infty'$ and $P\infty$, and let them intersect l and m respectively. These points of intersection already have coordinates assigned by rules 1) and 2). If these coordinates are $(x, 0)$ and $(0, y)$ respectively, assign to P the coordinates (x, y).

By this procedure, no coordinates are assigned to the points on the line $\infty\infty'$. We call this line the *line at infinity*. We call the points on it *ideal points* and we call all the other points of the plane ordinary points with respect to this system of non-homogeneous coordinates. Notice that any line of the plane may be chosen to serve as the line at infinity.

After non-homogeneous coordinates are introduced in this way, homogeneous coordinates may be defined for all points of the plane, including the ideal points, by the procedure outlined on page 305. In this way, the analytic geometry of the projective plane is introduced by purely geometric means.

Cross Ratio

Once a field structure is defined for the set of all points but one on a line in a projective plane, this field constitutes a number scale for the points on the line except the point ∞. The number attached to each point on the line except ∞ is itself. This assignment of a number to each point except ∞ provides the line with a system of non-homogeneous coordinates in which each point except ∞ is represented by a single coordinate x. Then, as on page 308, each point including ∞ can be represented by an ordered pair of homogeneous coordinates (X_1, X_2) related to the non-homogeneous coordinate x by equation (10) on page 308.

With the help of these homogeneous coordinates on a line it is possible to define an important number called the *cross ratio* for any ordered set of four points A, B, C and D on the line, at least three of which are distinct. If the points A, B, C and D have the homogeneous coordinates (a_1, a_2), (b_1, b_2), (c_1, c_2) and (d_1, d_2) respectively, the cross ratio of the points A, B, C, D taken in that order is the number

$$\frac{\begin{vmatrix} a_1 & a_2 \\ c_1 & c_2 \end{vmatrix} \begin{vmatrix} b_1 & b_2 \\ d_1 & d_2 \end{vmatrix}}{\begin{vmatrix} b_1 & b_2 \\ c_1 & c_2 \end{vmatrix} \begin{vmatrix} a_1 & a_2 \\ d_1 & d_2 \end{vmatrix}},$$

if the number exists. If the number does not exist we say the cross ratio is "infinity." For example, if the four points are on a line in the real projective plane and have homogeneous co-ordinates $(0, 1)$, $(2, 1)$, $(1, 1)$ and $(1, 0)$ respectively, the cross ratio is

$$\frac{\begin{vmatrix} 0 & 1 \\ 1 & 1 \end{vmatrix} \begin{vmatrix} 2 & 1 \\ 1 & 0 \end{vmatrix}}{\begin{vmatrix} 2 & 1 \\ 1 & 1 \end{vmatrix} \begin{vmatrix} 0 & 1 \\ 1 & 0 \end{vmatrix}} = \frac{(-1)(-1)}{(1)(-1)} = -1.$$

The cross ratio of four points has many interesting properties. We mention only two that are relevant to the use we shall make of the cross ratio concept later in this chapter: 1) The cross ratio for four points on a line is independent of the choice of coordinate system on the line. 2) The cross ratio of four points on a line is unchanged by projective transformations. That is, if a projective transformation moves A, B, C and D to A', B', C' and D' respectively, the cross ratio of the ordered quadruple of points A, B, C, D is the same as the cross ratio of the ordered quadruple A', B', C', D'.

Reconstructing the Euclidean Plane

Of all the many projective planes that are possible, the most important one is the real projective plane, because of its intimate relationship to the Euclidean plane. We now have the conceptual equipment for exploring this relationship in greater depth. We originally obtained the real projective plane from the Euclidean plane by adding to the Euclidean plane the line at infinity and simultaneously eliminating some properties such as the existence of parallel lines, order relations, and congruence relations. To understand better the relationship between the real projective plane and the Euclidean plane, we shall now reverse this procedure. Starting with the real projective plane, we shall use it to construct a Euclidean plane. To do so, we remove the line at infinity, and we find ways of restoring to what is left of the plane the properties that we had eliminated before.

Affine Geometry

We saw on page 341 that any line in the real projective plane can be cast in the role of "line at infinity." Simply choose any line n in the real projective plane, and any two other lines l and m, and follow the procedure outlined on page 340 for introducing coordinates in the plane. Then l and m will be the axes for the system of coordinates, and n will be the line at infinity. Since we are dealing with the *real* projective plane, the coordinates will be members of the field of *real numbers*.

Now we want to delete the line n from the real projective plane. But we must do so in a way that makes sense in terms of the concepts of projective geometry. Projective geometry is the study of those properties of configurations in the projective plane that remain unchanged when they are subjected to continuous projective transformations (projective collineations). The set of all such projective collineations is a group with respect to the operation of multiplication of transformations. We call this group the *projective group*. Among the members of this group of transformations we seek those transformations that can help us separate the line n from the rest of the plane. The transformations that serve this purpose are all those for which n is a fixed line. We call a transformation of this kind, that keeps n fixed, an *affine* transformation. The set of all affine transformations is also a group and is called the *affine group*. Because it is contained within the projective group, it is called a *subgroup* of the projective group.

An affine transformation moves every point on n to a new position on n, and it moves every point that is not on n to a new position that is not on n. This fact makes it possible for us to ignore the points of n, and pay attention only to what an affine transformation does to the points that are not on n. We call the set of all points in the real projective plane that are not on n the *affine plane*. By studying what affine transformations do in the affine plane we have in effect deleted the line at infinity from the projective plane.

To determine the structure of the affine plane, we look for properties of configurations that are unchanged by affine transformations. The study of these properties is called *affine geometry*. First we define an affine line to be what is left of a line in the projective plane after we delete from it the point at which it intersects the line at infinity. We call this deleted point the ideal point of the affine line. Two affine lines that have the same ideal point *do not intersect in the affine plane*. Consequently we can introduce the concept of parallelism in the affine plane by the following definition: Two affine lines are parallel if they have the same ideal point. If two affine lines are parallel, their images under an affine transformation are also parallel. So parallelism is a property that is preserved by affine transformations.

An affine line is obtained from a projective line by deleting its ideal point. A projective line in the real projective plane is like a closed loop, with its points arranged in circular order. Deleting the ideal point is like cutting the loop. The points that remain are arranged in linear order, like the points on an open necklace. In fact, the linear order on the affine line can be defined in terms of the circular order on the projective line in the following way: Let A, B and C be three points on an affine line, and let I be the ideal point on the line. We say that B is *between* A and C if A and

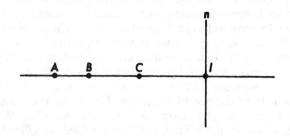

C separate B and I. With this definition we have introduced linear order relations in the affine plane. It can be shown that order relations of points are not changed by affine transformations. That is, if B is between A and C, and an affine transformation moves A, B and C to A', B' and C' respectively, then B' is between A' and C'.

Since linear order relations exist in the affine plane, we can define the concepts of segment and angle in the affine plane in the same way that we did in the Euclidean plane. We might be tempted to go one step further and try to introduce the concepts of length of a segment and measure of an angle by using equations (17), (18) and (22) of Chapter 4 to define them in terms of non-homogeneous coordinates in the plane. However, a segment length computed by equation (17) and an angle measure computed by equations (18) and (22) are not necessarily unchanged when the segment or the angle is subjected to an affine transformation. Consequently the concepts of length of a segment and measure of an angle do not belong to affine geometry, and the affine plane does not have all the properties of the Euclidean plane.

Although the concept of length does not exist in affine geometry, it is interesting that the concept of midpoint of a segment does. The midpoint of a segment can be defined without using the concept of length in the following way: Let I be the ideal point on the affine line determined by two points A and B. A point M on this line is called the *midpoint* of the segment AB if M

is the harmonic conjugate of I with respect to A and B. If M is the midpoint of AB, and an affine transformation moves A, M and B to A', M' and B' respectively, then M' is the midpoint of $A'B'$. (See exercise 32 on page 355.)

The existence of the concept of parallelism in affine geometry makes it possible to introduce the concept of equivalent directed segments, and the associated concept of vector: Two directed segments AB and DC are equivalent if the lines AB and DC are parallel and the lines AD and BC are parallel. The vector AB is the directed segment AB or any directed segment that is equivalent to it. On the basis of these definitions, and with the help of the system of non-homogeneous coordinates in the plane, it is possible to show that an affine plane is a vector space with properties I to VIII listed on pages 138 and 139. Therefore all the theorems proved in pages 143 to 144 are theorems of affine geometry. Affine geometry was first recognized as a separate branch of mathematics by Euler.

Similarity Geometry

We made the transition from projective geometry to affine geometry by reducing the size of the associated group of transformations from the projective group to the affine group. The result was a loss of generality and a gain of structure, through the reintroduction of parallelism and order relations. However, we have not yet restored the full structure of the Euclidean plane, because we have not yet succeeded in reintroducing the concepts of length and angle measure. To add these additional elements of structure to the affine plane, we have to reduce the size of the associated group of transformations some more. Instead of using the full affine group we shall use an appropriately chosen subgroup of the affine group.

We have already given the original real projective plane a system of coordinates, with the lines l and m as coordinate axes. Using the homogeneous coordinates introduced in this way, every point of the real projective plane is represented by families of ordered triples of real numbers. If we enlarge the system of homogeneous coordinates to include ordered triples of complex numbers, we are in effect embedding the real projective plane in a complex projective plane and thereby adding imaginary points to every line in the real projective plane. We single out for special attention the two points on the line at infinity that are represented by the coordinates $(1, i, 0)$ and $(1, -i, 0)$, and are known as the circular points at infinity. In the group of affine transformations there are transformations for which the circular points at infinity are fixed points. For example, the transformation given in the illustrative exercise on page 312 is such a transformation. An affine transformation for which the circular points at infinity are fixed points is called a *similarity* transformation. The set of all similarity transformations is a subgroup of the affine group, and is called the *similarity group*. The study of those properties of configurations in the affine plane that are unchanged by similarity transformations is called *similarity geometry*.

Using the smaller group of similarity transformations rather than the larger group of affine transformations, we can now introduce the concept of angle measure. This is now possible because it can be shown that the measure of an angle, as defined by equations (17), (18) and (22) of Chapter 4, is unchanged by a similarity transformation. Therefore angle measure is a valid concept in similarity geometry. With angle measure restored to the plane, we can also introduce the concept of congruence of angles by means of this definition: Two angles are congruent if they have the same measure. However, length of a segment, as defined by equation (17), is not itself an invariant under similarity transformations, so length of a segment is not a valid concept in similarity geometry.

Now that we have restored angle measure to the plane, it is interesting to inquire which pairs of lines are perpendicular. It is easy to verify that the lines l and m are perpendicular. (See exercise 33 on page 355.) This is especially interesting in view of the fact that the lines l and m were originally chosen as *any two lines* in the real projective plane other than the line already chosen for the distinction of being the line at infinity. The lines l and m are therefore any two lines in the affine plane. By choosing them to serve as axes for the coordinate system we introduced, we elevated them to the special distinction of being perpendicular lines as determined by equation (22). However, once we have given the pair of lines l and m this distinction, we cannot bestow the same distinction at will to other pairs of lines. Equation (22) determines which pairs of lines shall have this distinction and which pairs shall not. Each pair of lines that has this distinction inherits it from the pair l and m by virtue of the fact that there exists a similarity transformation that brings the pair of lines into coincidence with the pair l and m.

In our step-by-step construction of a plane in which angle measure exists, we have been playing at being God, in imitation of Bolyai. However, while Bolyai created a new universe from nothing (see page 213), our accomplishment is somewhat more modest. We have created the plane with angle measure out of something, namely, out of the real projective plane with which we started. In the course of doing so we have created a hereditary aristocracy among pairs of lines. The lines l and m were made perpendicular lines by divine decree. All other perpendicular lines acquire their aristocratic rank by inheritance, as described above.

There is an interesting geometric way of identifying which pairs of lines are perpendicular. There is a projective involution on the line at infinity whose fixed points are the circular points at infinity. (See page 312.) This involution is called the absolute involution. It can be shown that two lines are perpendicular if and only if their ideal points correspond to each other under the absolute involution.

Equiareal Geometry

Similarity geometry, by introducing the concept of angle measure, has taken us only part way in the direction of restoring the full structure of the Euclidean plane. We shall now start again with the affine group, and follow a different path that introduces the concept of area. This path, too, will take us only part way in the direction of restoring the full structure of the Euclidean plane.

In Chapter 4, equation (15) gave us a necessary and sufficient condition that a point with non-homogeneous coordinates (x, y) lie on the straight line that passes through two given points in a Euclidean plane. If the two points are P_2 and P_3, with non-homogeneous coordinates (x_2, y_2) and (x_3, y_3) respectively, the equation takes the form

$$\begin{vmatrix} x & y & 1 \\ x_2 & y_2 & 1 \\ x_3 & y_3 & 1 \end{vmatrix} = 0.$$

If we substitute for (x, y) the coordinates (x_1, y_1) of a point P_1, we see that P_1, P_2 and P_3 are collinear if and only if

$$\begin{vmatrix} x_1 & y_1 & 1 \\ x_2 & y_2 & 1 \\ x_3 & y_3 & 1 \end{vmatrix} = 0.$$

Therefore, if P_1, P_2 and P_3 are not collinear, and determine a triangle, the determinant on the left-hand side of this equation has a value that is different from zero. Let us denote by

$m(P_1, P_2, P_3)$ one-half of the absolute value of this determinant. That is,

$$m(P_1, P_2, P_3) = \tfrac{1}{2} \text{ absolute value of } \begin{vmatrix} x_1 & y_1 & 1 \\ x_2 & y_2 & 1 \\ x_3 & y_3 & 1 \end{vmatrix}.$$

Among the affine transformations there are some which leave the value of $m(P_1, P_2, P_3)$ unchanged when they are applied to any triple of points (P_1, P_2, P_3). We call the number $m(P_1, P_2, P_3)$, which is unchanged by such transformations, the *area* of triangle $P_1P_2P_3$, and the affine transformations which leave unchanged the areas of all triangles are called *equiareal transformations*. The set of all equiareal transformations is a subgroup of the affine group, and is called the *equiareal group*. The study of those properties of configurations in the affine plane that are unchanged by equiareal transformations is called *equiareal geometry*.

Euclidean Geometry

The similarity group and the equiareal group overlap. That is, there are some affine transformations that belong to both the similarity group and the equiareal group. For example, the transformation given in the illustrative exercise on page 312 is one of them. Such transformations which belong to both groups are called *Euclidean* transformations. The set of all Euclidean transformations is a subgroup of the similarity group and of the equiareal group. The study of those properties of the affine plane that are unchanged by Euclidean transformations is called *Euclidean* geometry.

Every Euclidean transformation, being a member of the similarity group, preserves angle measure. As a member of the equiareal group it also preserves the areas of triangles. It can be shown, too, that if the length of a segment is defined by equation (17) of Chapter 4, it also preserves the length of segments. Consequently the Euclidean transformations are the isometries we talked about in Chapter 6. Thus, by restricting ourselves to the Euclidean group of transformations, we have finally restored to the plane the concept of length of a segment. With segment-length restored to the plane we can introduce the concept of congruence of segments by means of this definition: Two segments are congruent if they have the same length. The affine plane, now equipped via the Euclidean group of transformations with the concepts of congruence of angles and congruence of segments, is called a *Euclidean plane*. It can be shown that it satisfies all the Hilbert axioms, so that it is really the same as the Euclidean plane we talked about in Chapter 3.

Klein's Erlangen Program

In the reconstruction of the Euclidean plane from the real projective plane we have been guided by a particular idea of what constitutes a geometry: A geometry is the study of those properties of configurations in a space that are invariant under a particular group of transformations of the space. To each group of transformations there corresponds a definite geometry. Thus it is possible to have many geometries in one and the same space. This fruitful idea on the nature of a geometry was first propounded by Felix Klein in 1872 in his inaugural address on the occasion of his installation as professor of mathematics at Erlangen. This famous address, now known as the "Erlangen Program," gave a great impetus to mathematical research by equipping mathematicians with a specific technique for inventing new geometries.

A Hierarchy of Geometries

The Euclidean group is a subgroup of the similarity group which is a subgroup of the affine group which in turn is a subgroup of the projective group. This fact establishes a hierarchy of the geometries defined by these groups. The more inclusive the group of transformations that defines a geometry, the more general the geometry is, and the less structure it has. If two geometries are defined by groups such that one group is a subgroup of the other, all the theorems of the more general geometry belonging to the more inclusive group are also valid in the less general geometry that belongs to the subgroup. Thus the theorems of real projective geometry are also valid in affine geometry, similarity geometry and Euclidean geometry; the theorems of affine geometry are also valid in similarity geometry and Euclidean geometry; and the theorems of similarity geometry are also valid in Euclidean geometry. Because the Euclidean group is a subgroup of the equiareal group, which is a subgroup of the affine group, which in turn is a subgroup of the projective group, there is a similar hierarchic arrangement of real projective geometry, affine geometry, equiareal geometry, and Euclidean geometry. The two interlocking hierarchies are shown in the diagram below.

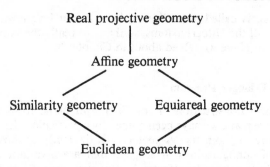

Because Euclidean geometry is at the bottom of the heap in
these two hierarchies, there are many different kinds of theorems
in Euclidean geometry: some are theorems that are valid in
Euclidean geometry and in no more general geometry; some are
theorems that are valid in similarity geometry and in no more
general geometry; some are theorems that are valid in equiareal
geometry and in no more general geometry; etc. In order to make
this distinction among theorems, we shall say that a theorem
belongs to a particular geometry or is a theorem of a particular
geometry if that geometry is the most general geometry in which
the theorem is valid. With this definition in mind, we can re-
examine the theorems we proved in Chapter 3 to see which
geometry they really belong to:

The theorem that the medians of a triangle are concurrent
is a theorem of affine geometry.

Pappus' theorem is a theorem of projective geometry.

Euclid's proposition 27 of Book I of the *Elements* is a theorem
of similarity geometry.

Euclid's proposition 32 of Book I, the angle sum theorem, is a
theorem of similarity geometry.

Euclid's proposition 47 of Book I, the Pythagorean Theorem,
is a theorem of Euclidean geometry.

The Non-Euclidean Geometries

Because Euclidean geometry belongs to the Euclidean group
which is a subgroup of the projective group, Euclidean geometry
may be thought of as a branch of projective geometry. This fact
was first discovered by Cayley. Klein extended this discovery
by showing that hyperbolic geometry and elliptic geometry are
also branches of projective geometry in exactly the same sense.

Hyperbolic geometry is the geometry defined by the group of

all projective transformations that leave a given *real point conic* fixed. The fixed point conic is called the *absolute*. The interior of the absolute is the hyperbolic plane. That part of a projective line that is in the interior of the absolute is a line in the hyperbolic plane. The lengths of segments and the measures of angles in the hyperbolic plane are defined in terms of the projective concept of cross ratio. On the basis of these definitions, it can be shown that the hyperbolic plane constructed in this way within the real projective plane satisfies all the axioms of hyperbolic geometry.

Two lines in the hyperbolic plane are parallel if the projective lines of which they are parts intersect on the absolute. Two lines in the hyperbolic plane are ultra-parallel if the projective lines of which they are parts intersect outside the absolute. The points inside the absolute are the ordinary points of the hyperbolic plane. The points on the absolute are the ideal points, and the points outside the absolute are the ultra-ideal points that are associated with the hyperbolic plane. Thus we see that the extended hyperbolic plane pictured in the diagram on page 233 is more than a mere visual aid. It represents an actual state of affairs. The extended hyperbolic plane is the real projective plane in which the hyperbolic plane is embedded.

Elliptic geometry is the geometry defined by the group of all projective transformations that leave a given *imaginary point conic* fixed. The entire projective plane, equipped with appropriate definitions of angle measure and arc measure, turns out to be what we have called the Cayley-Klein elliptic plane.

Because hyperbolic geometry and elliptic geometry are defined by groups of transformations that are subgroups of the projective group, we can include them in the diagram showing the hierarchies of geometries:

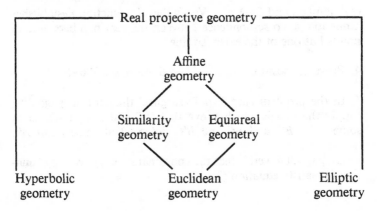

The diagram expresses clearly the remarkable fact that hyperbolic geometry, Euclidean geometry, elliptic geometry, similarity geometry, equiareal geometry, and affine geometry are all branches of projective geometry. This fact led Cayley to say that "Projective geometry is all of geometry." We shall see in Chapter 12 that his conclusion was premature.

EXERCISES FOR CHAPTER 10

1. Draw two distinct lines l and l' in a plane and let O be a point that is not on either line. If A is any point on l, find the point A' on l' that corresponds to it under the perspectivity with O as center. If B' is any point on l', find the point B on l that it corresponds to under this perspectivity.

2. Let P and Q be two distinct planes, let O be a point that is not on either of them, and consider the perspectivity T between P and Q with center O. Let A and B be two points in P. Let A' and B' respectively be their images under T. a) Prove that if C is any point on the line AB, its image C' under T is on the line $A'B'$. b) Prove that the correspondence between the points of AB and the points of $A'B'$ that associates with each point C of AB its image C' under T is a perspectivity in the plane OAB, as defined on page 291.

3. Put a real number scale on a Euclidean line. Add the point at infinity and attach to it the "number" ∞ (read as "infinity"). Let a be any real number. If x is any number on the extended line except a, define a function f as follows: $f(x) = \dfrac{1}{x - a}$ when x is a real number, and $f(\infty) = 0$. Verify that the function f establishes a one-to-one correspondence between the real numbers and all points but one of the extended line.

4. Prove propositions $1'$, $2'$, $3'$ and $6'$ on pages 296–297.

5. In the proof of case I of Desargues' theorem on page 299, supply the details which prove that AC and $A'C'$ intersect at some point B'', and BC and $B'C'$ intersect at some point A''.

6. On page 306, verify that the coordinates (x, y) given by equation (3) satisfy equation (2).

7. What are the old-style coordinates (x, y) represented by the homogeneous coordinates a) $(1, 2, .01)$; b) $(1, 2, .001)$; c) $(1, 2, .0001)$?

8. a) Write the equation in homogeneous coordinates for the line whose old-style equation is $y = mx + d$. b) If $m = \dfrac{b}{a}$, show that the homogeneous coordinates $(a, b, 0)$ satisfy this equation.

9. Write the equation in homogeneous coordinates for the conic sections whose old-style equations are

a) $y = x^2$.
b) $x^2 + y^2 = 25$.
c) $xy = 12$.
d) $(x - a)^2 + (y - b)^2 = r^2$.

10. Prove that $(i, 1, 0)$ is not proportional to any ordered triple of real numbers.

11. Prove that the point whose homogeneous coordinates are $(1, i, 1 + i)$ is an imaginary point.

12. Verify that the determinant

$$\begin{vmatrix} .6 & .8 & 1 \\ -.8 & .6 & 1 \\ 0 & 0 & 1 \end{vmatrix} \neq 0.$$

13. The circle with center at the origin and radius 5 has the equation $x^2 + y^2 = 25$.

a) Write its equation in homogeneous coordinates (X_1, X_2, X_3).
b) Verify that the homogeneous coordinates $(1, i, 0)$ and $(1, -i, 0)$ satisfy this equation.

14. Define a complete quadrilateral (the dual of a complete quadrangle).

15. Define a diagonal line of a complete quadrilateral.

16. Define a harmonic sequence of lines on a point.

17. Prove that if a projectivity between two distinct lines moves a point into itself, it must be a perspectivity.

18. a) Write in homogeneous coordinates the equation of the parabola $x = y^2$.
b) Verify that this equation is satisfied by the ideal point $(1, 0, 0)$ on the x axis.

19. a) Write in homogeneous coordinates the equation of the hyperbola $xy = 12$.
b) Verify that this equation is satisfied by the ideal point $(1, 0, 0)$ on the x axis, and the ideal point $(0, 1, 0)$ on the y axis.

20. a) Let A, B and C be the vertices of a triangle, and let l and m be two distinct lines that contain neither A nor B nor C. Construct a projectivity between the lines on A and the lines on C as follows: Let p be any line on A. Let P be its intersection with l. Draw PB and call it p'. Let Q be the intersection of p' and m. Draw QC and call it p''. Then p'' is the line on C that corresponds to the line p on A.

b) Determine the intersections of the corresponding pairs of lines, and observe that they generate a conic section.

21. Locate on a circle

a) two point pairs A, B and C, D that separate each other;
b) two point pairs A, B and E, F that do not separate each other.

22. Apply steps 1) to 4) for addition of points to show that $0 + X = X$.

23. Apply steps a) to d) for multiplication of points to show that $1X = X$.

24. Prove without using the commutative law that if $X \neq 0$ or ∞, there exists a point Y on l such that $YX = 1$.

25. Define $X + \infty$ via steps 1) to 4) on page 329, and show that $X + \infty = \infty$.

26. Define $\dfrac{1}{0}$ and $\dfrac{1}{\infty}$ by substituting 0 and ∞ for X in the construc-

tion of the reciprocal of X on page 336. Then show that $\frac{1}{0} = \infty$, and $\frac{1}{\infty} = 0$.

27. Use steps a) to d) for multiplication of points to verify that if X is not ∞, $0X = 0$.

28. Show that if steps a) to d) are used to try to multiply 0 by ∞, the result is indeterminate.

29. Prove that if a line in a finite projective plane contains exactly $N + 1$ points, every line in the plane contains exactly $N + 1$ points.

30. Identify all the finite projective planes which contain fewer than 100 points.

31. In the real projective plane, find the cross ratio of the ordered quadruples of points on a line with the following sets of homogeneous coordinates:

a) $(1, 1), (2, 1), (3, 1), (4, 1)$;
b) $(1, 0), (2, 1), (5, 1), (6, 1)$;
c) $(1, 3), (2, 1), (7, 1), (1, 3)$.

32. In the affine plane, let M be the midpoint of AB, and let A', M', and B' be the images respectively of A, M and B under an affine transformation T. Prove that M' is the midpoint of $A'B'$.

33. Using the system of non-homogeneous coordinates provided for the affine plane on page 340, the point $(0, 0)$ lies on both l and m, $(1, 0)$ lies on l, and $(0, 1)$ lies on m. Use equations (17), (18) and (22) of Chapter 4 to show that if θ is the angle between l and m, $\cos \theta = 0$.

11

Geometry in Relativity Physics

The Observer and the Observed

Every physical observation is a relationship between the observer and the physical phenomenon that is observed. The observer does not see directly what the phenomenon *is*. He sees what the phenomenon *looks like* to him, as seen from his position and measured by his measuring instruments. Thus his observation may be thought of as the product of two factors. One factor is contributed by the phenomenon being observed. The other factor is contributed by the special point of view of the observer. This fact poses a fundamental problem for the physical scientist: How can he eliminate from his observation that part of it that is contributed by his own special point of view, to obtain as a residue that part that is contributed by the phenomenon being observed. He tries to solve this problem by deriving from his observations a description of the observed phenomenon that will be the same for all observers. Newtonian physics succeeded in obtaining a partial solution to this problem. The theory of relativity of Albert Einstein (1879–1955) obtained a more complete solution of this problem that took into account nineteenth century discoveries in the study of electricity and magnetism. A consequence of this theory that is relevant to the subject matter of this book was a revolutionary change in our conceptions of physical space.

Newton's Absolute Space

Newton assumed the existence of a physical space that is independent of the phenomena that occur in it. He pictured this space as being empty, immovable, infinite, three-dimensional, and Euclidean. In this space, each point had a definite position that did not depend on the point of view of any observer, and hence was absolute. After a unit of length was chosen, the straight line segment between two points had a definite length that did

356

not depend on the point of view of any observer, and hence was absolute.

Newton also assumed that there was an absolute flow of time. After a unit of time was chosen, a one-to-one correspondence could be set up between the set of all instants of time, past, present and future, and the set of all real numbers. Then the interval between two instants of time had a definite length, represented by the difference of the real numbers that correspond to these instants. This length of the time interval was assumed to be independent of the point of view of any observer, and hence was absolute.

Newton assumed, too, that interactions between material particles were transmitted instantaneously across the space that separated them.

Frames of Reference

Let us picture an observer in Newton's absolute space, making some observations. The observer is equipped with a frame of reference, consisting of a set of axes, a yardstick for measuring distances from the axes, and a clock for measuring the passage of time. Let us assume first that the axes are not moving through space, or, if they are moving, that we consider only a specific instant of time when the axes are in a definite position. To simplify the picture, let us restrict our attention to observations made in a particular plane with respect to rectangular axes OX and OY in the plane. To give a specific example of the problem the physicist faces when he tries to draw from his observations conclusions that are valid for all observers, let us see what he can do when he observes two particles in the plane, one at point A and the other at point B. Using the yardstick, he can measure the coordinates of both points. Suppose that the coordinates of A are (x_1, y_1), and the coordinates of B are (x_2, y_2). Obviously the values of these coordinates depend on the position of the axes. If the observer used axes OX' and OY' (shown in the diagram) instead of OX and OY, he would observe different sets of coordinates for the same two points A and B. Consequently the coordinates of points are *relative* measurements, expressing a relationship between the points and the axes from which distances are being measured.

Now consider the directed segment from A to B. This directed segment can be decomposed into two components AC and CB that are parallel to OX and OY respectively. Let us call them the x and y components of AB. The lengths of AC and CB are the absolute values of $x_2 - x_1$ and $y_2 - y_1$ respectively. If the ob-

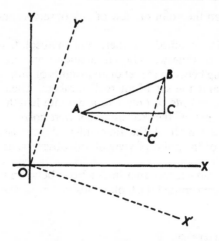

server used axes OX' and OY' instead of OX and OY, he could determine in a similar manner the lengths of components AC' and $C'B$, which are parallel to OX' and OY' respectively. Notice that the lengths of the two components depend on the choice of axes; consequently the lengths of the x and y components of a directed segment are *relative* measures.

Now, using the components AC and CB of the directed segment AB, the observer can compute the length of AB by using the formula $d = \sqrt{(x_2 - x_1)^2 + (y_2 - y_1)^2}$. If he used other axes that were in a different position, although the coordinates of A and B would be different, and the x and y components of AB would be different, this computation would yield the same value for the length of AB. Consequently, in Newtonian physics, the length of a directed segment is an *absolute* measure that is independent of the choice of axes by the observer. If there are many observers, each equipped with his own frame of reference, then while each may observe different x and y components for the directed segment AB, depending on the orientation of his axes, all will compute the same length for the segment.

Inertial Frames of Reference

In the preceding paragraph we dealt with measurements made at a fixed instant of time, so that it might be assumed that the frames of reference were stationary in absolute space. However, as soon as we begin to observe events that occur as time passes, we are not free to make this assumption any more, because it is possible that the frame of reference of an observer may be moving through absolute space.

Among all possible frames of reference, there are some that have the special property that observations made from them are subject to Newton's First Law of Motion, which says, "Every body continues in its state of rest, or of uniform motion in a straight line, unless it is compelled to change that state by forces impressed on it." Frames of reference that have this property are called *inertial* frames. Newton assumed that frames of reference that are stationary in absolute space are inertial. It follows from this assumption that any frame of reference that has a uniform speed through absolute space is also inertial. Conversely, any frame of reference that is inertial has a uniform speed through absolute space, and in the special case where the speed is zero, the frame is at rest in absolute space.

The frame of reference usually used for the observation of events in a laboratory is the ground on which we stand. For small intervals of time, during which the effects of the earth's rotation are negligible, the ground as a frame of reference is approximately inertial. A train moving at a uniform speed across the ground is also approximately inertial.

Inertial frames of reference have a further common property that distinguishes them from all other frames of reference: *The laws of mechanics are the same for all inertial frames of reference.* This fact is known as the Newtonian principle of relativity. The meaning of the principle of relativity may be stated in two ways. Mathematically, it means that the equations in which a law of mechanics is expressed have the same form for all inertial frames of reference. Experimentally, it means that a mechanical experiment carried out in the same way will have the same outcome in any inertial frame of reference. This is the underlying reason why we can write, drink water, or play ball on a train that is moving uniformly over the ground just as we do on the ground itself.

The assumption that there is an absolute space implies that an inertial frame of reference has an absolute speed through this space. However, the principle of relativity implies that there is no way of detecting this speed by mechanical means.

The Addition of Velocities

Suppose an observer on the ground watches a railroad train go by at a speed of 60 miles per hour. If a man on the train walks forward on the train at a speed of 2 miles per hour with respect to the floor of the train, the observer on the ground sees him move over the ground with a speed of 62 miles per hour. If the man were walking toward the rear of the train, his speed with respect to the ground would be 58 miles per hour. In general, in Newtonian mechanics, if the velocity of the train over the ground is represented by a vector, and the velocity of the walking man with respect to the train is also represented by a vector, then his velocity over the ground is the sum of these two vectors, where the sum is obtained by the triangle rule for addition of vectors given on page 138.

Waves Through the Ether

During the nineteenth century a crisis developed in Newtonian mechanics as a result of progress in the study of optics, electricity and magnetism. It was well known that light, on its way from the sun to the earth, travels through empty space. The phenomenon of interference of light, discovered in 1802 by Thomas Young (1773–1829), showed that light must be understood as a vibration traveling in waves through space. The equations of Clerk Maxwell (1831–1879) for the electromagnetic field predicted that vibrations in an electromagnetic field also travel in waves through space with the speed of light. This prediction was confirmed by Heinrich Hertz (1857–1894) in his laboratory, and is reconfirmed every day by millions of people who receive in their radio sets electromagnetic waves sent out from broadcasting studios. This discovery led to the recognition that light is only one of many kinds of electromagnetic wave that differ from each other in wavelength.

Physicists at first found it difficult to think of vibrations in space without picturing something that was vibrating. They therefore postulated that space was permeated by a substance they called the *ether*, and that electromagnetic waves were vibrations moving through the ether. With the assumption of the existence of the ether, physical space ceased to be a void and became a plenum again.

The Michelson-Morley Experiment

If the ether really existed, then it was a frame of reference that was stationary with respect to absolute space. As the earth moved through absolute space, the ether would drift past it in an ether wind, just as the air drifts past a moving car in a wind that you can feel when you put your hand out of the car window. Then the relationship between light, the ether, and the earth would be comparable to the relationship between the walking man, the train, and the ground described on page 360. The velocity of light with respect to the earth would be the sum of the velocity of light with respect to the ether and the velocity of the ether drift past the earth. Moreover, the value of this sum would vary with the direction of the light's path with respect to the path of the earth. In particular, two light rays traveling along paths that are perpendicular to each other would have a detectable difference in speed with respect to the earth. In 1887, Albert Michelson (1852–1931) and E. W. Morley (1838–1923) performed an experiment with instruments capable of detecting such a difference if it existed. They found that there was no difference at all. The Michelson-Morley experiment showed that the speed of light with respect to an observer is independent of the motion of the observer through the ether, even though this fact is contradicted by the triangle rule for the addition of velocities.

The Special Theory of Relativity

In order to eliminate this contradiction, Einstein proposed in 1905 a new theory of mechanics now known as the *Special Theory of Relativity*. The foundation of the theory is an extension of the principle of relativity. While the Newtonian principle of relativity asserts that the laws of *mechanics* are the same for all inertial frames of reference, the Einstein principle of relativity says that the laws of *mechanics and of electromagnetism* are the same for all inertial frames of reference. Einstein's theory also rejects the Newtonian assumption that interactions between material particles are transmitted instantaneously through space. Electromagnetic interactions are transmitted with the speed of light, as shown by Maxwell, and in accordance with the extended principle of relativity, this speed should be the same for all observers. Starting with these assumptions, Einstein drew some startling conclusions: If two observers whose frames of reference are in relative motion measure the length of a segment that is parallel to the direction of their relative motion, they obtain different lengths, and the ratio of these two lengths depends on the speed of their relative motion. If they measure the duration of time between two events, they obtain different measures, and the ratio of these measures, too, depends on the speed of their relative motion. Moreover, they do not agree on whether or not two events at different places are simultaneous. Events that are simultaneous as seen from one frame of reference may be consecutive as seen from the other. Thus, in the theory of relativity, the length of segments, the duration of time intervals, and simultaneity have meaning only relative to a definite frame of reference. In this respect they are like the coordinates of a point, or like the x and y components of a segment in a Euclidean plane.

Space-Time

The Special Theory of Relativity led to a new conception of space, formulated by Einstein's former teacher Minkowski (1864–1909). Minkowski said that space and time should not be viewed as two separate continua but as components of one four-dimensional space-time continuum. A point in space-time *is a particular point in space at a particular time*, and is called an *event*. In any given frame of reference, an event has four coordinates, (x, y, z, t), where the coordinates (x, y, z) indicate its position in space, and the coordinate t indicates its position in time. If two events have coordinates (x_1, y_1, z_1, t_1) and (x_2, y_2, z_2, t_2) respectively, the distance d between the places where the events occur, and the length of time T between the instants when they occur are given by the formulas $d = \sqrt{(x_2 - x_1)^2 + (y_2 - y_1)^2 + (z_2 - z_1)^2}$ and $T = t_2 - t_1$. Although d and T are not the same for all inertial frames of reference, there is a quantity related to them that is. This quantity, called the *interval s* between the two events, is given by the formula $s^2 = d^2 - c^2T^2$, where c is the speed of light. The quantities d and T are the space and time components of the interval s in space-time just as the segments AC and CB are the x and y components of the segment AB in the Euclidean plane in the diagram on page 120. Different frames of reference in space-time resolve an interval into different space and time components just as different sets of axes in a Euclidean plane resolve a line segment into different x and y components. Consequently there is no absolute space and absolute time. Each frame of reference has its own private space and time.

The General Theory of Relativity

The Special Theory of Relativity deals with the motions of material particles in physical space as observed from inertial frames of reference. In 1915 Einstein extended the theory so that it could be applied to any frame of reference, non-inertial as well as inertial. The extended theory, now known as the General Theory of Relativity, is based on the observed fact that the two kinds of mass that enter into physical theory, *inertial mass* and *gravitational mass*, are numerically equal. The inertial mass of a material particle is a measure of its tendency to resist changes of motion. When a force on a particle causes it to accelerate, the inertial mass of the particle is the ratio of the force to the acceleration. The gravitational mass of a particle is a measure of its tendency to be attracted toward other particles by a force of gravitation, and thus to be accelerated toward them. In Newtonian mechanics, the numerical equality of these two masses appears as an unexplained but fortunate accident. Einstein proposed that this was not an accident, and that inertial mass and gravitational mass are intrinsically equivalent. On the basis of this assumption, accelerated motions in a non-inertial frame of reference can be treated as motions in an inertial frame of reference in the presence of a gravitational field. The General Theory of Relativity is essentially a new theory of gravitation that replaces Newton's theory of gravitation.

In the General Theory of Relativity, the space-time continuum is a four-dimensional Riemannian space. Its three-dimensional space-component relative to a particular frame of reference is a Riemannian space whose curvature varies from point to point with the density of matter. Thus, in this theory, space cannot be considered apart from the matter that is in it. In Newtonian physics, space is a pre-existing stage on which material particles are the characters acting out the drama of physical events. In relativity physics, however, as the mathematician Sir Edward Whittaker points out, the *characters create the stage as they walk about on it.* Since the properties of space in relativity theory are inseparable from the matter that is in it, relativity theory represents a return to the Aristotelian view that space is a plenum and not a void.

The Geometry of Physical Space

The surface of the earth is curved. Because of the mountains and valleys on it, its curvature varies from point to point. However, we can imagine the surface of the earth smoothed out by cutting off the mountain tops and filling in the valleys. On this smoothed surface the curvature is approximately the same at all points. Consequently the geometry of the surface of the earth must be one of the three geometries discussed in earlier chapters, namely, Euclidean, hyperbolic or elliptic. Since the constant curvature of the surface of the earth is positive, its geometry is in fact elliptic.

Similarly, although the curvature of space varies from point to point with the density of matter, we can imagine physical space smoothed out by a uniform distribution of its matter. In this smoothed-out physical space, which is an idealized and simplified version of actual physical space, the curvature of space is constant. Consequently we may legitimately ask whether this constant curvature is zero, negative or positive. This is equivalent to asking which geometry, Euclidean, hyperbolic, or elliptic, correctly describes the relations in this smoothed-out space. Before we examine ways in which this question may be answered, we must first explore the meaning of the question itself.

Euclidean, hyperbolic and elliptic space are abstract mathematical systems in which certain assertions are made about undefined terms such as *point* and *line*, and undefined relations such as *congruence*. Some of these assertions are assumed as the axioms of the system, and others are proved as the theorems of the system. When we ask which system correctly describes physical space, we are asking which set of axioms and theorems gives us true statements about the points, lines and congruence relations in physical space. But this question assumes that we have already decided in advance which configurations in physical space are to be called points and lines, and which relations are to be called congruence relations. That is, in order to answer the question, we must first set up a one-to-one correspondence between the terms and relations of abstract geometry and certain physical configurations and relations that will be their counterparts in physical space.

Poincaré pointed out that since there may be more than one way of choosing the physical counterparts of the undefined terms and relations of geometry, the answer to the question is indeterminate because it is conditional on the choice that is made. That is, there may be one answer if one choice is made, but there may be another answer if a different choice is made. For example, the region inside a circle in a Euclidean plane has a Euclidean geometry if we interpret *line* and *congruence* in the usual sense. However, if we use the word *line* to mean arc of a circle perpendicular to the boundary of the region, use the word *angle* to mean the angle between the tangents to two such arcs at their point of intersection, and use the terms *congruent segments* and *congruent angles* to mean segments with equal length and angles with equal measure respectively, where length is defined by the formula given on page 250, then the same region has a hyperbolic geometry.

However, once a choice has been made of the physical configurations and relations that are to be represented by the words *point, line,* and *congruence,* there is a determined answer to the question, "Is the curvature of physical space zero, positive or negative?" Then finding out what the answer is becomes a problem for the experimental physicist.

Point, Line and Congruence in Physical Space

We begin therefore by specifying the meaning of the words *point, line segment,* and *congruence* when used in reference to physical space. We do so by appealing to some intuitive ideas that are part of our everyday experience.

The fundamental idea on which all conceptions of physical space rests is that of a *rigid body.* We classify some bodies, such as a rock, or a block of wood, or a stiff sheet of metal, as rigid. All these bodies are rigid only under certain specified conditions of temperature, pressure, etc., but these details need not concern us here.

We shall understand a *point* to be the position of a small particle whose size is negligible in the context in which it is being examined.

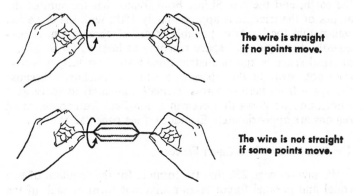

The wire is straight if no points move.

The wire is not straight if some points move.

We shall understand a *segment of a straight line* to be an axis of rotation of a rigid body, where rotation, and axis of rotation are defined as follows: We say that a rigid body has undergone a rotation if it is moved in such a way that two of its points do not move while at least one of its points does move. The axis of the rotation is the set of all the points in the body that do not move during the rotation. In accordance with this definition, if we hold a stiff wire of negligible thickness between the thumb and forefinger of each hand, and then roll the wire between our fingers, then, if no part of the wire changes its position, the wire represents a segment of a straight line.

We shall understand two line segments to be *congruent* if a line segment drawn on a rigid body can be brought into coincidence with each of them. Similarly, two angles are said to be congruent if an angle drawn on a rigid body can be brought into coincidence with each of them. These definitions underlie our everyday use of a yardstick to see if two line segments are congruent, and our use of a protractor to see if two angles are congruent.

The definitions given above are crude. But the refinements that are needed would alter only the details and not the principle, so they need not concern us here.

The Angle Sum of a Triangle

Theoretically, we should be able to determine whether physical space is Euclidean, hyperbolic or elliptic by measuring the sum of the angles of a triangle. Gauss performed this experiment using a terrestrial triangle whose vertices are at Inselsberg, Brocken, and Hoher Hagen. Lobatschewski performed a similar experiment, using astronomical data for a triangle whose vertices are the sun, the earth, and the star Sirius. Both found that the sum of the angles of the triangle is approximately 180°, with the deviation well within the limits of possible errors of observation. Consequently, for regions of space reaching at least as far as Sirius, physical space is approximately Euclidean. This fact, however, does not assure us that space as a whole is Euclidean, because the space from here to Sirius is small compared to space as a whole, and we know that even in a non-Euclidean space, small regions are approximately Euclidean. (See page 252.)

The Resultant of Two Equal Forces

We saw on page 256 that the formula for the resultant of two equal and parallel forces takes a different form in each of the three different geometries. So, theoretically, we should be able to determine which geometry is true for physical space by seeing which formula best fits the facts of experiment. If each of the forces has magnitude F, the resultant has magnitude $2F$ if space is Euclidean; it has magnitude $2F \cosh \dfrac{b}{k}$ if space is hyperbolic; and it has magnitude $2F \cos \dfrac{b}{r}$ if space is elliptic. In these formulas, b is half the distance between the lines of action of the forces, and k and r are constants that are characteristic of the space. Lobatschewski's astronomical calculations showed that if space is non-Euclidean, the numbers k or r are very large. For values of b that are very small compared to k or r, both $\cosh \dfrac{b}{k}$ and $\cos \dfrac{b}{r}$ are almost equal to 1. Consequently for small values of b the three formulas are approximately the same, and therefore cannot be used to distinguish one geometry from the other. Unfortunately, any experiment in which we try to measure two equal forces and their resultant must be limited to circumstances in which the value of b is small.

Finding the Curvature of Space

Measuring the angle sum of a triangle or the resultant of two forces are necessarily local experiments in a small region of physical space. They cannot help us determine whether space is Euclidean, hyperbolic or elliptic because the local differences among the three kinds of space are smaller than the possible experimental error. In order to determine the curvature of space we shall have to use some criterion which is not local, but relies on some significant property of space as a whole. There is such a criterion that is provided by the general theory of relativity. It relies on two properties of space as a whole: the average density of matter in space, and the "red shift" constant.

The "Red Shift" and Its Meaning

The stars in the observable universe are gathered in great assemblages called *galaxies*, that are scattered throughout space in all directions. Analysis of the light spectra of distant galaxies shows that the light of a galaxy is shifted to the red or low frequency end of the spectrum compared to light from a stationary source. According to the theory of the Doppler effect, such a shift to the red end of the spectrum indicates that the galaxy is receding from us.* If light whose frequency is ω_0 when sent out by a stationary source has a lower frequency ω when sent out by a galaxy, the ratio by which the frequency of the light has changed is given by the formula $(\omega - \omega_0)/\omega_0$. This ratio, according to the theory of the Doppler effect, is a measure of the speed with which the galaxy is receding. The measurement of this ratio for many

---

* For an explanation of the Doppler effect, see *The Stars: Stepping-stones into Space*, by the same author, The John Day Company, 1958.

galaxies whose distances were known led to a remarkable discovery: the more distant a galaxy is from us, the faster it recedes. The connection between the red shift and the distance l of a galaxy is given by Hubble's Law, formulated in 1928:

$$(1) \qquad \frac{\omega - \omega_0}{\omega_0} = -\alpha l.$$

From astronomical measurements, the red-shift constant that occurs in this formula has been found to have the following value:

$$(2) \qquad \alpha = 5.6 \times 10^{-26} \text{ per centimeter.}$$

This value of α corresponds to an increase in the velocity of recession of 160 kilometers per second for each million light-years of distance.

The Expanding Universe

To account for the observed fact that the galaxies seem to be receding from us, and that the speed of recession increases with distance, physicists have postulated that the universe is expanding. In an expanding universe, if every unit of distance increases by an amount d in one second, then a distance of l units increases by an amount ld per second. Then a galaxy whose distance from us is l would be receding from us at a speed of ld per second. Thus the theory of an expanding universe successfully accounts for the observed fact that the speed of recession of a galaxy is proportional to its distance l.

The Expanding Universe in Relativity Theory

Relativity theory enters the picture by supplying models of the universe that fit the assumption that the universe is now expanding. If it is assumed that the density of matter in space is uniform, so that the curvature of space is constant, it is possible to derive from the equations of relativity theory some simple models of an expanding universe. In one of these models the universe is described as expanding ever since an initial "explosion." In another model the universe is described as vibrating by alternately expanding and contracting, and the present moment is located in one of the periodic phases of expansion. In the equations associated with these models, the red-shift constant α is related to λ, a measure of the curvature of space, and to μ, the density of matter, by the following formula:

$$(3) \qquad \lambda = \frac{8\pi k}{3c^2} \mu - \alpha^2,$$

where c is the speed of light, and k is the gravitational constant, with these values:

(4) $c = 3 \times 10^{10}$ centimeters per second;

(5) $k = 6.67 \times 10^{-8}$ cubic centimeters per gram per (second)^2.

From equation (3) we see that $\lambda = 0$ if

$$(6) \quad \mu = \mu_0 = \frac{3c^2\alpha^2}{8\pi k} = 5 \times 10^{-24} \text{ grams per cubic centimeter.}$$

Moreover, λ is positive if $\mu > \mu_0$, and λ is negative if $\mu < \mu_0$. Consequently, according to the expanding universe models derived from the theory of relativity, and the assumption that the density of matter in space is uniform, physical space is Euclidean, elliptic or hyperbolic according as the average density of matter in the universe is equal to, greater than, or less than 5×10^{-24} grams per cubic centimeter. Thus we can determine the nature of the geometry of physical space by measuring the average density of matter in the universe. Unfortunately our present knowledge about the density of matter in the universe is not accurate enough to allow us to determine whether it is indeed equal to or greater than or less than 5×10^{-24} grams per centimeter. However, with better telescopes, and with the new techniques of radio astronomy, X-ray astronomy, and neutron astronomy, there is hope that in the future the density of matter may be measured accurately enough to decide the question. So, while the problem of identifying which of the three geometries is true of physical space has not yet been solved, it is in principle solvable.

Is Space Finite or Infinite?

The answer to this question is linked to the determination of the kind of geometry that is true for physical space. If physical space is elliptic, then it is finite, though unbounded, just as the surface of a sphere is finite and unbounded. However, if space is either Euclidean or hyperbolic, it is infinite in extent.

If we modify the question by asking, "Is the *observable* universe finite or infinite?" then we have to modify the answer. When we look at a distant galaxy, we are looking back in time as well as out into space, because we see the galaxy as it was when the light by which we see it left the galaxy. If the galaxy is one billion light years away, it means that the light took one billion years to reach us. Consequently we see the galaxy as it was one billion years ago. Now, if the universe is expanding as the result of an initial explosion, the theory shows that only a finite amount of time has elapsed since the expansion began. Similarly, if the universe is vibrating, only a finite amount of time has elapsed since the present expansion phase began. If this amount of elapsed time is T years, then we cannot see farther back in time than T years. Consequently the farthest out that we can possibly see in space is the distance that light can travel in T years. This distance is finite. Hence, on the basis of the theory of either the expanding or the vibrating universe, even if the geometry of space is Euclidean or hyperbolic, the observable part of the universe is finite.

There is some merit to the argument that it makes no sense to talk of the existence of anything that is in principle not observable. If this argument is valid, the observable universe is the whole universe in any meaningful sense of the latter term. It would follow then, that under the theories of the expanding or vibrating universe that we have described, the universe is finite no matter what its curvature may be. However, this conclusion is valid only if we assume that the equations of these particular theories accurately describe the state of the universe in the remote past. There is no experimental evidence that they do. There are other rival theories of an expanding universe that allow for a past of infinite duration, and hence for a space of infinite extent. We do not yet know enough to choose rationally from among these different theories.

12

Geometry Generalized

Topology

In spite of Cayley's statement quoted on page 352, projective geometry is *not* all of geometry. Just as there are geometries that are more specialized than projective geometry and which are associated with subgroups of the projective group of transformations, there is a more general geometry associated with a group of transformations that contains the projective group as a subgroup. The projective group in a projective space is the group of all continuous collineations of the space into itself. We obtain a larger group of transformations that contains the projective group if we drop the requirement that each of the transformations be a collineation, and merely require that both the transformation and its inverse be continuous. A reversible transformation that is continuous in both directions is called a *topological transformation*. The topological transformations of a projective space form a group called the *topological group* of the projective space. The study of those properties of configurations in a projective space that are invariant under all transformations of the topological group is called the *topology* of that space.

Topological Space

Besides widening the group of transformations to include all reversibly continuous transformations, we can also generalize the space to which the transformations are applied. Reversibly continuous transformations exist in any space in which the concept of continuity can be defined. Continuity can, in fact, be defined in a very general class of spaces called *topological spaces*, which includes projective spaces and also many others that are not projective spaces.

A set of objects called "points" is called a topological space if it contains a special class of subsets that has the three properties that are listed next on page 372.

1) The entire space and the empty set are both members of this special class of sets. (The empty set is the set with no members in it.)

2) The set obtained by combining the memberships of any number of sets that belong to this special class is also in the special class.

3) For any two sets that belong to this special class, the set consisting of the points they have in common is also in the special class.

The sets that belong to this special class of subsets in a topological space are called the *open sets* of the space, or the *neighborhoods* of the space. Points that belong to the same open set are said to be *neighbors*, and neighbors are said to be *near* each other. (Note that this general concept of *nearness* is defined without invoking at all any concept of distance between points.)

Under this definition. *any collection of objects can be converted into a topological space,* usually in more than one way. For example, let us consider the set of four symbols {x, 2, +, *}. Its elements were chosen arbitrarily, so they have no relationship to each other beyond the fact that they happen to have been thrown together in the same set. The set is a loose aggregation, and has no structure. However, the set acquires a structure, and the elements become related to each other as neighbors, as soon as we single out certain subsets that will be called neighborhoods or open sets.

For example, we might specify that these four sets should constitute the class of open sets: {x, 2, +, *}, {x, 2}, {+, *}, and {   }, where the latter symbol represents the empty set. This class satisfies the three requirements listed above. 1) The whole space, {x, 2, +, *}, and the empty set, {   }, are members of the class. 2) The set formed by combining the memberships of any number of sets in the class is also in the class. For example, by combining the memberships of the sets {x, 2} and {+, *} we obtain the set {x, 2, +, *}, which is in the class. 3) The set consisting of the common membership of any two sets in the class is also in the class. For example, the set consisting of the common membership of {x, 2} and {+, *} is the empty set {   }, which is in the class. Since the requirements are met, this class of "open sets" defines a topological structure for the set {x, 2, +, *} and converts it into a topological space. The way the topological structure relates the elements of the space to each other as members of neighborhoods is indicated in the diagram below, where each open set that is not empty is represented by a loop enclosing its members.

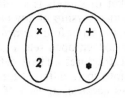

We can give the same set another, different topological structure by picking out other subsets to use as "open sets." We might, for example, decide that we want the class of open sets to include all those listed above, and in addition the set $\{x, +\}$. The inclusion of this one additional set among the open sets compels us to include more sets, in order to meet the three requirements. If we combine the memberships of $\{x, +\}$ and $\{x, 2\}$ we obtain the set $\{x, 2, +\}$. To meet requirement 2), we must include it among the open sets. The set containing the common membership of $\{x, +\}$ and $\{x, 2\}$ is the set $\{x\}$. To meet requirement 3), we must classify it as an open set. For similar reasons, we have to include $\{x, +, *\}$ and $\{+\}$ among the open sets. We find that all three requirements are met now by the enlarged collection consisting of these nine sets:

$$\{x, 2, +, *\}, \{x, 2\}, \{+, *\}, \{ \ \},$$
$$\{x, +\}, \{x, 2, +\}, \{x, +, *\}, \{x\}, \{+\}.$$

The enlarged collection of open sets gives the space a more complicated structure of interlocking neighborhoods, as shown in the diagram below:

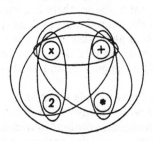

A third topological structure can be defined for the same set by specifying that *every* subset shall belong to the class of open sets. This definition meets the three requirements, because the whole set and the empty set are both subsets; the set obtained by combining any number of subsets is a subset; and the set containing the common membership of any two subsets is a subset. Under this definition, the set of four elements becomes a topological space containing sixteen open sets, including the empty set. The three topological structures we have defined convert the same set of four elements into three different topological spaces. They are different as spaces because, in each of the topological structures, the elements hang together differently as members of interlocking neighborhoods.

General Topology

The concept of continuity is defined for any topological space in the following manner: A one-to-one transformation is continuous if any set whose image under the transformation is an open set is also an open set. Since the concept of continuity has meaning in any topological space, then so does the concept of topology: Topology is the study of the invariants of the group of reversibly continuous transformations of *any topological space*. Consequently topology is a very general kind of geometry that has a wider range of application than projective geometry. That is why Cayley was wrong when he said that projective geometry is all of geometry.

Topology in Euclidean Space

In order to give a Euclidean space of three dimensions a topological structure, it is first necessary to specify which non-empty sets of points in the space are open sets. This is usually done as follows: A non-empty set in Euclidean space is called open if for every point in the set there is a sphere with that point as radius such that the entire interior of the sphere is also in the set. If the empty set is also taken as an open set, it can then be verified that the open sets in Euclidean three-space satisfy requirements 1), 2) and 3) on page 372, and so they convert it into a topological space.

A property of a figure in this space is called *topological* if it remains unchanged when the figure is subjected to a topological transformation (a reversibly continuous transformation). For example, the properties *orientability* and *non-orientability* of a surface are topological. (See page 99.)

As another example of a topological property in Euclidean

three-space we define the *genus* of a surface. The genus of a surface is the largest number of non-intersecting simple closed curves that can be drawn on the surface without separating it into two or more parts. (A curve is called simple if it does not intersect itself. Thus, an oval is simple, but a figure eight is not.) The genus of a simple polyhedron and of a sphere is 0, but the genus of a torus, which is the surface of a doughnut, is 1.

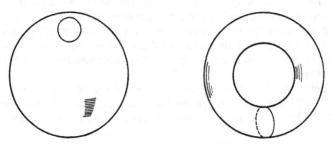

One closed curve separates a sphere but need not separate a torus.

Topological Equivalence

Two configurations are said to be *topologically equivalent* if each can be transformed into the other by a topological transformation. For example, a cube is topologically equivalent to a sphere, because it can be converted into a sphere by a reversible continuous deformation, and every reversible continuous deformation is a topological transformation. Similarly, a coffee cup is topologically equivalent to a doughnut, because each, if made of modeling clay, can be obtained from the other by a reversible continuous deformation. For this reason, mathematicians jestingly define a topologist as somebody who doesn't know the difference between a coffee cup and a doughnut.

Topologically equivalent figures

Closed Surfaces

If two closed surfaces are topologically equivalent, they are either both orientable or both non-orientable. This follows from the fact that *orientability* and *non-orientability* are topological properties. Similarly, if two closed surfaces are topologically equivalent, they both have the same genus, since genus, too, is a topological property. Conversely, it can be shown that if two closed surfaces are either both orientable or both non-orientable, and they have the same genus, then they are topologically equivalent. Consequently the topological properties of a closed surface are determined once we specify its genus and whether or not it is orientable.

Simple models of all closed surfaces of finite genus can be constructed from a sphere in the following way: If p handles are attached to a sphere, where each handle is like the ear of a cup,

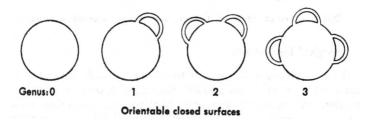

Genus: 0   1   2   3

Orientable closed surfaces

the resulting surface is an orientable surface of genus p. If p holes are punched in a sphere, and each hole is sealed with a Möbius strip in the manner described on page 249, the resulting surface is a non-orientable surface of genus p. (Each non-orientable closed surface in Euclidean three-space crosses itself. It can be

The common pretzel

drawn without crossing itself only in a space of four dimensions or more.) A torus is topologically equivalent to a sphere with one handle, so its genus is 1. The surface of the common pretzel is topologically equivalent to a sphere with three handles, so its genus is 3. Both the torus and the surface of the common pretzel are orientable surfaces. The Cayley-Klein elliptic plane (see page 249) and the real projective plane are non-orientable and are topologically equivalent to a sphere with one hole plugged by a Möbius strip, so each has genus 1.

Euler's Formula

If V, E and F are the number of vertices, edges and faces of a simple polyhedron, they are related by Euler's formula, $V - E + F = 2$. If the simple polyhedron is converted by a topological transformation into another orientable surface of genus 0, the network of edges of the polyhedron is transformed into a network of lines drawn on the surface, dividing the surface into many pieces. Each of these pieces corresponds to a face of the polyhedron, each line between adjacent pieces corresponds to an edge of the polyhedron, and each point where three or more of these lines meet corresponds to a vertex of the polyhedron, and vice versa. Thus, F, E and V are respectively the number of separate pieces of the surface formed by the network of lines on the surface, the number of lines between adjacent pieces, and the number of points where three or more lines meet. Consequently Euler's formula is a topological property of networks drawn on a closed orientable surface of genus 0.

There are generalizations of Euler's formula that apply to networks on any closed surface of any finite genus. If the genus is p, Euler's formula takes the following two forms:

For orientable surfaces, $V - E + F = 2 - 2p$.
For non-orientable surfaces, $V - E + F = 2 - p$.

(See the exercises on page 380.)

Measure Theory

Just as topology arises from generalizing the concept of nearness, another branch of geometry known as *measure theory* arises when we generalize the concepts of length, area and volume. The study of measure theory was initiated in 1901 by Henri Lebesgue (1875–1941).

On a Euclidean line equipped with a real number scale of coordinates, we associate with every segment on the line a real number called the length of the segment, by using the following rule: If the ends of the segment have coordinates a and b with $a \leq b$, then the length of the segment is $b - a$. A segment is a set of points. So when we assign a length to a segment we are attaching a number to a set of points. It is possible to generalize the concept of length by defining it for some sets of points that are not segments as well as for those sets that are segments. For example, if a set of points consists of all the points of two segments which do not overlap and have lengths m and n respectively, it is natural to define the length of this set as $m + n$. Measure theory arose from the attempt by Lebesgue to define a length for as many sets as possible on the line, no matter how complicated they look as compared to a segment. The generalized concept of length introduced by Lebesgue is known as *Lebesgue measure*. Similarly, the concept of area has been generalized by the introduction of an area-like measure that is defined for an extensive class of sets of points in a Euclidean plane. The concept of volume has been generalized by the introduction of a volume-like measure that is defined for an extensive class of sets of points in Euclidean three-space. There are also other measures that are not mere extensions of the concepts of length, area or volume. *Probability* is an important example of such a measure.

Properties of a Measure

A measure such as length cannot be generalized by merely assigning to a set that is more complicated than a segment any arbitrary real number. The assignment of numbers must be made in such a way that certain familiar properties of segment-lengths are preserved. The most significant of these properties is that if a segment is subdivided into a finite number of pieces that do not overlap, the length of the segment is the sum of the lengths of these pieces. Because the length of segments has this property, we say that segment length is *finitely additive*. Segment length also has a somewhat stronger property. If a segment is subdivided into a denumerably infinite number of pieces (see page 50) that do not overlap, the length of the segment is the sum of the lengths of these pieces. Because the length of segments has this property, we say that segment length is *denumerably additive*. Another significant property of segment-length is that congruent segments have the same length.

The length of a segment is never a negative number. If we wish to be able to speak of the "length" of the whole Euclidean line, we assign to it the length ∞, where ∞ obeys the addition rule $\infty + m = \infty$, where m is any other length.

With these considerations in mind, we define the general concept of a *measure* in a space as follows: A system of measure in a space is a function that assigns to each set of a certain specified class of sets in the space a number called its measure such that 1) the measure of a set is a non-negative real number or ∞; 2) the measure of the empty set is 0; 3) the system of measure is either finitely additive or denumerably additive; 4) the class of sets to which measures are assigned has the properties described in the next paragraph. If the measure is intended to be an extension of the concept of length, or area or volume, we impose two more requirements: 5) congruent sets have the same measure; 6) the measure of a unit segment, or unit square or unit cube respectively is equal to 1.

Measurable Space

The measures of sets are numbers that can be added and subtracted. Analogous to the addition and subtraction of numbers there is a kind of addition and subtraction of sets defined as follows: The union of any collection of sets is the set obtained by combining their memberships. If E and F are sets, we designate by $E - F$ the set consisting of those members of E that are not in F.

In order to make possible the addition and subtraction of measures, we require that the class of sets to which measures are assigned have these properties: a) if E and F belong to the class, then so does $E - F$; b) If $E_1, E_2, E_3, \ldots$ is a denumerable sequence of sets in the class, then their union is also in the class.

A set of objects that contains a special class of sets satisfying conditions a), b) and c) is called a *measurable space*. The sets that belong to this special class of subsets in a measurable space are called *measurable sets*. A measurable space is the natural domain for the study of measure theory just as a topological space is the natural domain for the study of topology.

Sets That Have No Volume

In attempts to extend the concept of volume, it is natural to try to define a volume for as many sets as possible in Euclidean three-space. This inevitably leads to the question, "Is it possible to define a volume measure for *every* set in three-space?" A startling discovery known as the *Banach-Tarski paradox* shows that the answer to this question is "No." Banach and Tarski showed that if S and S′ are the interiors of any two spheres of unequal radii, it is possible to divide S into a finite number n of non-overlapping sets $s_1, s_2, \ldots, s_n$, and to divide S′ into an equal number of non-overlapping sets $s_1', s_2', \ldots, s_n'$, such that s_1 and s_1' are congruent, s_2 and s_2' are congruent, etc. If the sets into which S and S′ are subdivided all had volume measures assigned to them, it would follow from conditions 3) and 5) for systems of measure that S and S′ have the same volume. However, this is impossible, because S and S′ have unequal radii, and the volume of a sphere is related to its radius by the formula $V = \frac{4}{3}\pi R^3$. Consequently it is not possible to have a system of finitely additive volume measure in which a measure is assigned to every set in Euclidean three-space. A similar theorem holds for the analogue of volume measure in Euclidean spaces of more than three dimensions.

Area in the Euclidean Plane

It is interesting that there is no Banach-Tarski paradox in the plane. It has been proved that there are finitely additive systems of area measure in the Euclidean plane that assign an area to every set in the plane without exception. Evidently Euclidean spaces of three or more dimensions are far more complicated things than a Euclidean space of two dimensions.

EXERCISES FOR CHAPTER 12

1. Draw a cubical network on a sphere and remove two opposite faces from the sphere. Then plug the two holes by joining them with a handle so that four "parallel" lines drawn on the handle join pairs of vertices on the boundaries of the two holes. Verify from the resulting network that on an orientable closed surface of genus $p = 1$, $V - E + F = 2 - 2p$.

2. A Möbius strip can be made by drawing a rectangle $AB'BA'$ and joining opposite edges AB and $A'B'$ so that A coincides with A' and B coincides with B'. Let M be any point on AB' between A and B. Draw BM and $A'M$. The network of lines on the Möbius strip (including AM, MB' and $A'B$) has 3 vertices (since A coincides with A', and B coincides with B'), 6 edges (since AB coin-

cides with $A'B'$), and 3 faces. Draw a tetrahedral network on a sphere so that one of its faces is a circular disc. Remove the disc and replace it by the Möbius strip. By using the resulting network, verify that on a non-orientable closed surface of genus $p = 1$, $V - E + F = 2 - p$.

13

Geometry and Algebra

The Evolution of Geometry

In the preceding chapters we have traced the course followed by geometry as it has evolved over the centuries. We shall now summarize the main features of this evolution, in order to see more clearly the nature of geometry and its relationship to the rest of mathematics.

Geometry began as the study of the spatial properties of material bodies. Then there was a conceptual separation of material bodies from the physical space that the bodies occupy, and geometry became the study of properties of physical space. In the next stage of development, mathematical space, in the form of Euclidean space, emerged as an abstraction from physical space. It was studied deductively on the basis of a set of axioms, and presumably was the only true mathematical model of physical space.

The unique position of Euclidean space as a model of physical space was undermined by the study of the theory of parallels and by the use of the methods and concepts of the calculus for the development of differential geometry. The theory of parallels led to the discovery of two more geometric structures capable of serving as models of physical space, namely hyperbolic space and elliptic space. Differential geometry identified the three rival geometric structures as spaces of constant curvature, distinguished from each other by the fact that in Euclidean space, elliptic space and hyperbolic space the curvatures are zero, positive and negative respectively. It also established the fact that there are infinitely many mathematical spaces known as Riemannian spaces, of which the spaces with constant curvature are only special cases.

The emergence of many different geometric structures, each consistent within itself, but inconsistent with the others as a model of physical space, led to a sharp distinction between pure mathematics and applied mathematics. Pure mathematics came to be understood as the study of the implications of any arbitrary consistent set of axioms. Applied mathematics is the construction and use of particular mathematical systems designed to serve as mathematical models of other systems being investigated, such as physical systems, biological systems, economic systems, etc. From the point of view of pure mathematics, Euclidean space, hyperbolic space, and elliptic space are all equally valid systems. It is only from the point of view of applied mathematics that one may ask which of these spaces is the best model of physical space. Finding the answer to this question is a problem for the physicist and not the mathematician.

While the theory of parallels and differential geometry led to a proliferation of geometric structures, projective geometry, which arose as an abstraction from Euclidean geometry, led to the reunification of some of them. Euclidean geometry, hyperbolic geometry, and elliptic geometry were all found to exist within the body of projective geometry as the study of the invariants of certain subgroups of the group of all projective collineations in real projective space.

A trend toward greater and greater abstraction was a feature of the evolution of geometry from the very beginning. The concept of physical space was developed as an abstraction from material bodies. Mathematical space is an abstraction from physical space. Projective space is an abstraction from Euclidean space, obtained by considering only some of the properties of Euclidean space and simplifying these properties. The process of abstraction proceeded in many directions. In one direction it led to the development of spaces of more than three dimensions that have all the essential properties of Euclidean space. In another direction it took the form of a generalization of geometry by the construction of many spaces that have only some of the properties of Euclidean space, or properties that are analogous to those of Euclidean space. Spaces in this category are the non-Euclidean spaces, the projective spaces, topological spaces, measurable spaces, and many others. The trend toward abstraction and generalization was accentuated by the recognition that a mathematical system can be built on the basis of any arbitrary set of consistent axioms. Then new spaces could be obtained from any given space by merely omitting, modifying or replacing one or more of its axioms.

While mathematical space was going through the process of proliferation, unification, abstraction, and generalization, our conceptions of physical space also went through many changes. Ancient thinkers oscillated between the views that space was either infinite and a void or finite and a plenum. The development of Newtonian physics tended to favor the former view, but relativity physics, which has displaced Newtonian physics, tends to favor the latter. Relativity theory definitely requires that physical space be a plenum and that the properties of space depend on the matter that is in it. The most popular cosmologies derived from relativity theory also imply that the observable universe is finite. However, there are other cosmologies that are consistent with the known facts of physics, and the question "Is physical space finite or infinite?" still remains essentially unanswered.

The Evolution of Algebra

The etymology of the word *geometry* (geo–metry = earth measurement) reminds us that at the beginning there was an intimate connection between the study of space and number. However, space and number soon became separate subjects of study, and out of them emerged the two principal divisions of mathematics. The study of space developed into geometry, and the study of number developed into algebra (in which we include arithmetic). While geometry evolved through a process of proliferation, unification, abstraction, and generalization, algebra did, too, in its own way.

The first algebraic structure known was the system of natural numbers. But the requirements of practical life and mathematical theory led to the development of the system of integers, the rational number system, the real number system, and the complex number system. These five number systems are nested within each other like a set of bowls of different sizes: the complex number system includes the real number system, which includes the rational number system, which includes the system of integers, which includes the system of natural numbers.

Additional algebraic structures have been obtained by constructing mathematical systems that have only some of the properties of these number systems, or properties that are analogous to those of the number systems. The most important of these abstract algebraic structures are groups, rings, fields, and vector spaces.*

The axiomatic method, originally developed in geometry, was ultimately introduced into algebra, so that now all mathematical structures, algebraic as well as geometric, are defined by means of systems of axioms.

---

* For an elementary introduction to the principal algebraic structures, see *The New Mathematics*.

Relations Between Geometry and Algebra

After the initial separation of geometry and algebra they developed as separate branches of mathematics. Then they were reunited as partners in the analytic geometry of Descartes. But their mutual relations soon entered a new phase, transcending that of partnership, when each of the partners swallowed the other. First algebra swallowed geometry, when it was shown that Euclidean plane geometry could be developed as a branch of algebra where points are defined as ordered pairs of numbers, lines are defined as sets of points satisfying equations of the first degree, and the concepts of the distance between two points and the measure of an angle are defined by appropriate algebraic formulas. Then geometry swallowed algebra when it was shown that algebra could be developed as a branch of geometry by using geometric constructions to give a field structure to the set of points that remain on a projective line after one of the points has been removed. The complete fusion of the geometric structure of a projective plane with the algebraic structure of the field on any of its lines shows that geometry and algebra, at least in their classical forms, study essentially the same subject matter, but study it from different points of view. The fact that algebra and geometry are two sides of the same coin is symbolized very well by the theorem that the Pappus property of the projective plane is equivalent to the commutative property of the multiplication of points on a line in the plane. (See page 337.)

Geometric and Algebraic Structures

The evolution of mathematics has produced many geometric structures and many algebraic structures. Both types of structures are abstract mathematical systems whose properties are derived from axioms. How then, do they differ from each other? What kind of mathematical system is classified as a geometric structure, and what kind is classified as an algebraic structure? We can answer this question by examining the typical structures of each kind described in this book.

If we look at the axioms for a projective space, given on page 295, we see that a projective space is a collection of objects which has certain special subsets called "lines" whose basic properties are described in the axioms. If we look at the axioms for a topological space, given on page 372, we see that a topological space is a collection of objects which has certain special subsets called "open sets" whose basic properties are described in the axioms. If we look at the axioms for a measurable space, given on page 379, we see that a measurable space is a collection of objects which has certain special subsets called "measurable sets" whose basic properties are described in the axioms. These examples suggest the following definition of a geometric structure: *A mathematical system is called a geometric structure if it is characterized by a special class of subsets whose basic properties are described in the axioms of the system.* The word "space" is usually used only for a geometric structure, although there are some exceptions to this rule. The elements of a geometric structure are usually called "points."

If we look at the axioms for a group, given on page 39, we see that a group is a collection of objects in which there is an operation called multiplication whose basic properties are described in the axioms. If we look at the axioms for a field, given on page 32, we see that a field is a collection of objects in which there are operations called addition and multiplication whose basic properties are described in the axioms. If we look at the axioms for the algebraic structure called a vector space, given on pages 138–9, we see that a vector space is a collection of objects in which there are operations called addition and scalar multiplication whose basic properties are described in the axioms. These examples suggest the following definition of an algebraic structure: *A mathematical system is called an algebraic structure if it is characterized by one or more operations whose basic properties are described in the axioms of the system.*

Thus the essential difference between a geometric structure and an algebraic structure is that one is concerned with special subsets and the other is concerned with operations in the system.

There are some mathematical systems that have a hybrid character: they are both geometric and algebraic structures. This is true, for example, of any structure studied in topological algebra, where it is required that the structure be both an algebraic structure and a topological space and that the algebraic operations define continuous functions in the topological space. The real number system is an example of such a hybrid structure. The branch of mathematics called "analysis" studies such hybrid structures by using simultaneously their algebraic and topological properties.

Fusion in Modern Mathematics

The fusion of algebra and geometry that we encountered in classical mathematics is characteristic of modern mathematics as well. It arises as a result of the typical modern technique for studying mathematical structures: The modern mathematician often studies a structure by examining the functions on this structure into itself or into another structure of the same kind or even of a different kind.

For example, we can study a geometric structure by studying functions which assign to each element of this structure an element of the same structure. Of particular importance are those reversible functions or transformations that preserve the special subsets which determine the structure. In a topological space, the important reversible transformations are those which preserve open sets. In a measurable space, they are those which preserve measurable sets. In a projective space they are those which preserve lines. In each case the set of all reversible structure-preserving transformations is a group with respect to the operation of multiplication of transformations. Thus there is associated with every geometric structure a significant algebraic structure, the group of reversible structure-preserving transformations. Moreover, as Klein pointed out in his Erlangen program, every subgroup in this group determines a separate geometry in the geometric structure, where a geometry is defined as the study of the properties of configurations that are left unchanged by all the transformations in a group.

Similarly, we can study an algebraic structure by studying functions which assign to each element of this structure an element of the same structure. The set of all such functions can be given a topological structure, which is a kind of geometric structure, by specifying in an appropriate way which of its subsets shall be open sets. Thus there can be associated with every algebraic structure a related geometric structure. The properties of this geometric structure often reveal significant properties of the underlying algebraic structure.

It is common in modern mathematics to build structure upon structure in this way. As a result there is an interweaving of geometric and algebraic structures and a mingling of geometric and algebraic methods that reveal the essence of a subject being studied by alternately exposing its geometric and its algebraic side. Geometry and algebra are two complementary aspects of mathematical systems just as wave and particle are two complementary aspects of the physical world.

Bibliography

I. Adler, *Magic House of Numbers*, John Day, New York, 1957.
——, *The New Mathematics*, John Day, New York, 1958.
C. F. Adler, *Modern Geometry*, McGraw-Hill, New York, 1958.
Aleksandrov, Kolmogorov, & Lavrent'ev, *Mathematics, Its Content, Methods and Meaning*, M.I.T. Press, Cambridge, 1963.
F. Bachmann, *Aufbau der Geometrie aus dem Spiegelungsbegriff*, Springer, Berlin, 1959.
R. H. Bing, *Elementary Point Set Topology*, Mathematical Association of America, Buffalo, 1960.
G. Birkhoff & S. MacLane, *A Survey of Modern Algebra*, Macmillan, New York, 1951.
L. M. Blumenthal, *A Modern View of Geometry*, Freeman, San Francisco, 1961.
W. Bonnor, *The Mystery of the Expanding Universe*, Macmillan, New York, 1964.
R. Bonola, *Non-Euclidean Geometry*, Dover, New York, 1955.
E. Borel, *Space and Time*, Dover, New York, 1960.
L. Brand, *Vector and Tensor Analysis*, Wiley, New York, 1947.
J. L. Coolidge, *A History of Geometrical Methods*, Dover, New York, 1963.
H. S. M. Coxeter, *Introduction to Geometry*, Wiley, New York, 1961.
L. P. Eisenhart, *Riemannian Geometry*, Princeton, 1949.
W. T. Fishback, *Projective and Euclidean Geometry*, Wiley, New York, 1962.
H. G. Forder, *The Foundations of Euclidean Geometry*, Dover, New York, 1958.
A. Grünbaum, *Philosophical Problems of Space and Time*, Knopf, New York, 1963.
T. L. Heath, *Euclid's Elements*, Dover, New York, 1956.
D. Hilbert, *Foundations of Geometry*, Open Court, La Salle, 1962.
D. Hilbert & S. Cohn-Vossen, *Geometry and the Imagination*, Chelsea, New York, 1956.

F. Klein, *Elementary Mathematics from an Advanced Standpoint, Geometry*, Dover, New York, 1939.

——, *Vorlesungen Uber Nicht-Euklidische Geometrie*, Chelsea, New York.

M. Kline, *Mathematics in Western Culture*, Oxford, New York, 1953.

A. S. Kompaneyets, *Theoretical Physics*, Dover, New York, 1962.

L. Landau & E. Lifshitz, *The Classical Theory of Fields*, Addison-Wesley, Reading, 1951.

H. P. Manning, *Geometry of Four Dimensions*, Dover, New York, 1956.

——, *Introductory Non-Euclidean Geometry*, Dover, New York, 1963.

S. F. Mason, *Main Currents of Scientific Thought*, Schuman, New York, 1953.

J. R. Newman, *The World of Mathematics*, Simon & Schuster, New York, 1956.

H. Poincaré, *Science and Hypothesis*, Dover, New York, 1952.

W. Prenowitz, *A Contemporary Approach to Classical Geometry*, Mathematical Association of America, Buffalo, 1961.

Plato, *Timaeus*, and *The Republic*.

B. Riemann, *Ueber die Hypothese welche der Geometrie zu Grunde liegen, Gesammelte Mathematische Werke*, Dover, New York, 1953.

B. Russell, *Foundations of Geometry*, Dover, New York, 1956.

D. E. Smith, *History of Mathematics*, Dover New York, 1958.

D. J. Struik, *A Concise History of Mathematics*, Dover, New York, 1948.

O. Veblen & J. W. Young, *Projective Geometry*, Ginn, New York, 1910.

C. F. von Weizsäcker & J. Juilfs, *The Rise of Modern Physics*, Braziller, New York, 1957.

A. Weber & R. B. Perry, *History of Philosophy*, Scribners, New York, 1925.

H. Weyl, *Space, Time and Matter*, Dover, New York, 1950.

E. Whittaker, *From Euclid to Eddington*, Dover, New York, 1958.

H. E. Wolfe, *Non-Euclidean Geometry*, Dryden, New York, 1945.

I. M. Yaglom, *Geometric Transformations*, Random House, New York, 1962.

J. W. Young, *Fundamental Concepts of Algebra and Geometry*, Macmillan, New York, 1925.

J. W. Young et al., *Monographs on Modern Mathematics*, Dover, New York, 1955.

Answers to Exercises

Chapter 2

4. 4, 9, 16, 25, 36, 49, 64, 81, 100, etc. are square numbers.

8. $(2n + 1)^2 = 4n^2 + 4n + 1 = 2(2n^2 + 2n) + 1$.

9. a) $-2, -5, 3, 0$; b) $\frac{3}{2}, \frac{1}{5}, -1, 1$.

11. c) If r is real, r^2 is positive or zero, and hence $r^2 + 1$ is positive.

12. a) 2; b) 1; c) 1; d) 2.

13. 49.

16. a) No; b) Yes.

17. a) If no woman is a bigamist;
    b) If no woman is unmarried.

18. a) finite; b) finite; c) infinite; d) infinite.

19. a) 12; b) c; c) 12 is finite and c is infinite.

20. a) $\frac{1}{2}$; b) No.

21. a)

	A	B	C	D
A	B	C	D	A
B	C	D	A	B
C	D	A	B	C
D	A	B	C	D

   b) D;

   c) C, B, A, D;

   d) Yes.

Chapter 3

1. b) No; no. c) No; no.

Chapter 4

2. $BA + AC = BC$. Therefore $AC = -BA + BC = AB + BC$.

3. a) $y = 5$; b) $x = -4$; c) $y = 0$; d) $x = 0$.

4. a) 3; b) $y - 1 = 3(x - 1)$; c) Yes; d) No.

5. a) 0; b) No slope; c) $-\frac{3}{4}$; d) 4.

6. a) Yes, because the lines are distinct and $3(-8) - 6(-4) = 0$.
   b) No, because $3(7) - 5(-4) \neq 0$.
7. a) No, because $3(2) + (-4)(1) \neq 0$.
   b) Yes, because $3(8) + (-4)(6) = 0$.
8. a) parallel; b) coincide; c) intersect at the point whose coordinates are $(7, -3)$.
9. a) 7; b) 17.
10. a) $\begin{vmatrix} x & y & 1 \\ 3 & 1 & 1 \\ 2 & 2 & 1 \end{vmatrix} = 0$, or $-x - y + 4 = 0$.
    b) $y - 1 = -1(x - 3)$. c) Both yield $y = -x + 4$.
11. 13.
12. $l = 5/13, m = 12/13$.
13. 4 units.
14. $\dfrac{3 + 4\sqrt{3}}{10}$.
15. a) parabola, since $B^2 - 4AC = 0$;
    b) hyperbola, since $B^2 - 4AC = 1$;
    c) ellipse, since $B^2 - 4AC = -144$
16. $(x + 2)^2 + (y - 3)^2 = 64$.
17. $N_0 = 16, N_1 = 32, N_2 = 24, N_3 = 8; 16 - 32 + 24 - 8 = 0$.

Chapter 5

1. $\mathbf{b} - \mathbf{a}, \mathbf{c} - \mathbf{b}, \mathbf{a} - \mathbf{c}$.
2. a) $\mathbf{c}$; b) $\mathbf{d} - \mathbf{c}$; c) $\mathbf{p} = \mathbf{c} + r(\mathbf{d} - \mathbf{c})$.
3. $\mathbf{p} = \dfrac{4\mathbf{a} + 3\mathbf{b}}{7}$.
4. $\mathbf{p} = \tfrac{1}{4}(\mathbf{a} + \mathbf{b} + \mathbf{c} + \mathbf{d})$.
5. $\overrightarrow{OL} + \overrightarrow{OM} + \overrightarrow{ON} = \mathbf{l} + \mathbf{m} + \mathbf{n} = \tfrac{1}{2}(\mathbf{a} + \mathbf{b}) + \tfrac{1}{2}(\mathbf{b} + \mathbf{c}) + \tfrac{1}{2}(\mathbf{c} + \mathbf{a}) = \mathbf{a} + \mathbf{b} + \mathbf{c} = \overrightarrow{OA} + \overrightarrow{OB} + \overrightarrow{OC}$.
6. a) 5; b) 0; c) 25.
8. 13.   9. $-\dfrac{1}{\sqrt{5}}$ 10. c).
11. a) $\overrightarrow{AB} = \mathbf{b} - \mathbf{a}, \overrightarrow{BC} = \mathbf{c} - \mathbf{b}$.
    b) $|\overrightarrow{AB}| = \mathbf{b} \cdot \mathbf{b} - 2\mathbf{b} \cdot \mathbf{a} + \mathbf{a} \cdot \mathbf{a}$;
    $|\overrightarrow{BC}| = \mathbf{c} \cdot \mathbf{c} - 2\mathbf{c} \cdot \mathbf{b} + \mathbf{b} \cdot \mathbf{b}$.

c) $\mathbf{a} \cdot \mathbf{a} - \mathbf{c} \cdot \mathbf{c} + 2\mathbf{b} \cdot (\mathbf{c} - \mathbf{a}) = \mathit{0}$

d) $\mathbf{m} = \frac{1}{2}(\mathbf{a} + \mathbf{c})$.

e) $\overrightarrow{BM} = \frac{1}{2}\mathbf{a} + \frac{1}{2}\mathbf{c} - \mathbf{b}$.

f) $\overrightarrow{AC} = \mathbf{c} - \mathbf{a}$.

g) $\overrightarrow{AC} \cdot \overrightarrow{BM} = (\mathbf{c} - \mathbf{a}) \cdot (\frac{1}{2}\mathbf{a} + \frac{1}{2}\mathbf{c} - \mathbf{b})$
$= -\frac{1}{2}\mathbf{a} \cdot \mathbf{a} + \frac{1}{2}\mathbf{c} \cdot \mathbf{c} - \mathbf{b} \cdot (\mathbf{c} - \mathbf{a})$
$= -\frac{1}{2}[\mathbf{a} \cdot \mathbf{a} - \mathbf{c} \cdot \mathbf{c} + 2\mathbf{b} \cdot (\mathbf{c} - \mathbf{a})] = -\frac{1}{2}0 = 0$.

Chapter 6

1. 360.   2. 260.

3. $B^{-1}A^{-1}$.   4. R_A; R_a.

5. a) $AT = B$; b) $PR_a = Q$.

6. If A coincides with one of the other two points, say B, then since $A = B$, we have $A' = B'$, and so A' lies on $B'C'$. If A coincides with neither B nor C, one of the three points lies between the other two, say B lies between A and C. Then $AB + BC = AC$. Since T is an isometry, $AB = A'B'$, $BC = B'C'$, and $AC = A'C'$. Hence $A'B' + B'C' = A'C'$. This is possible only if B' is on $A'C'$.

7. A.

8. a) Let T be the translation that moves the plane through a distance equal to segment PP' in the direction from P to P'. Let k be the image of h under T. Let R be the rotation about P' that moves k to h'. Let a' be the line on which h' lies, and denote by $R_{a'}$ the reflection across a'. Then either TR or $TRR_{a'}$, is the desired isometry. b) Let U and V be isometries that move P to P', h to h', and S to S'. Let a and a' be the lines on which h and h' lie respectively. Let A' be the image of A under U, and let A'' be the image of A under V.

Case I. Assume A is on a. Then by hypothesis, A' and A'' are on a', and they are on h' or not on h' according as A is on h or not on h. Since U and V are isometries, $PA = P'A'$, and $PA = P'A''$. Consequently $P'A' = P'A''$, and therefore $A' = A''$.

Case II. Assume A is not on a. Let B be any point other than P on a, and let B' be its image under U and V. (These images are the same by case I.) By hypothesis, A' and A'' are both on S' or not on S' according as A is on S or not on S. Then triangles $P'B'A'$ and $P'B'A''$ lie on the same side of a'. A' is the intersection of the lines $P'A'$ and $B'A'$. A'' is the intersection of the lines $P'A''$ and $B'A''$. Since the segments $P'B' = P'B'$, $P'A' = P'A''$, and $B'A' = B'A''$, triangles $P'B'A'$ and $P'B'A''$ are congruent. Consequently angle $P'B'A' =$ angle $P'B'A''$, and angle $B'P'A' =$ angle $B'P'A''$. Therefore the lines $P'A'$ and $P'A''$ coincide, and the lines $B'A'$ and $B'A''$ coincide. Hence $A' = A''$.

9. a) B; b) C; c) A; d) C; e) D.

10. It suffices to prove that P' is not on PQ. If P' is on PQ, then A, which is the midpoint of PP', is also on PQ, which is contrary to the hypothesis.

11. Since $b \neq a$, there is a point P on b that is not on a. a is the perpendicular bisector of PP'. So if P' is on b, a is perpendicular to b, which is contrary to the hypothesis. Hence P' is not on b, and therefore bR_a is not b.

12. Several cases arise, depending on the relative positions of P, A and B on g. We give the proof for one case. The proofs for the other cases are similar. Suppose A is between P and B, and $PA < AB$. Then Q is between A and B, and B is between Q and P'. That is, the points are arranged on g in the order $PAQBP'$. Moreover $PA = AQ$, $QB = BP'$, and $AQ + QB = AB$. Then $PP' = PA + AQ + QB + BP' = 2(AQ + QB) = 2AB$.

13. a) $R_a R_b R_c = R_a R_a R_a = (R_a R_a) R_a = 1 R_a = R_a$.

b) $R_a R_b R_c = R_a R_a R_c = (R_a R_a) R_c = 1 R_c = R_c$.

c) If a and b intersect, draw d through their intersection so that angle $(a, b) =$ angle (d, a). If a is parallel to b, draw d parallel to b so that distance $(a, b) =$ distance (d, a). In either case, $R_a R_b = R_d R_a$. Multiplying on the right by R_a, we get $R_a R_b R_a = R_d$. Since $c = a$, $R_a R_b R_c = R_d$.

14. Through A draw c perpendicular to b, and draw a perpendicular to c at A. Then $R_A R_b = (R_a R_c) R_b = R_a (R_c R_b) = R_a R_C =$ a glide reflection, as shown on page 169.

15. Suppose P is the only fixed point of T. Then PS is a fixed point of $S^{-1}TS$. Suppose B' is also a fixed point of $S^{-1}TS$. Since S is onto, there is a point B such that $B' = BS$. Then, since BS is a fixed point of $S^{-1}TS$, B is a fixed point of T. Hence $B = P$, and $B' = BS = PS$.

16. a) $SR_PS = R_Q$; b) $CS = D$; c) $SR_hS = R_k$;
d) $fS = g$; e) $R_b R_Q R_b = R_Q$.

17. *Given:* M is the midpoint of AC and BD.

Prove: $AB = DC$ and AB is parallel to DC.

Proof: Since M is the midpoint of AC and BD, $R_A R_M = R_M R_C$ and $R_B R_M = R_M R_D$. Multiplying each equation by R_M on the right, we get $R_A = R_M R_C R_M$ and $R_B = R_M R_D R_M$. Multiplying the two equations, we get

$$R_AR_B = (R_MR_CR_M)(R_MR_DR_M) = R_MR_C(R_MR_M)R_DR_M$$
$$= R_MR_C1R_DR_M = (R_MR_CR_D)R_M = (R_DR_CR_M)R_M$$
$$= (R_DR_C)(R_MR_M) = (R_DR_C)1 = R_DR_C.$$

Therefore $AB = DC$, and AB is parallel to DC.

18. a) R_aR_P. b) If P is on a, draw b through P perpendicular to a. $R_PR_a = R_bR_aR_a = R_b \neq 1$. If P is not on a, draw b through P perpendicular to a. Then $P(R_PR_a) = (PR_P)R_a = PR_a \neq P$ since P is not on a. Hence $R_PR_a \neq 1$. c) By entry 1 on page 176, if P is on a, then $R_PR_a = R_aR_P$. Since R_aR_P is the inverse of R_PR_a, $(R_PR_a)(R_aR_P) = 1$, and hence $(R_PR_a)(R_PR_a) = 1$. Then, since $R_PR_a \neq 1$, R_PR_a is an involution. Therefore, if P is on a, R_PR_a is an involution. Conversely, if R_PR_a is an involution, $(R_PR_a)(R_PR_a) = 1$. By a), we also have $(R_PR_a)(R_aR_P) = 1$. Hence $(R_PR_a)(R_PR_a) = (R_PR_a)(R_aR_P)$, and therefore $R_PR_a = R_aR_P$. Then, by entry 1, P is on a.

Chapter 8

1. If two intersecting lines are cut by a transversal that does not pass through their point of intersection, then the sum of the interior angles that are on the same side as this point of intersection is less than two right angles. *Proof:* These two angles are in a triangle. Hence their sum is less than two right angles, by Euclid's proposition 17, Book I.

2. Side-angle-side.

3. Four right angles minus less than two right angles = more than two right angles.

4. a) $x + (y + r) + t = S$.
   b) $S = 2$ right angles, and $z + s = 2$ right angles.

5. The four right angles at T exhaust the $360°$ at T. Angle $TSU +$ angle $TSV = 180°$. Therefore USV is a straight line segment equal to $2RT$. There are four such straight line segments enclosing a quadrilateral. Each angle of this large quadrilateral is an angle of one of the four small quadrilaterals.

6. a) Side-angle-side; b) side-side-side.

7. In a plane, there is only one perpendicular bisector of a segment, and by the theorem proved on page 206, the line joining the midpoints of the base and summit of a Saccheri quadrilateral is the perpendicular bisector of the base and the summit.

8. Let $ABDC$ be a Saccheri quadrilateral with the base BD and summit AC. Let M and N be the midpoints of AB and CD re-

spectively. Since $AB = CD$, $MB = ND$. Therefore $MBDN$ is a Saccheri quadrilateral with base BD and summit MN. The line joining the midpoints of AC and BD is the perpendicular bisector of BD. Therefore it is also the perpendicular bisector of MN, by exercise 7.

9. If Ω' is the other ideal point on $B\Omega$, a half-line through A intersects $\Omega'B\Omega$ if and only if it lies inside angle $\Omega'A\Omega$. Therefore AB is inside angle $\Omega'A\Omega$. Hence AP, which is inside angle $BA\Omega$, is inside angle $\Omega'A\Omega$. Therefore AP intersects $\Omega'B\Omega$. AP cannot intersect $B\Omega'$, since they are in separate half-planes on opposite sides of the line AB. Therefore AP intersects $B\Omega$.

10. Draw $AB\Omega$ and a line EF that does not pass through A, B or Ω, but intersects $A\Omega$ at P. Draw BP. A half-line on EF with vertex at P lies either inside angle APB or inside angle $BP\Omega$. In the first case EF intersects AB. (This is a consequence of the Pasch axiom.) In the second case EF intersects $B\Omega$, by the result of exercise 9.

11. Draw $AB\Omega$ and a line EF that does not pass through A, B or Ω, but intersects AB at P. Draw $P\Omega$. A half-line on EF with vertex at P lies either inside angle $AP\Omega$ or inside angle $BP\Omega$. In the first case EF intersects $A\Omega$, and in the second case it intersects $B\Omega$, by the result of exercise 9.

12. Either $AB = FE$ or $AB < FE$ or $AB > FE$. If $AB = FE$, then angle F = angle A, by the theorem on page 206. If $AB < FE$, then angle $F <$ angle A, by theorem III on page 224. Both of these consequences are contrary to the hypothesis. Therefore $AB > FE$.

13. Use the diagram on page 222 for Saccheri quadrilateral $ABDC$ with base BD, with E and F the midpoints of the base and summit respectively. In quadrilateral $ABEF$, angles BEF and AFE are right angles. Hence theorem III of page 224 applies. Since angle ABE is a right angle and angle BAF is acute, angle $ABE >$ angle BAF. Therefore $AF > BE$. Consequently AC (which is $2AF$) $> BD$ (which is $2BE$).

14. a) $10°$; b) $27°$.

15. If $AB \neq A'B'$, then one of them is greater than the other, say $AB > A'B'$. Draw AE on AB and DF on DC so that $AE = A'B'$ and $DF = D'C'$. Then triangles AED and $A'B'D'$ are congruent. Consequently triangles DEF and $D'B'C'$ are also congruent. It follows that angles AEF and DFE are right angles. Then angles BEF and CFE are also right angles. Then quadrilateral $EBCF$ has four right angles, which is impossible.

16. Since $AB = A'B'$, by exercise 15, then triangles ABD and $A'B'D'$ are congruent, and triangles BDC and $B'D'C'$ are congruent.

17. Triangle PQR = quadrilateral $QMNR$ + triangle PMT + triangle PNT. Quadrilateral $GQRH$ = quadrilateral $QMNR$ + triangle GMQ + triangle NHR. Quadrilateral $QMNR \cong$ quadrilateral $QMNR$. Triangle $PMT \cong$ triangle GMQ. Triangle $PNT \cong$ triangle NHR.

18. Let AB and $A'B'$ be the equal sides of triangles ABC and $A'B'C'$. Let $S = A + B + C = A' + B' + C'$. Use the construction of theorem IV on page 224 to make a Saccheri quadrilateral with summit AB and a Saccheri quadrilateral with summit $A'B'$. They will have equal summit angles each equal to $\frac{1}{2}S$. Then, by exercise 16, the two Saccheri quadrilaterals are congruent, and hence they are equivalent. By exercise 17, each of triangles ABC and $A'B'C'$ is equivalent to one of these Saccheri quadrilaterals. Hence the triangles are equivalent.

19. Triangle $TQZ \cong$ triangle RSM by side-angle-side. Then triangle $QZU \cong$ triangle SMN by side-angle-side.

20. Angle MPK = angle $LP''M$, angle LMP'' = angle PMK, and $PM = P''M$.

21. $P\Omega$ and $P'\Omega$ make equal angles with PP' because these are angles of parallelism that correspond to equal distances.

22. Let M and N be the midpoints of PP' and $P'P''$ respectively. Let the perpendicular bisectors of PP' and $P'P''$ meet h at A and B respectively. Then MA and NB are perpendicular to AB. Draw PS, $P'S'$, and $P''S''$ perpendicular to h. Triangles PSA and $P'S'A$ are congruent. Hence $PS = P'S'$. Similarly, triangles $P'S'B$ and $P''S''B$ are congruent. Hence $P'S' = P''S''$.

23. a) The step, "Through A draw DE parallel to BC" is not valid in elliptic geometry.
b) The statement, "When two parallel lines are cut by a transversal, the alternate interior angles are equal" is not valid in hyperbolic geometry.

Chapter 10

2. a) OA and OB determine a plane R. The line $A'B'$ is the intersection of R and Q. C, being on AB, is in R. Therefore OC is in R. C', being on OC, is in R, but it is also in Q. Therefore C' is on $A'B'$, the intersection of R and Q.

b) All the lines OC lie in plane OAB, and C and C' have a common projection from O.

3. $f(x)$ exists and is a real number for all values of x except $x = a$; and for any real number $r \neq 0$, $f(x) = r$ if and only if $x = a + \dfrac{1}{r}$; and $f(\infty) = 0$, by definition.

4. a) 1 and 2 imply 1'. b) Let P be any point. 1, 2 and 3 imply that there is a line that is not through P. (See c below.) There are three points, say Q, R and S, on this line. Then PQ, PR and PS are distinct lines on P. c) Suppose all lines lie on P. By 3 and 4, there are at least two distinct lines l and m. By 2 there is another point A on l and another point B on m. Because of 4, A is not on m, and B is not on l. If AB is on P, then A is on BP, which is m, and we have a contradiction. d) Let P be any point, a any line not on P. Join P to every point on a. This establishes a one-to-one correspondence between lines on P and points on a, and all points but one of the latter are in one-to-one correspondence with the real numbers.

5. AC and $A'C'$ meet because they are coplanar. BC and $B'C'$ meet because they are coplanar.

6. $\dfrac{b}{X_3} = \dfrac{b}{a} \cdot \dfrac{a}{X_3}$.

7. a) (100, 200);   b) (1000, 2000);   c) (10000, 20000).

8. a) $X_2 = mX_1 + dX_3$.   b) $b = \dfrac{b}{a} \cdot a + d \cdot 0$.

9. a) $X_2 X_3 = X_1^2$.   b) $X_1^2 + X_2^2 + X_2^2 = 25X_3^2$.
   c) $X_1 X_2 = 12X_3^2$.   d) $(X_1 - aX_3)^2 + (X_2 - bX_3)^2 = r^2 X_3^2$.

10. If r, s and t are real, and $(i, 1, 0)$ are proportional to (r, s, t) then there exists a complex number k such that $i = kr$ and $1 = ks$. Then $k = 1/s$ is real, and then i is real, a contradiction.

11. If $(1, i, 1 + i)$ is real, there exists a complex number k and real numbers (r, s, t) such that $1 = kr$, $i = ks$, and $1 + i = kt$. Then $k = 1/r$ is real, and i is real, a contradiction.

12. $d = 1$.

13. a) $X_1^2 + X_2^2 = 25X_3^2$.
    b) $1^2 + (i)^2 = 25(0)^2$; $1^2 + (-i)^2 = 25(0)^2$.

14. A complete quadrilateral is the configuration consisting of four lines, no three of which are concurrent, and the six points determined by pairs of these lines.

15. The line that lies on a pair of opposite vertices of a quadrilateral is called a diagonal line.

16. Four lines a, b, c and d on a point are said to form a harmonic sequence if a and b are diagonal lines of a complete quadrilateral, and c and d are on the vertices of the quadrilateral that are on the third diagonal line.

17. Suppose a projectivity T between distinct lines l and m moves P into itself. Then P is the intersection of l and m. Let A and B be two other points on l, and let A' and B' be their images respectively under T. Let O be the intersection of AA' and BB'. Both the projectivity T and the perspectivity with center O move P, A and B to P, A' and B' respectively. Therefore, by the fundamental theorem of projective geometry, T is the perspectivity with center O.

18. a) $X_1 X_3 = X_2{}^2$. b) $1(0) = 0$.

19. a) $X_1 X_2 = 12 X_3{}^2$. b) $1(0) = 12(0)^2$; $0(1) = 12(0)^2$.

22. Draw m and n through ∞, and choose A on n.
    $0A$ meets m at B.
    $0B$ meets n at A.
    XA meets m at D.
    AD meets l at $0 + X$. But AD meets l at X.
    Therefore $0 + X = X$.

23. Draw m through 0 and n through ∞. Choose A on n but not on m. $1A$ meets m at B.
    $1B$ meets n at A.
    XA meets m at D.
    AD meets l at $1X$. But AD meets l at X.
    Therefore $1X = X$.

24. Draw m through 0 and n through ∞. Choose A on n but not on m. $1A$ meets m at B.
    XA meets m at D.
    $1D$ meets n at C.
    CB meets n at Y.
Then steps a) to d) show that $YX = 1$.

25. $0A$ meets m at B.
    XB meets n at C.
    ∞A meets m at ∞.
    $C\infty$ meets l at ∞.

26. a) $1A$ meets m at B.
    $0B$ meets n at C.
    $1C$ meets m at C.
    AC meets l at $1/0$.
    But AC meets l at ∞.
    Therefore $1/0 = \infty$.

    b) $1A$ meets m at B.
    ∞B meets n at ∞.
    1∞ meets m at 0.
    $A0$ meets l at $1/\infty$.
    But $A0$ meets l at 0.
    Therefore $1/\infty = 0$.

27. $1A$ meets m at B.
   OB meets n at C.
   XA meets m at D.
   CD meets l at OX.
   But CD meets l at 0.
   Therefore $0X = 0$.

28. $1A$ meets m at B.
   OB meets n at C.
   ∞A meets m at C.
   CC does not determine a
   line. Therefore its intersection with l is indeterminate.

29. Suppose a has $N + 1$ points. Let b be any other line, and O any point not on a or b. The perspectivity between a and b with center O sets up a one-to-one correspondence between the points of a and the points of b. Therefore b has $N + 1$ points.

30. There are seven such planes as follows:

$p^n = N$	$N^2 + N + 1$ Number of points in the plane	$N + 1$ Number of points on a line
$2^1 = 2$	7	3
$2^2 = 4$	21	5
$2^3 = 8$	73	9
$3^1 = 3$	13	4
$3^2 = 9$	91	10
$5^1 = 5$	31	6
$7^1 = 7$	57	8

31. a) $4/3$; b) $4/3$; c) ∞.

32. Let Ω be the ideal point on AB, and Ω' the ideal point on $A'B'$. Then the images of A, M, B and Ω under T are respectively A', M', B' and Ω'. Since M is the midpoint of AB, M is the harmonic conjugate of Ω with respect to A and B. Consequently M' is the harmonic conjugate of Ω' with respect to A' and B'. Therefore M' is the midpoint of $A'B'$.

33. For the segment joining $(0, 0)$ and $(1, 0)$ on l, $d_1 = 1$, $l_1 = 1$, amd $m_1 = 0$. For the segment joining $(0, 0)$ and $(0, 1)$ on m, $d_2 = 1$, $l_2 = 0$, and $m_2 = 1$. $\cos \theta = 1(0) + 0(1) = 0$.

Chapter 12

1. On the sphere we had $V = 8$, $E = 12$, and $F = 6$. When the two opposite faces were removed and replaced by the handle, the

number of vertices was not changed, the number of edges was increased by 4, and the number of faces was increased by 2. The resulting network on the sphere with one handle has $V = 8$, $E = 16$, and $F = 8$. Then $V - E + F = 8 - 16 + 8 = 0 = 2 - 2p$, since $p = 1$.

2. On the sphere we had $V = 4$, $E = 6$, and $F = 4$. When the disc is replaced by the Möbius strip, the three vertices on the boundary of the disc are replaced by the three vertices on the Möbius strip; the three edges on the boundary of the disc are replaced by the six edges on the Möbius strip; and the single face of the disc is replaced by the three faces on the Möbius strip. The resulting network on the projective plane has $V = 4$, $E = 9$, and $F = 6$. Then $V - E + F = 4 - 9 + 6 = 1 = 2 - p$, since $p = 1$.

Index

Index

Index